ADVANCES IN

Applied
Microbiology

VOLUME 50

ADVANCES IN

Applied Microbiology

Edited by

ALLEN I. LASKIN
Somerset, New Jersey

JOAN W. BENNETT
New Orleans, Louisiana

GEOFFREY M. GADD
Dundee, Scotland

VOLUME 50

ARCHAEA

Ancient Microbes, Extreme Environments,
and the Origin of Life

Edited by

PAUL BLUM
Lincoln, Nebraska

ACADEMIC PRESS

San Diego   New York   Boston
London   Sydney   Tokyo   Toronto

Academic Press
A division of Harcourt, Inc.
525 B Street, Suite 1900, San Diego, California 92101-4495, USA
http://www.academicpress.com

Academic Press
Harcourt Place, 32 Jamestown Road, London NW1 7BY, UK
http://www.academicpress.com

International Standard Serial Number: 0065-2164

International Standard Book Number: 0-12-002650-3

PRINTED IN THE UNITED STATES OF AMERICA
01 02 03 04 05 06 MM 9 8 7 6 5 4 3 2 1

CONTENTS

DNA Recombination and Repair in the Archaea

ERICA M. SEITZ, CYNTHIA A. HASELTINE, AND
STEPHEN C. KOWALCZYKOWSKI

Basal and Regulated Transcription in Archaea

JÖRG SOPPA

Protein Folding and Molecular Chaperones in Archaea

MICHEL R. LEROUX

Archaeal Proteasomes: Proteolytic Nanocompartments of the Cell

JULIE A. MAUPIN-FURLOW, STEVEN J. KACZOWKA,
MARK S. OU, AND HEATHER L. WILSON

Archaeal Catabolite Repression: A Gene Regulatory Paradigm

ELISABETTA BINI AND PAUL BLUM

PREFACE

Biologists divide living organisms into prokaryotes, represented by the bacteria, and eukaryotes such as ourselves. In this scheme, bacteria are considered to be the most primitive form of life. One popular idea about how life arose supposes that humans evolved from bacteria. In this age of genomics, however, DNA sequence comparisons indicate that our genes are unlike those of bacteria. Where, then, did we come from? An answer to that question comes from a truly unexpected source, life in extreme environments.

It was generally thought that boiling acid hot springs or saturating saline lakes were sterile, however, life is both present and highly successful. Many of the resident organisms are still microbes, just not bacteria. We call them archaea. Whole-genome DNA sequences of five archaeal species reveal remarkable gene matches to human genes and those of other eukaryotes. These matches occur in the most essential of the subcellular processes carried out by all organisms, the synthesis and repair of DNA, RNA, and protein. This suggests that eukaryotes evolved from archaea or perhaps that archaea and eukaryotes derive from a common ancestor.

Gene sequences aside, archaea lay additional claim to the title of ancient organism based on geologic and taxonomic considerations. The early Earth (Archean age) was a time of elevated surface temperatures. Fossil dating indicates the presence of microbes at the close of this period, suggesting that earlier forms of life from which these fossils would have derived must have been adapted to temperature extremes. Microbes called hyperthermophiles with just these abilities are still found on the Earth in geothermal springs and hydrothermal ocean vents. These extreme locations exhibit geochemistries most like that of early Earth. To understand how such organisms relate to other forms of life, taxonomic methods based on ribosomal RNA sequences have been used to create phylogenetic "trees" of life. These hyperthermophilic microbes, dominated by the archaea, exhibit the deepest phylogenetic branches, suggesting that they have undergone the longest period of evolution among extant organisms. Taken together, these ideas suggest that hyperthermophilic archaea may represent a form of the earliest type of life.

The intent of this book is to expand the general understanding of the archaea. As simple organisms with sequenced genomes, they present unique opportunities to understand better our own origins and, indeed, the origin of life. As prokaryotes they provide powerful experimental systems for genetic and molecular experimentation.

Acknowledgments

I am most grateful for the support and interest of many colleagues, including Rolf Bernander, Mike Dyall-Smith, Peter Kennelly, John Leigh, William Metcalf, Kenneth Noll, Frank Robb, Richard Shand, Dieter Soll, and William Whitman. The pioneering interests of Thomas Brock, Richard Morita, Norman Pace, Carl Stetter, Carl Wose, and Wolfram Zillig helped create my interest in extreme environments and the evolutionary implications of life native to such habitats.

INTRODUCTION

Ribosomal RNA gene sequence comparisons separate life into three distantly related groups or domains (Fig. 1). Eukaryotes constitute one domain, encompassing single- and multicelled organisms such as plants and animals. Despite their morphologic simplicity and apparent similarity, the other two domains are both prokaryotic. These domains are represented by the bacteria and the archaea. Differences in rRNA sequence are but one feature leading to this classification system. Most scientists are well versed about the evidence which distinguishes bacterial from eukaryotic life. This includes a diversity of mechanisms governing all aspects of basic cell biology, from gene organization to gene expression, from signal transduction to metabolism. What, then, beyond differences in rRNA sequence supports the separate classification of the archaea from bacterial prokaryotes and eukaryotes? As predicted by their rRNA sequence divergence and by analogy to the dissimilarity between bacteria and eukaryotes, archaea are likely to employ archaeal-specific subcellular mechanisms for conducting essential processes. While the existence of archaeal-specific subcellular mechanisms is supported by the occurrence of a unique group of archaeal orthologous genes, the functions of these genes remain unknown. The truly remarkable discovery which is the focus of this book concerns the finding that archaea use eukaryotic-like mechanisms and not bacterial mechanisms for much of their information processing functions.

The literature on bacterial prokaryotes is both extensive and diverse, however, such is not yet the case for archaea. As a recognized group, archaea are relative newcomers to the prokaryotic world. Despite their recent entry into the research arena, studies on archaea are blossoming, largely in response to the availability of whole-genome DNA sequences. Three distinctive biotypes of archaea are best known: the methanogens, the halophiles, and the hyperthermophiles. Due to the radical nature of their respective niches, these organisms have been dubbed extremophiles. Studies on life in extreme environments portend many exciting areas in biology research including studies on the origins of life and the possibility of extraplanetary organisms. Archaea have contributed, and will continue to contribute, greatly to these areas. However, phylogenetic studies of marine, dirt, and other environmental samples indicate that some archaea are not extremophilic but mesophilic members of microbial communities. More importantly, these types of

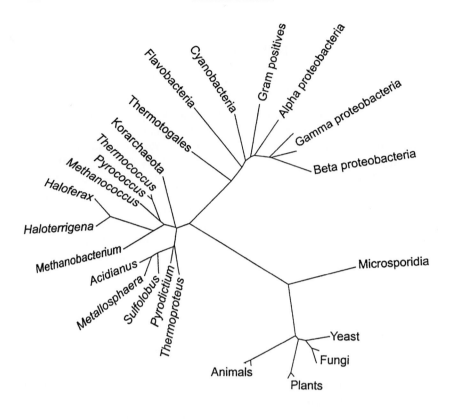

FIG. 1. Archaeal, bacterial, and eukaryotic sequences were aligned with ClustalW. The programs DNADIST, NEIGHBOR, CONSENSE, and FITCH, of the Phylip package, were used to build the tree. The alignment was bootstrapped 100 times with SEQBOOT.

archaea appear to constitute a significant proportion of the total planetary microbial biomass.

This book provides an overview of key aspects of the archaea including chapters on their genomics and phylogeny, micropaleontology, cell biology, and molecular genetics. This information should be useful to microbiology students at both the graduate and the undergraduate levels. More importantly, this text should focus the attention of researchers on the importance of archaea as model systems to address fundamental biological questions. As prokaryotes, archaea can and should be used to wield the power of haploid genetics providing cost-efficient systems for scientific investigations into life.

Section I. Extremophiles, Fossils, and the Search for Exobiology

The first chapter in this book, by Sherry Cady, provides a geologic perspective to questions about the evolution of life. Paleontological consideration of the microbial fossil record reveals a convergence between phylogenetic trees and the geologic record. The early Archean geologic age comprises a time when Earth surface temperatures were episodically elevated above those of today. New microbial fossils dating to this period suggest that early life was thermophilic. This observation fits nicely with the positioning of thermophiles and hyperthermophiles in the 16S rRNA phylogenetic tree. Of all organisms this prokaryotic biotype occurs most closely to the "root" or last common ancestor at the base of the tree, providing molecular evolutionary evidence consistent with the fossil record. Sherry Cady provides an introduction to the fossil record and its role in understanding the evolution of early life.

Paleobiology of the Archean

SHERRY L. CADY

Department of Geology
Portland State University
P.O. Box 751
Portland, Oregon 97207-0751

I. Introduction

Four billion years ago our planet began accumulating a rock record, a geological event that heralded the beginning of the Archean Eon (Fig. 1). Though the Earth itself is approximately 4550 million years old, the oldest known rocks are the 4030 million-year-old Acasta gneisses of northwestern Canada (Bowring and Williams, 1999). These rocks, and others of comparable age, indicate that the ancient Earth harbored a number of environments that could have supported life. The presence of even older minerals, 4400 million-year-old detrital zircons from the Yilgarn Craton, Western Australia (Wilde *et al.*, 2001), provides evidence

3

ADVANCES IN APPLIED MICROBIOLOGY, VOLUME 50

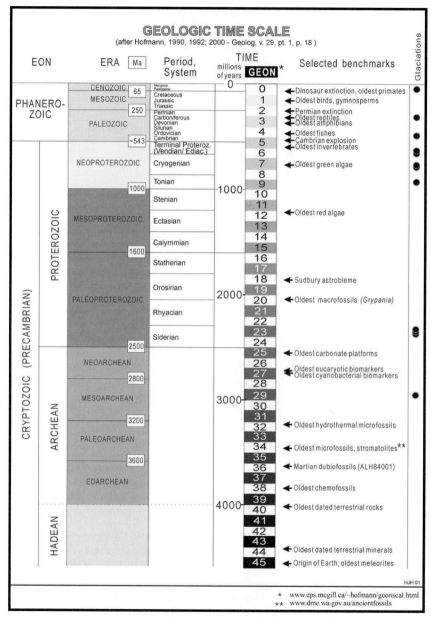

FIG. 1. Geological time scale. Geologic time is represented here by numerically designated bins of geologic time, each 100 million years long. The intervals, called geons (from
GEOlogical EON), are counted from the present. The geon scale allows for Earth history
(and that of other planets) to be viewed in a succession of large, simply named, equal time
units, in the same way that centuries and millennia are used for familiar smaller intervals.

that the Earth was tectonically active shortly after it formed. Without plate tectonic activity commencing early and persisting throughout our planet's history, it is unlikely that life would exist today. It is yet a mystery when life began; its origin is not revealed in the paleobiological record. Chemical fossil evidence in rocks from West Greenland indicate, however, that microbial life had evolved by 3700 Ma (mega-annum; meaning million years) (Rosing, 1999). The possibility exists that life emerged within a few hundred million years after crustal rocks began accumulating and began to leave traces in the paleobiological record. The Archean Eon, by definition, ends at precisely 2500 Ma.

Fossil evidence indicates that once life emerged it evolved rapidly and occupied new ecological niches as soon as they became available (e.g., hydrothermal systems, shallow coastlines, continental seas). The Archean paleobiological record is preserved in geological units that consist primarily of chemical precipitates and volcanogenic sediments. Such deposits accumulated on the edges of accreting continental cratons and along the rims of protocontinents. Although only meager bits of evidence have been discovered to date, new and significant discoveries regarding Archean paleobiology and paleoecology continue to be made.

In general, paleobiological information comes from a potentially wide range of fossil data known collectively as biosignatures. Evidence for the existence of the Archean biosphere is preserved in the fossil remains of microorganisms (i.e., microfossils), in the sedimentary structures microorganisms helped construct (i.e., stromatolites), and in the biomarker compounds, isotopic signatures, and biominerals life leaves behind (i.e., chemofossils). Bona fide microfossils, carbonaceous microorganisms preserved in three dimensions, contain structural and chemical rem- nants of cellular and extracellular components (e.g., cell walls, sheathes, exopolysaccharides), which together can provide direct evidence for life. Also, microbially influenced fabrics and structures that form in ecosystems where authigenic minerals precipitate and rock detritus accumulate provide indirect evidence of microscopic life forms. Microbially influenced accretionary growth structures and fabrics include nonlaminated microbialites and laminated biogenic stromatolites. Although both types of biogenic structures are found throughout the Archean, they usually lack microfossils. Chemical fossils ("chemofossils"), such as biomarkers and isotopic signatures, also provide indirect

---

The geon scale complements the traditional, history-based scale with complex names and unequal geologic periods, much like dynasties in history overlap centuries. The Archean Eon is defined as that interval of time that began with the formation of the oldest geological record and that ends at 2500 Ma (mega annum; meaning million years), which is the beginning of the Proterozoic Eon. [Reproduced with permission from Hofmann (1990, 1992, 2000).]

evidence for life. Biomarker compounds consist of organic molecules highly diagnostic for their parent organisms. Elements that can be isotopically fractionated by biological or biochemical processes include carbon, sulfur, and nitrogen.

Not only have a number of types of fossil biosignatures shown that life emerged and spread globally during the early Archean, but evidence for life has been reported from all of the major Archean cratons. The paleobiological record has shown that life occupied several niches during the Archean and, within a few hundred million years, diversified and became metabolically sophisticated. Although fossil evidence cannot provide precise data as to when a metabolic innovation evolved, it can reveal that a particular metabolic capability existed. Chemofossil evidence demonstrates that many major metabolic pathways evolved during the Archean. Though the data for these evolutionary innovations are scattered throughout the ancient rock record, the continuous and relatively robust paleontological record of the Proterozoic Eon (2500 to 543 Ma) has confirmed that prior to the end of the Archean, analogues of modern microbial ecosystems, including those considered extreme, were established.

Several lines of evidence suggest that the earliest microbial inhabitants occupied sediments and colonized newly formed minerals that precipitated around hydrothermal vents, environments characterized by high temperatures and an abundance of dissolved metal ions (Nisbet, 1995). As shown in Fig. 2 molecular phylogenetic analysis of the small-subunit ribosomal RNA of extant life indicates that hyperthermophiles in the bacterial and archaeal domains are the closest living relatives of the hypothetical last common ancestor (Pace, 1997). Whether life originated at high temperatures (Shock and Schulte, 1998) is not known, and evidence for a hyperthermophilic ancestry has been challenged (Doolittle, 2000; Galtier et al., 1999). The possibility exists that hyperthermophiles emerged during that brief but finite period of time after the collision that presumably formed the Earth–Moon system; after the collision (~4450 to 4500 Ma), when the Earth began to cool, surface temperatures could have hovered around 100°C for 100 to 200 million years (Sleep et al., 2001). The catastrophic conditions created by large impactor events during the latter part of the heavy bombardment period (~3900 to 3800 Ma) would also have created and sustained high-temperature habitats that could have supported hyperthermophilic communities; large impactors (>400 km in diameter) could have vaporized the ocean, while smaller impactors (>200 km in diameter) might have been able to heat the ocean above 100°C (Sleep and Zahnle, 1998). Alternatively, if life originated at lower temperatures, and then radiated and adapted to high-temperature environments, hyperthermophiles living in the subsurface portions of deep-sea

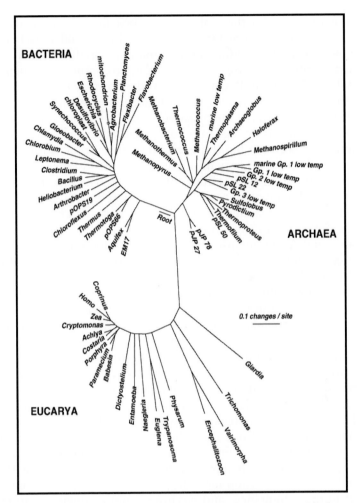

FIG. 2. Universal phylogenetic tree based on SSU rRNA sequences. The largest formal unit of the tree consists of domains that include the Archaea, Bacteria, and Eukarya. The tree can be considered a rough map of the evolutionary distance between the organisms shown. Although the nature of the hypothetical last common ancestor is not known, hyperthermophilic and thermophilic microbial lineages are clustered around the root of the microbial domains. Constructed by comparison of 16S small-subunit ribosomal RNA sequences, the tree includes 64 rRNA sequences representative of all known phylogenetic domains. [Reproduced with permission from Pace (1997).]

hydrothermal systems may have been the only microbial communities that could have survived the impact-induced ocean sterilizing events (Maher and Stevenson, 1988).

The Earth has always been a volcanically active planet, and hydrothermal activity prevailed after liquid water became stable at the Earth's surface. Estimates of a threefold greater heat flux on the early Earth suggest that there was a greater amount of hydrothermal activity during the Archean (Turcotte, 1980). The combination of hydrothermal activity and life can produce macroscopically and microscopically recognizable biosignatures. Where molten magma intrudes within a few kilometers of the Earth's surface, it heats subsurface fluids that rise and erupt through hydrothermal effluents at deep-sea vents on the ocean floor. In continental settings, fluids emerge into hot springs and geysers. In any case, the interaction between hot fluids and the crustal rock through which it passes alters the chemistry of the fluid. Dissolved ions and gases become concentrated in the hot fluids that boil upon ascent, and minerals precipitate in the open fractures through which the boiling fluids pass. Reduced gases, if separated from boiling hydrothermal fluids, find alternative pathways to the surface, where they form acidic fumaroles (i.e., gas vents). The remaining dissolved ions precipitate as metal-rich hydrothermal deposits when subaqueous oceanic hydrothermal fluids mix with overlying seawater. If the fluids emerge from subaerial hot spring and geyser effluents, hydrothermal minerals precipitate as sinter deposits when the fluids cool and evaporate. Hyperthermophilic and thermophilic microbial communities are likely to have thrived in these types of environments since life originated.

Paleontological interpretations of biosignatures rely upon comparisons with modern counterparts. Comparisons with modern-day organisms assessed within a taphonomic framework that considers how microbial cells become degraded and altered with time places paleobiological evidence in a phylogenetic and functional framework. Paleoenvironmental interpretations place fossil biosignatures in a geological context characterized by physical and chemical constraints. Because rocks and the organic remains they contain become altered with time as a result of diagenetic and metamorphic processing, it becomes more difficult to draw biological and environmental analogies from biosignatures found in older deposits. Furthermore, it is unlikely that extant microbial communities or their environments are exact homologues of their ancient counterparts, given the expanse of time and the secular changes that have occurred on Earth since ancient microbial lineages originated. Even so, the physical and chemical processes that operated in ancient ecosystems continue to do so today. By applying what is known about how microbial life influences its environment, and vice versa, we can continue to extract meaningful paleobiological information from the ancient rock record.

The geological record indicates that the Earth evolved in a number of ways during the Archean. On a global scale, the crust differentiated from the core and solidified. The oceans were established; the atmosphere stabilized. The geological record indicates the present style of plate tectonics ensued by the end of Archean. The combination of rock weathering, crustal generation, and crustal recycling throughout the Archean produced a variety of environments that rapidly became inhabited as microbial communities spread worldwide. As illustrated in Fig. 3, hydrothermal systems would also have been in existence since crustal rocks began forming in the Archean.

II. Historical Review

A. Main Phases of Archean Paleobiological Studies

An analysis of the nature and timing of publications related to Archean paleobiology reveals that research in the field has undergone three main phases since the midnineteenth century (Schopf and Walter, 1983). As discussed recently by Hofmann (2000), each phase was "defined by a different tempo of significant contributions to our knowledge of the most ancient fossil record."

Archean paleobiology began with the published description of *Eozoön Canadense* (Dawson, 1875), laminated structures of putative biological origin. The *Eozoön Canadense* structures were subsequently dismissed as biogenic structures, but the controversy associated with their origin reflects the nature of paleontological research. In fact, debate regarding the origin of the earliest reported structures is similar to present-day controversies associated with the claims of putative microfossils in extraterrestrial materials (McKay *et al.,* 1996) and the origin of stromatolite structures in general (Grotzinger and Rothman, 1996). The first phase ended just after the beginning of the twentieth century with the discovery of the first real stromatolites in Archean rocks at Steep Rock Lake, Ontario (Lawson, 1912; Walcott, 1912).

During the second phase, between 1912 and the early 1960s, the first Archean microfossils were reported from iron-bearing chert formations in northern Minnesota (Gruner, 1923, 1924). These early reports met with skepticism (Hawley, 1926), and over 30 years passed until the first unequivocal microfossils were reported from the $\sim$2100 million-year-old chert of the Gunflint Formation (Tyler and Barghoorn, 1954). A second discovery of microfossils in Proterozoic-age chert deposits (Barghoorn and Schopf, 1965) compelled most paleobiologists to study rocks of similar age and composition, a sound strategy that continued for several decades. Convincing Archean stromatolites were discovered during this period, and again, initial reports were questioned, in some cases even by the scientists who originally discovered the objects (for a complete list of references see Hofmann, 2000). The first actively

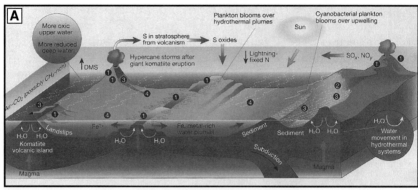

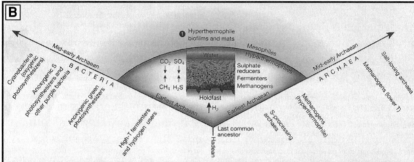

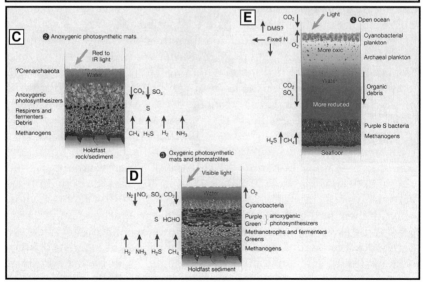

FIG. 3. Map of microbial ecology established during the Archean Eon. (A) Numbers refer to typical settings in the habitat model: (1) mid-ocean ridges (chemotrophic communities, including hyperthermophiles); (2) shallow coastal waters (oxygenic and

forming stromatolites were discovered during this period along the shoreline of Shark Bay, Western Australia (Logan, 1961).

Widespread interest in the history of early life increased exponentially during the third phase of study, which began in the mid 1960s and continues to the present. Detailed investigations of stromatolites in modern ecosystems led to the development of actualistic models for Archean and Proterozoic paleobiology (e.g., Hofmann, 1973; Walter 1976). The effects of diagenesis on the fidelity of microbial biosignatures were also systematically addressed during this period (e.g., Knoll and Golubic, 1979). Numerous discoveries were made of Archean and Proterozoic microfossils and stromatolites, and of their nonbiological mimics (Schopf, 1999; Hofmann, 2000). Decades of intensive geological and paleobiological studies of Archean sediments, and the carbonaceous matter they contain, have refined our views of the emergent biosphere (e.g., Lowe and Byerly, 1999). The nature of early microorganisms, as revealed by molecular phylogenies and hypotheses regarding early Earth environments, has stimulated new theories of life's origins. A new understanding of the types of habitats in which life could have thrived during the Archean has emerged (e.g., Buick and Dunlop, 1990; Bengtson, 1994; Mojzsis and Harrison, 2000; Nisbet and Sleep, 2001).

The search for the oldest evidence of terrestrial life has intensified as a result of the search for life beyond Earth. Three critical areas where paleobiological research can be expanded has arisen from extraterrestrial search strategies for Mars (Cady, 1998): (1) recent discoveries of novel ecosystems have increased the number of potential paleobiological repositories that may harbor evidence of early microbial communities on Earth and elsewhere—these ecosystems must be systematically investigated to extract the maximum amount of paleobiological and paleoenvironmental information from ancient analogue deposits; (2) a fundamental understanding of the physical, chemical, and biological forces that contribute to the preservation of biogenic signatures in systems whose geochemical context is well understood is needed if we are to have any chance of deciphering fossilized evidence of life from

---

anoxygenic photosynthesis); (3) oxygenic photosynthetic mats and stromatolites around oceanic komatiite shields and continental andesitic island arc volcanoes; (4) open ocean [photosynthetic plankton, muds in the seafloor (methanogens and organisms dependent upon the sulfur cycle)]. The arrows in the individual habitats show the direction of fluid movement. (B) Evolutionary heritage follows the standard evolutionary model based on the universal phylogenetic tree shown in Fig. 2. (C–E) Columns show possible mat communities and biofilms along with chemical products; numbers refer to typical settings in the habitat model. [Reproduced with permission from Nisbet and Sleep (2001).]

small samples returned from other planets, whose geological context will be vague at best; and (3) research regarding the origins of biosignatures requires characterization at the submicroscopic scale for a number of reasons—to understand the geomicrobiological processes that led to the lithification of microbial mats and biofilms, to identify the mechanisms by which microbial cells are fossilized, and to refine and expand the chemofossil database. Understanding which extrinsic environmental factors and intrinsic biochemical attributes are responsible for the preservation of biosignatures will provide the cornerstone needed to distinguish microfossils and stromatolites from their nonbiological mimics (Cady, 2001). The challenges faced in detecting ancient life on Earth are the same challenges faced in our search for life on other planetary bodies (i.e., Mars and Europa). These challenges are forcing us to refine our ability to detect evidence of life in ancient rocks (e.g., Conrad and Nealson, 2001).

B. Reviews of Archean Paleobiology

A number of seminal papers on Archean paleobiology have been published during the past two decades. In the book *Earth's Earliest Biosphere,* edited by Schopf (1983), all aspects of Archean paleobiology are thoroughly reviewed, including paleobiological aspects of the origin of life (Chang *et al.,* 1983), the Archean organic geochemical (Hayes *et al.,* 1983) and carbon, sulfur, hydrogen, and nitrogen isotopic records (Schidlowski *et al.,* 1983), Archean biochemistry (Gest and Schopf, 1983; Chapman and Schopf, 1983), Archean stromatolites (Walter, 1983), microfossils (Schopf and Walter, 1983), and environmental (Walker *et al.,* 1983) and ecological (Hayes *et al.,* 1983; Schopf *et al.,* 1983; Schopf, 1999) evolution. The book *The Proterozoic Biosphere,* edited by Schopf and Klein (1992), includes updates on Archean paleobiology (Schopf and Klein, 1992), paleobiological aspects of origin of life (Kasting and Chang, 1992), Archean microfossils (Schopf, 1992), and geological and environmental evolution (Lowe, 1992). Major implications of the Archean microfossil stromatolite (Walter, 1994) and organic geochemical (Hayes, 1994) records are discussed in the Nobel Symposium Volume *Early Life on Earth,* edited by Bengtson (1994). Additional noteworthy publications include a recent review by Hofmann (2000) of Archean stromatolites; a review by Knoll (1996) of paleontological evidence for Archean fossils, with special emphasis on their paleoecology, biostratigraphy, and evolution; an overview of theories and paleobiological evidence related to the origin of life by McClendon (1999); and critical reviews of the criteria by which the biogenicity of ancient microfossils and stromatolites are assessed by Buick (1981, 1990).

III. Evidence for Life—Microbial Biosignatures

Biosignatures indicative of ancient life include cellularly preserved microfossils and carbonaceous cellular and microbial mat remains, molecular fossils known as biomarkers, biologically fractionated isotopic signatures, and stromatolites and microbialites. Each fossil type provides a unique indicator for some aspect of life. When found together in the geological record, the different types of biosignatures can provide mutually reinforcing lines of evidence; multiple biosignatures strengthen the case for life. Bona fide microfossils, characterized by chemical and structural biogenic attributes, provide direct evidence of life.

A. Bona Fide Microfossils

Bona fide microbial fossils retain a portion of the biochemical and structural remains of their cellular components. The remains are recognizable either by their physical structure (as biomarker compounds) or by their degree of complexity (higher-order structures such as layered cell walls). Bona fide microfossils include permineralized and nonmineralized cellular remains (Schopf and Walter, 1983). Permineralized cells retain carbonaceous compounds composed of complex organic biopolymers that, by definition, retain enough morphological fidelity to be recognizable (Schopf, 1975). Newer instruments that allow the detection and characterization of structural and chemical evidence for life from the same object, particularly when that object looks like the remains of extant life, will have a significant impact on the ability to identify taxonomic affinity.

B. Microbialities and Biogenic Stromatolites

A long-standing controversy exists over the definition and meaning of the term stromatolite. The debate centers on whether a genetic or descriptive definition should be used. The potential for stromatolites to form in a variety of environments, including extreme ecosystems, via various contributions from different types of microorganisms (phototactic and nonphototactic) demonstrates the restrictions imposed by genetic definitions of stromatolites. A nongenetic definition for stromatolites recommended by Semikhatov and colleagues (1979) states that a stromatolite is "an attached, laminated, lithified sedimentary growth structure, accretionary away from a point or limited surface of initiation." This description allows for the exploration of the various roles of biological and nonbiological processes in stromatolite accretion. Even though stromatolites may have multiple or even indeterminate origins, their basic geometric and textural properties are included in Semikhatov's definition. Another definition by Hofmann (2000) acknowledges the potential for biology to be involved in the morphogenesis of the structures without compromising the nongenetic physical description. Hofmann (2000) asserts that "stromatolites are morphologically circumscribed accretionary growth structures with a primary lamination that is, or may be, biologically influenced (biogenic)." With either definition, the aim is to determine which components of stromatolites or microbialites indicate the involvement or former presence of life during their construction. The potential for stromatolites to harbor evidence of early microbial life is a matter of particular relevance to the Archean. Stromatolites can preserve evidence of the early biosphere and, more specifically, of benthic ecosystems on the early Earth (Walter, 1994). Because stromatolites are macroscopic structures, they are potentially recognizable by remote imaging on planetary rovers and can assist in determining locations to search for past extraterrestrial life.

C. CHEMOFOSSILS—BIOMARKERS AND ISOTOPES

Biomarker compounds, organic molecules that are highly diagnostic for their parent organisms, are particularly revealing with regard to metabolic innovations. As our understanding of the amount of biodiversity in modern ecosystems improves and our database of unique biomarker compounds attributed to individual taxa expands, biomarker compounds will continue to reveal much about early life. Stable isotopes, when found in abundance, are generally metabolically fractionated. Elemental isotopic indicators of life, which can sometimes be linked to rather specific metabolic processes, are relatively resistant to alteration by geological processes such as thermal metamorphism. Subsequent discoveries of reliable chemofossils throughout Earth history are needed to establish a continuous record for assessing biomarker compounds (e.g., Summons and Walter, 1990).

IV. Criteria for Assessing Biogenicity

Since microfossils have the potential to provide definitive evidence for life, robust criteria for assessing their validity have been established. In general, acceptable evidence of past life must meet five stringent criteria (Schopf and Walter, 1983) that include demonstrating the following: (1) the source (i.e., provenance) of the rock sample that contains the putative biological features must be capable of preserving paleobiolog-ical information, (2) the age of the rock sample must be known with appropriate precision, (3) the putative biological features must be in-digenous to the rock sample within which they were found, (4) the putative biological features must be syngenetic with (i.e., formed at the same time as) the primary mineral phases of the rock sample, and (5) the features of the putative fossils must be biogenic in origin. Proving the biogenicity of fossil evidence is often the most challenging criterion.

A number of criteria for assessing the biogenicity of microstructures found in ancient rocks have been proposed (e.g., Muir, 1978; Cloud and Morrison, 1979; Awramik *et al.,* 1983; Schopf and Walter, 1983; Buick, 1990). The most rigorous classification system for microfossils, as pro-posed by Buick (1990), states that structures can be considered biogenic if they (1) occur in a petrographic thin section, (2) occur in sedimen-tary or low-grade metasedimentary rock, (3) are larger than the small-est extant free-living microorganisms ($\sim$0.01 μm^3), (4) are composed of kerogen, (5) occur with others of similar morphology, (6) are hollow, and (7) show cellular elaboration. This classification system differen-tiates pseudofossils that do not meet criteria 1 through 4, dubiofossils (Hofmann, 1972) that satisfy only criteria 1 through 4, possible micro-fossils that meet criteria 1 through 5, and probable microfossils that fulfill criteria 1 through 6. Existing criteria for demonstrating microfos-sil biogenicity continue to be modified, refined, and challenged (e.g., Westall, 1999).

A parallel topic for paleobiological study includes experiments de-signed to determine the mechanisms by, rates at, and conditions un-der which pseudofossils form and become preserved in the geological record. Objects not accepted by all specialists in the field as bona fide evidence of life continue to be reported in peer-reviewed publications. But as noted by Hofmann (2000), such objects should not be discounted completely; more precise criteria will be forthcoming as new instru-ments for detecting and characterizing life's biosignatures are devel-oped. Until such time, though, only the most stringent criteria should be used when assessing the biogenicity of evidence for ancient terres-trial life or, as in the case for Mars, extraterrestrial life.

V. Biogeochemical Interactions and Biosignature Formation

Microorganisms interact with their environment in numerous ways that alter the chemistry and physical characteristics of their environ-ment. The environment also creates circumstances that favor the preser-vation of microbial biosignatures. The strategy used by paleobiologists to explore the ancient fossil record of microbial life on Earth is based upon locating specific rock types and paleoenvironments thought to have the highest potential for capturing and preserving fossil biosignatures. The most informative assemblages of organically preserved microfossils are found in mid to late Proterozoic sediments deposited in marine and lacustrine environments. Microbial fossils in these deposits occur either as three-dimensional cellularly preserved (permineralized) forms embedded in a relatively stable mineral matrix or as two-dimensional compressed or flattened forms (acritarchs) preserved in fine-grained, typically clay-rich detrital sediments. Microbial permineralization occurred primarily in shallow evaporative peritidal environments of elevated salinity (e.g., Knoll, 1985), whereas acritarchs were preserved primarily in fine-grained siliciclastic sediments that formed in a wide range of marine environments. Although there are probably no bona fide acritarchs known from Archean deposits (H. Hofmann, personal communication), permineralized fossils, albeit rare, have revealed much about the antiquity of life and the nature of the early Earth.

In recent years, the discovery of microbial life in a number of extreme environments, along with the realization that extreme ecosystems have existed throughout Earth's history, has expanded the search for microbial biosignatures. Exopaleobiological and astropaleobiological search strategies also focus on extreme ecosystem deposits, especially hydrothermal deposits, as potential extraterrestrial paleobiological repositories (Walter and Des Marais, 1993; Walter et al., 1996, 1998; Farmer and Des Marais, 1999). Ancient sedimentary chert deposits have produced permineralized microfossils with a high cellular fidelity. A wide range of rock types has revealed evidence of life in the form of carbon and sulfur isotopic signatures. Microbial biosignatures are likely to have been preserved in any type of rock that was altered by sedimentary processes (e.g., infiltration of mineral-charged fluids in cracks and fissures in igneous and metamorphic rocks). Indeed, the voluminous subsurface biosphere could harbor biosignatures in any crevice where fluids have passed.

VI. Extreme Ecosystems

Extreme environments often represent modern analogues for ancient ecosystems. The environmental extremes of such ecosystems exclude higher-order taxa that would normally compete with or graze upon microbial mats or biofilms. Much remains to be learned regarding the detection and proper interpretation of paleobiological and paleoecological information from ancient analogues of modern extreme environments— a reliable and continuous database of microbial biosignatures from sequentially older deposits has yet to be established. A continuous fossil record is important because of the deleterious effects of diagenetic and taphonomic processes on the fidelity of structural and chemical biosignatures. Modern extreme ecosystems provide an opportunity to study the processes by which paleobiological information is preserved in the fossil record. For example, a study of modern hot spring deposits in Yellowstone National Park (Cady *et al.*, 1995; Cady and Farmer, 1996) revealed that hyperthermophilic biofilms can influence the microstructural development of high-temperature siliceous sinter deposits previously considered to have formed abiologically. The ways in which microbial communities interact with their environment and impact global biogeochemical cycles can also be studied in real time in extreme ecosystems. Extreme ecosystems also provide geochemical and physical constraints to refine experimental studies.

VII. Geological and Paleobiological History of the Archean (4000–2500 Ma)

Discoveries from the geological and paleobiological records that provide insight regarding the potential to find evidence of Archaea in the Archean paleobiological record are discussed below. Unfortunately, little direct evidence has been discovered regarding the distribution and occurrence of Archaea in the rock record, and future systematic searches that involve extreme ecosystem analogue studies are clearly warranted.

A. Oldest Dated Terrestrial Rocks and Minerals

Isotopic data from meteorites and the Moon place the age of the solar system at approximately 4600 Ma and the age of the Earth at approximately 4560 Ma (Patterson, 1956; Papanastassiou and Wasserburg, 1971). Evidence that crustal formation and weathering processes were active within the first few hundred million years after the Earth accreted is found in the oldest minerals (~4300 Ma), detrital zircon crystals from the Yilgarn Craton, Western Australia. Although these detrital crystals are the oldest indigenous objects found on Earth (Wilde *et al.*, 2001), they reside in younger sedimentary deposits, having been eroded out of the rocks in which they crystallized. Fortunately, the oxygen isotopic signatures of the detrital zircon crystals reveal that they crystallized from magmas containing a significant component of reworked continental crust that formed in the presence of water near the Earth's surface (Mojzsis *et al.*, 2001). Even though all crustal rocks of Hadean age were recycled, isotopic evidence from the oldest minerals has revealed much about the way in which the early Earth operated.

B. OLDEST TYPES OF CRUSTAL TERRANE

Three types of rocks make up the oldest crustal terranes: gneissic terranes that have been highly metamorphosed, greenstone terranes that consist of volcanic rock overlain by metasediments, and cratonal margin deposits (Lowe and Ernst, 1992). Recall that the oldest rocks are from gneissic terranes (e.g., Mojzsis *et al.*, 2001; Bowring *et al.*, 1989). Greenstone terranes harbor the oldest microbial biosignatures.

The oldest known greenstone belt, the 3800 million-year-old Isua supracrustal belt in West Greenland, consists of thick units of volcanics, conglomerates, banded iron formation, chert, and detrital sediments. As discussed below, the oldest putative chemofossils have been found in Isua supracrustal rocks. A younger generation of approximately 3500 to 3200 million-year-old greenstone belts includes the Barberton Greenstone Belt and related rocks in the Kaapvaal Craton, South Africa, and greenstone belts in the eastern Pilbara Block, Western Australia. The lower units of these rock complexes consist primarily of volcanic rocks; the upper units include sedimentary rocks, such as silicified or carbonated volcanic tuffs, cherts, and evaporites. These sedimentary rocks, as discussed below, contain the oldest reported stromatolites and carbonaceous microfossils. The absence of large stable cratons with extensive shallow continental shelves during most of the Archean limited the distribution of ancient continental margin and terrestrial microbial ecosystems. The ages of crustal blocks formed during the Archean Eon indicate that nearly 90% of the crust formed during the Neoarchean, between 2700 and 2500 Ma (e.g., Lowe, 1994).

While it is apparent that the crustal lithosphere expanded dramatically during the Archean, secular changes also altered the composition of the Archean atmosphere, hydrosphere, and biosphere (e.g., reviewed by Lowe, 1994; Des Marais, 1994; Des Marais, 1997; Mojzsis and Harrison, 2000; Nisbet and Sleep, 2001). Although details are still forthcoming, it is apparent that the Archean hydrosphere was characterized by a lower pH (i.e., a higher [Fe(II)]), and the Archean atmosphere was characterized by a higher partial pressure of CO_2 and a significantly lower partial pressure of O_2 (Holland, 1984). Without an oxygen-rich atmosphere, the Earth would have experienced a higher ultraviolet (UV) flux during the Archean. The sun was also 30% less luminous at 3800 Ma than it is today (Kuhn *et al.*, 1989). These conditions at the Earth's surface would have restricted the number of suitable environments for microbial inhabitants. Given that photosynthesis directly powers >99% of the modern biosphere's productivity (e.g., Des Marais, 1997), it is clear that microbial life during the Archean differed in significant ways from that of today.

C. CHEMOFOSSILS, WEST GREENLAND

Chemofossil evidence indicative of autotrophic carbon dioxide fixation has been found in ~3700 million-year-old seafloor sediments from the Isua supracrustal belt in West Greenland (Rosing, 1999). Even older chemofossil evidence for autotrophy has been found in ~3850 million-year-old banded iron formations from Akilia Island, West Greenland (Mojzsis et al., 1996). As noted by Mojzsis and Mark (2000), the age of the sediments within which the oldest chemofossils are found is significant. These rocks would have been deposited during the period of intense meteor bombardment as recorded on the Moon at 3800 to 3900 Ma, which coincided with the period of time when liquid water existed on the surface of Mars. Whether these chemofossils represent the oldest evidence of life on Earth is still debated. The carbon isotopic signatures were obtained from carbon blebs that occur as inclusions in apatite. A more detailed look at the occurrence could not substantiate the previous claims of the age and origin of the host rock (Myers and Crowley, 2000). The discovery by Rosing (1999), however, of carbon of possible biogenic origin in >3700 million-year-old schists interpreted as metasedimentary marine rocks in the Isua greenstone belt provides more plausible evidence of life. These rocks are less altered by metamorphic processes, and absolute ages have been obtained from all of the tectonostratigraphic units that occur in association with the metasediments. It is interesting to note that the carbon isotopic values for the carbonaceous inclusions in apatite from ~3770-Ma Isua sediments ($\delta^{13}C = -30$ per mil) and the ~3850-Ma banded iron formation from Akilia island ($\delta^{13}C = -37$ per mil) closely match those from a variety of younger, less metamorphosed rock. Mojzsis et al. (1996) propose that the simplest interpretation indicates the presence of diverse photosynthesizing, methanogenic, and methanotrophic bacteria on Earth before 3850 Ma (Mojzsis and Arrhenius, 1998; Mojzsis and Harrsion, 2000). However, as discussed below, a more significant ^{13}C depletion (up to −60 per mil) in the organic fraction of sediments is attributed to the combination of phototrophs, methanogens, and methanotrophs. To date, chemofossils provide the only evidence consistent with life in Eoarchean-age (4000- to 3600-Ma) rocks.

D. OLDEST MICROFOSSILS

The oldest microbial fossils and stromatolites discovered to date have been found in greenstone belts of two Archean terranes of nearly the same age in the Pilbara Craton in Western Australia and in the Kaapvaal

Craton of South Africa. The nature of the microfossils indicates that they represent distinctly different types of microbial communities.

Microfossils from the Pilbara Block of northwestern Western Australia consist of cyanobacterium-like filaments from the ~3465-Ma Apex Formation (Fig. 4) and sheath-enclosed colonial unicells of the Towers Formation (Schopf and Packer, 1987; Schopf, 1993). The cellular level

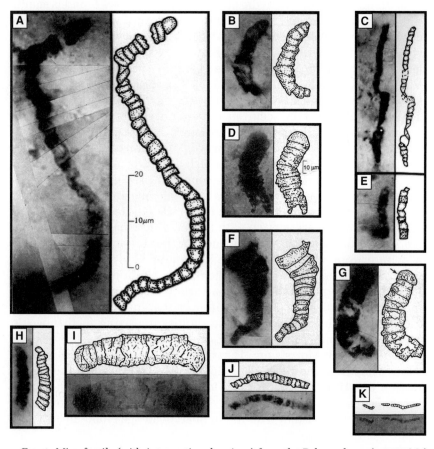

Fig. 4. Microfossils (with interpretive drawings) from the Paleoarchean (~3465-Ma) Apex Chert Formation, Pilbara Block, Western Australia. All of the microfossils, photographed under plane light in thin sections, are the representative holotypes of microbial taxa. n. gen., new genera; n. sp., new species, (A) *Primaevifilum amoenum* Schopf (1992). (B) *Primaevifilum conicoterminatum* Schopf (1992). (C) *Eoleptonema apex*, n. sp. (D) *Archaeoscillatoriopsis maxima*, n. gen., n. sp. (E) *Primaevifilum minutum*, n. sp. (F) *Primaevifilum attenuatum*, n. sp. (G) *Primaevifilum laticellulosum*, n. sp. (H) *Archaeoscillatoriopsis disciformis*, n. gen., n. sp. (I) *Archaeoscillatoriopsis grandis*, n. gen., n. sp. (J) *Primaevifilum delicatulum* Schopf (1992) (holotype). (K) *Archaeotrichion septatum*, n. sp. [Reproduced with permission from Schopf (1993).]

of organization, morphological complexity, and similarity to younger microorganisms support a biogenic origin for the objects (Schopf, 1993). All of the Apex fossils occur in subangular to rounded siliceous sedimentary clasts that are less than 1 mm to a few millimeters in diameter. As shown in Fig. 4, 11 taxa of filamentous microfossils have been described from the Apex chert. Though the microfossils were reported as being carbonaceous, hence considered bonafide, this claim has recently been challenged (Brasier *et al.*, 2001). The irregular distribution and random orientation of the solitary filaments in the clasts preclude speculation as to whether the microfossils represent a benthic or planktonic community. The filaments are surrounded by fine-grained particles of kerogen (i.e., carbonaceous matter) hypothesized to be the remains of mucilaginous extracellular components. Like most ancient fossil finds, the Apex microfossil assemblage is poorly preserved. Schopf (1993) noted that less than 1% of the filaments detected in the deposit warrant detailed study and formal description. Although most of the Apex microorganisms resemble trichomic oscillatoriacean cyanobacteria, their specific taxonomic affinity cannot be established because of the profound gap (865 million years) in the fossil record between the Apex fossil assemblage and younger, morphologically similar microfossils. An unbroken fossil record of diverse cyanobacterial families begins in the Neoarchean with the ~2600-Ma Campbell Group fossils (Altermann and Schopf, 1995).

Microfossils of the Kromberg Formation in the ~3259-Ma uppermost Hooggenoeg chert from Kaapvaal Craton of South Africa include two distinct populations of narrow, nonseptate bacterium-like filaments (Fig. 5). The most abundant filaments are thread-like ones reported by Walsh and Lowe (1985), ranging in diameter from 0.1 to 0.6 μm and in length from 10 to 150 μm. The other population of tubular filaments has diameters of 1.4 to 2.2 μm and lengths of from 10 to 150 μm. Walsh (1992) noted that the fossiliferous filaments, especially the tubular ones, display morphologies similar to those of modern filamentous bacteria. Different types of filaments co-occur with one another, and the filaments often extend from one lamina to another, as do motile filamentous bacterial and cyanobacteria in layered microbial mats (Walsh, 1989, 1992). Preservation of the cellular morphology of many of the filaments has been attributed to the presence of pyrite crystallites that encrust the cellular remains. Walsh (1992) assessed the biogenicity of the filamentous objects using the criteria discussed previously as proposed by Buick. For example, although the filament shown in Fig. 5D is interpreted as a sheathed set of trichomes (i.e., Fig. 5E), the filament is classified as a dubiofossil since it does not occur with others of similar morphology. Solid filamentous objects, like the ones shown in Figs. 5A and B, are classified as possible microfossils. Only hollow filaments, like

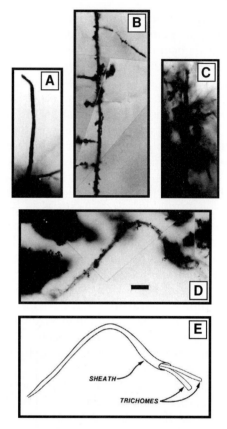

FIG. 5. Filamentous microfossils (with interpretive drawing) from the Paleoarchean (∼3416-Ma) Onverwacht Group, Barberton Mountain Land, South Africa. (A) Solid thread-like filament (possible microfossil). (B) Solid filament with hair-like projections and apparent branching (microfossil). (C) Hollow cylindrical filament (bone fide microfossil). (D) Solitary large hollow filament with trichomes extending from the flaring end (dubiofossil). (E) Interpretive sketch of D. Scale bar equals 5 μm for A and 12 μm for B–E. [Reproduced with permission from Walsh (1989, 1992).]

the one shown in Fig. 5C, are classified as bone fide microfossils. The filamentous microfossils are associated with fossilized evidence of microbial mats or biofilms preserved as various types of carbonaceous matter (Fig. 6). These varieties of carbonaceous matter are relatively common compared to the microfossil-like objects. The presence of layered microbial communities, however, suggests that taxis behavior (phototaxis and chemotaxis) cannot be distinguished in fossilized mats from the rock record. Therefore it is not yet known whether the organisms occupied shallow- or deep-water environments, and the exact details of their occurrence remain unresolved.

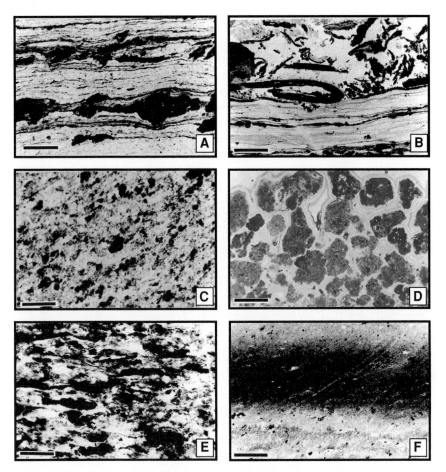

FIG. 6. Variety of carbonaceous matter in cherts from the Paleoarchean (~3416-Ma) Onverwacht Group, Barberton Mountain Land, South Africa. Plane-light photomicrographs of thin sections of rock ground to a 30-μm thickness. (A) Fine carbonaceous microbial mat-like laminations with scattered simple and composite grains in a fossiliferous sample. Scale bar = 200 μm. (B) Mat-like laminations folded over on themselves and loose detrital fragments of carbonaceous laminations. Scale bar = 1 mm. (C) Layer of simple carbonaceous grains that overlies mat-like laminations in a fossiliferous sample. Scale bar = 200 μm. (D) Layer of composite carbonaceous grains with a botryoidal coating of silica cement. Scale bar = 1 mm. (E) Partially flattened carbonaceous wisps in massive black chert. Scale bar = 200 μm. (F) Layer of very fine grained cloudy carbonaceous matter in a fossiliferous sample. Scale bar = 200 μm. [Reproduced with permission from Walsh and Lowe (1999).]

E. MICROFOSSILS IN HYDROTHERMAL DEPOSITS

The oldest putative hydrothermal fossils were recently discovered in a 3235 million-year-old deep-sea volcanogenic massive sulfide deposit that formed in a submarine thermal spring system found preserved in the Pilbara Craton of Australia. Pyrite crystals located along their entire length encrust the filaments. Morphologically, the mineralized filaments appear to be thread-like, unbranched, and of uniform thickness along their length (0.5–2 μm in diameter, up to 300 μm long) (Fig. 7). The filaments range from straight to sinuous and sharply curved, and they are intertwined when found in high numbers. Rasmussen (2000) proposed that their biogenicity is demonstrated by their sinuous morphology, lengthwise uniformity, intertwined habit, and morphological attributes comparable to those of Archean and younger microfossils. Their mode of occurrence indicates that the microorganisms were probably chemotrophs that inhabited the pores and crevices of rocks at shallow depths below the seafloor. The presence of mineralized filaments preserved in hydrothermal silica within a volcanic massive sulfide deposit is consistent with a thermophilic community preserved in an ancient subseafloor hydrothermal system. The probable absence of dissolved oxygen in the deep Archean ocean and the abundance of reduced chemical species in seafloor hydrothermal fluids suggest that chemical pathways for metabolic processes were probably anaerobic. The abundance of reduced sulfur in the hydrothermal fluids, as evidenced by the encrustation of filaments by pyrite and iron sulfide, suggest that sulfur may have been central to energy-yielding reactions for microbial growth. Although there is no direct evidence that the mineralized filaments are the fossilized remains of Archaea, the geological setting is consistent with such an interpretation. If the mineralized filaments are subsequently deemed to be compelling evidence of microbial life, the discovery will extend the known range of submarine hydrothermal biota by more than 1595 million years. The oldest known hydrothermal fossil microorganisms are found in the Early Proterozoic (1640-Ma) McArthur River (HYC) lead–zinc–silver deposit of northern Australia (Oehler, 1976; Oehler and Logan, 1977). Hydrocarbon biomarkers from the ore and associated sediments indicate the former presence of sulfide oxidizing bacteria (Logan *et al.*, 2001).

F. OLDEST STROMATOLITES

Nine occurrences of stromatolites from the Paleoarchean are known (e.g., Hofmann, 2000) and their biogenicity has been challenged (Buick *et al.*, 1981; Buick, 1990; Lowe, 1994). While it has been argued that

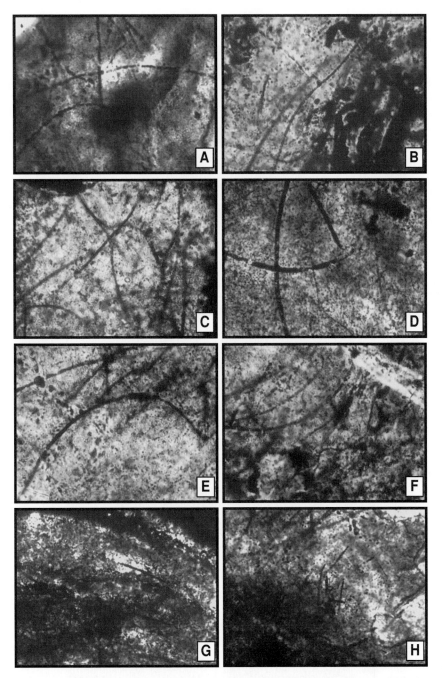

FIG. 7. Filaments from the Paleoarchean (∼3235-Ma) volcanogenic massive sulfide deposit from the Pilbara Craton of Australia. Plane-light photomicrographs of thin sections. (A–F) Straight, sinuous, and curved morphologies, some densely intertwined. (G) Filaments parallel to concentric layering. (H) Filaments oriented subperpendicularly to banding. Scale bar = 10 μm. [Reproduced with permission from Rasmussen (2000).]

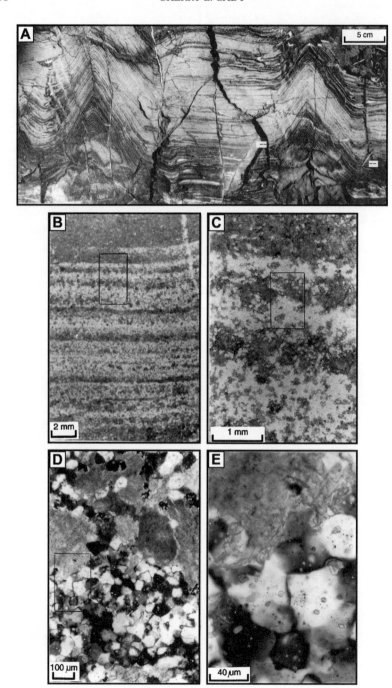

similar structures could be formed by inorganic precipitation and soft-sediment deformation (Lowe, 1994), others (Buick *et al.*, 1981; Buick, 1990, Hofman *et al.*, 1999) have argued that at least some structures are neither deformational nor precipitational and are likely to have had a biological origin. Lowe (1995) noted that the diagnostic criteria for biogenicity could be equally applied to "abiogenic geyserites" and speleothems. Interestingly, geomicrobiological studies of "abiogenic structures" in modern ecosystems where columns and spicules of geyserite form around modern hot springs (Cady *et al.*, 1995; Cady and Farmer, 1996) and columnar speleothems ("pool fingers") form in submerged pools in caves (Boston *et al.*, 2001) indicate that microbial biofilms have contributed, in part, to the microstructural and morphological development of these structures.

A recent reexamination of conically laminated stromatolites from the Warrawoona group of the North Pole Dome by Hofmann and colleagues (1999) provides an illustrative example of ancient stromatolite structures that record a biogenic contribution to their construction (Fig. 8). Evidence that biofilms contributed to the formation of the structures includes the nature of the laminations of the conical pseudocolumns (Fig. 8B), their microstructure (Fig. 8C), the distribution of similar forms over tens of square kilometers, and their similarity to younger ancient and modern biogenic stromatolites. As proposed by Hofmann (2000), until a known genesis for a stromatolitic structure is demonstrated, "Structures that are possibly partly biogenic should continue to receive appropriate scrutiny deserving of laminated dubiofossils [whose origins are unknown], because they may have relevance in elucidating the Archean biosphere."

Fɪɢ. 8. Conically laminated stromatolites from the Paleoarchean (∼3450-Ma) Warrawoona Group, North Pole, Western Australia. (A) Outcrop exposure of a section oriented approximately perpendicularly to bedding and showing two pseudocolumns composed of uniform and wavy conical layers. Horizontal layers in between the conical stromatolites display a more irregular layer thickness. (B–E) Examples of the microstructure of the conical stromatolites when viewed in thin section under cross-polarized plane light. Enlarged views circumscribed by rectangles are shown sequentially. The microstructure of these ancient stromatolites is secondary due to recrystallization of the primary minerals and silicification from a later influx of silica-rich fluids. However, even at the microscopic scale, detailed characteristics of the primary banded layers are preserved. Laminations consist of alternating light (silica) and dark (carbonate) layers with irregular contacts. The lower contact of the dark carbonate layers is more evenly developed than the top contact, a feature observed in some younger stromatolites with preserved organic matter. [Reproduced with permission from Hofmann *et al.* (1999).]

G. Oldest Isotopic Evidence of a Specific Metabolic Pathway in Archean Evaporite Deposits

Chemofossil evidence supports the hypothesis that microorganisms were metabolically diverse by the Paleoarchean (3600–3200 Ma). Convincing evidence of microbial sulfate reduction has recently been found in microscopic sulfides from ~3470 million-year-old barites from North Pole, Australia (Shen *et al.,* 2001). The sulfides display maximum isotopic fractionations (21.1 per mil) that could not have been produced by volcanogenic processes. The barites came from layers of gypsum crystals replaced by barite, a process that requires the infiltration of relatively low-temperature barium-rich solutions. The microscopic sulfide crystals, localized along the growth faces of the original gypsum crystals, provide a means to determine the relative fractionation of cogenetic sulfides and sulfates. Although dissimilatory sulfate reduction occurs in both the archaeal and the bacterial domains, the only known archaeal sulfate reducer is a single genus, *Archaeoglobus,* a hyperthermophile characterized by an optimal growth temperature higher than 80°C. Consideration of the processes by which the original gypsum would have become baritized indicates barium transport by relatively cool hydrothermal fluids with temperatures below 60°C. The find indicates a minimum age of 3470 Ma for the emergence of sulfate-reducing microorganisms in the bacterial domain. This find extends the geological record of microbial sulfate reduction by more than 750 Ma and represents the oldest evidence of a specific metabolic pathway. The preservation of evaporitic sediments from several Paleoarchean cratons indicates that shallow marine and terrestrial conditions may have prevailed over a considerable portion of the primeval Earth (Buick and Dunlop, 1990). These environments could have supported diverse communities of microbial life (Groves *et al.,* 1981).

H. Evidence of Hydrothermal Oil Generation during the Archean

Hydrocarbon droplets preserved in hydrothermally silicified kerogeneous sediments have been reported from the Warrawoona Group (>3458 Ma) (Buick *et al.,* 1998). Although textural evidence indicates that the hydrocarbons have not migrated from their source rock, thermal maturation of organic matter does not appear to have resulted from burial but, rather, from hydrothermal heating by convecting fluids at shallow depths. No other fossil biosignatures indicative of an indigenous or fluid-transported microbial community are associated with this ancient hydrocarbon occurrence. On the other hand, evidence of hydrocarbons trapped in fluid inclusions and bituminous residues in cavities from the Pilbara craton of Australia have been found in a Paleoarchean deep-sea volcanogenic massive sulfide deposit ($\sim$3235 Ma) (Rasmussen and Buick, 2000). In addition to providing the oldest evidence for hydrothermal oil generation in a deep-sea volcanogenic setting, the deposit, as discussed above, has yielded the oldest probable fossil remains of thermophiles (Rasmussen, 2000). The presence of pyrobitumen within the sulfide suggests that oil was a significant component of the hydrothermal fluids. Rasmussen and Buick (2000) have proposed the geological occurrence and isotopic signatures of the oil indicate that a granitic intrusion produced a subsurface heating event which formed the volcanogenic massive sulfide deposit and inadvertently generated oil within a younger, highly carbonaceous shale unit. The migrating oil became entrained in the convectively circulating hydrothermal fluids and trapped within the sulfide deposit while it precipitated. This discovery indicates that subseafloor hydrothermal petroleum generation was active during the Paleoarchean and that some of the sulfur might have been derived by hydrothermal reduction of sulfate using hydrocarbons as a reductant.

I. Oldest Compelling Chemofossil Evidence of Methanogenic Archaea

Although direct evidence of methanogenesis has not been found in the geological record, data from ancient carbon and sulfur isotopes of kerogens and sulfide minerals (Schidlowski and Aharon, 1992; Grassineau et al., 2001) and from ancient carbon isotopes of sedimentary organic carbon and carbonates (Hayes et al., 1983) require the presence of methanogens as early as 2800 Ma. Biogeochemical considerations indicate that methanogens could have played an important role in the carbon cycle from the time when autotrophy arose. In the absence of respiratory CO_2 production, they would have provided the only biochemical means of remobilizing carbon that had been fixed by autotrophs (Hayes, 1994). Though methane reaching the atmosphere would be oxidized by photochemically driven reactions in the absence of methanotrophy, no ^{13}C-depleted organic material would form. Also, if the environment was anaerobic and methanogens were not geographically or environmentally restricted, ^{13}C-enriched carbonates would not form. However, the combination of oxygenic photosynthesis, methanogenesis, and methanotrophy is hypothesized to have produced a globally detectable anomalous isotopic signal during the Neoarchean (2800–2500 Ma).

Isotopically light (heavily ^{13}C-depleted) primary biomass (sedimentary organic carbon) characterizes most of the Archean isotopic carbon record. As detailed by Hayes (1994), the most plausible scenario is that a single mechanism of primary production has been globally dominant since 3500 Ma (3800), and that some process was added to the global carbon cycle about 2750 million years ago. This process would have produced highly depleted ^{13}C sedimentary organic carbon (to -60 per mil). By about 2000 Ma, the impact of the process apparently waned and no longer exerted an important influence on the ^{13}C content of sedimentary organic matter. As noted by Hayes (1994), the only process proposed to date that is consistent with the observed isotopic variation involves the recycling of methanogenic-produced methane by methanotrophs, the latter converting the ^{13}C depleted methane into biomass and biological debris that is the precursor of sedimentary organic carbon. The global distribution and simultaneous appearance of isotopically light sedimentary organic carbon are interpreted as indicating that methanotrophy, hence methanogenesis and oxygenic photosynthesis, was of global significance (Hayes, 1994).

J. OXYGENIC PHOTOTROPHY

Though numerous hypotheses that describe the sequence of events that led to the emergence of oxygenic phototrophy have been proposed, constraints discovered to date from the paleobiological and geological record include the following. An anomalous isotopic shift caused by the coupled biogeochemical cycling between Archaea and Bacteria, accompanied by the apparent onset of methane cycling by at least 2750 million years ago, caused a steady decline in the isotopic difference between sedimentary organic materials and cogenetic carbonates as recorded in the rock record (Schidlowski, 1988). Microfossil evidence consistent with the emergence of oxygenic phototrophy has been discovered in rocks at least as old as ~2600 Ma (Altermann and Schopf, 1995). The possibility exists that some of the most ancient Archean microfossils were oxygenic phototrophs (e.g., Fig. 4).Walter (1983) and Buick (1992) proposed that some Neoarchean stromatolites must be cyanobacterial in origin. As discussed by Buick (1992), Neoarchean stromatolites required oxygenic photosynthesis to develop abundantly in environmental settings that lacked evidence of hydrothermal activity. The morphological and isotopic fossil evidence is consistent with the presence of cyanobacteria, yet none of these lines of evidence require that cyanobacteria existed in the Archean.

Convincing evidence for the emergence of oxygenic phototrophy prior to the end of the Archean has been discovered in biomarker compounds; molecular fossil evidence (i.e., biomarkers) for cyanobacteria (2-methylhopanes) (Summons *et al.*, 1999) has been discovered in rocks as old as 2700 Ma (Brocks *et al.*, 1999). Neoarchean rocks (2700 Ma) also contain steranes (sedimentary molecules derived from sterols), which are biomarkers indicative of eukaryotic organisms. Free atmospheric oxygen would have been needed for eukaryotic sterol synthesis. The geological record supports the hypothesis that oxygenic photosynthesis evolved prior to 2700 Ma. The accumulation of massive banded iron formations before 2500 Ma is consistent with an abundant source of free oxygen. And as noted by Des Marais (1997; 2001), vast sedimentary deposits of organic carbon, reduced sulfide, ferric iron, and sulfate on continental platforms and along coastal margins are among the most prominent and enduring legacies of billions of years of oxygenic photosynthetic activity. Clearly, the end of the Archean was a time of significant change on Earth, and the reader is referred to Schopf and Klein (1992) and Knoll (1996) for a thorough discussion of the implications of these changes on Earth's biosphere, atmosphere, hydrosphere, and lithosphere.

VIII. Conclusion

The sparseness of the Archean fossil record prevents an accurate reconstruction of the sequence of evolutionary events that occurred on the early Earth. A more effective means of assessing the order of appearance of different physiological pathways is better evaluated on the basis of comparative molecular phylogenetic techniques (e.g., Pace, 1997; Xiong *et al.*, 2000). Evidence from the geological record and from molecular phylogenetic analyses of modern organisms, however, indicates that hydrothermal systems and the associated subterranean parts of these environments were likely to have been one of the earliest habitats for life. As shown in Fig. 3, a variety of Archean hydrothermal settings could have sustained microbial ecosystems (e.g., Nisbet and Sleep, 2001). Mid-ocean ridge volcanic systems would have been more active than they are today and would have delivered fluids rich in reduced manganese, iron, copper, zinc, and sulfur. Subaerial hot springs would have formed relatively early on broad komatiite shield volcanoes and would have been characterized by solutions rich in reduced magnesium. Volcanic island arcs would have hosted subaqueous and subaerial hydrothermal systems characterized by solutions rich in reduced copper, molybdenum, and zinc. The recent discovery of ancient microfossils and oil in Archean hydrothermal deposits serves as a bellwether for the types of biosignatures that could be found in ancient hydrothermal paleobiological repositories. The sediments around hydrothermal systems likely supported another type of ecosystem whose primary producers included methanogens and sulfate reducers that exploited the oxidation contrast between the air–water system and the more reduced rock-derived fluids (Nisbet, 1995). The Archean Eon also witnessed the origin of photosynthesis and evolution of Eukarya, major events in Earth history, second only to the origin of life. An overview of the various hypotheses proposed for the evolution of photosynthesis and origins of Eukarya are discussed by Nisbet and Sleep (2001) (see references therein).

Despite the miniscule amount of information preserved in the Archean paleobiological record, it is clear that life displays a remarkable degree of morphological conservatism. While the effects of postfossilization alteration processes reduce the fidelity of cellular structures and compositions as a function of time, the presence of three-dimensional carbonaceous microfossils in ancient Archean rocks demonstrates that microbial cells can be preserved intact for billions of years. The presence of stromatolite laminations preserved in the fossil record has shown that microbial life evolved the ability to maintain its position on surfaces or near the sediment–water interface. Either the rates at which nonmotile organisms replicated were faster than the rates of sediment accumulation/mineral precipitation or motility emerged during the Archean. The occurrence of ancient stromatolites in shallow-water ecosystems that were constructed in the absence of a UV-shielding atmosphere indicates that the microorganisms contributing to their construction had developed some minimum resistance to high-energy solar radiation. Evidence that different metabolic pathways emerged during the Archean is recorded in the carbonaceous matter dispersed in some Archean sedimentary sequences that have not been heavily metamorphosed (e.g., <300°C).

Numerous advances in Archean paleobiology will certainly occur over the next few decades as technological innovations catalyzed by the search for life beyond Earth are applied to the search for terrestrial life's origins. The present trend for multidisciplinary teams to collaborate in the search for evidence of ancient life will also escalate the rate at which new details about early life are forthcoming. There can be no doubt that the search for early life, whether here or on other planets, will continue to attract new generations of scientists who seek to answer the age-old question, "Are we alone?"

REFERENCES

Altermann, W., and Schopf, J. W. (1995). *Precambr. Res.* **75**, 65–90.
Awramik, S. M., Schopf, J. W., *et al.* (1983). *Precambr. Res.* **20**, 357–374.
Barghoorn, E. S., and Schopf, J. W. (1965). *Science* **150**, 337–339.
Bengtson, S. (ed.) (1994). "Early Life on Earth," Nobel Symposium Volumes. Columbia University Press, New York.
Boston, P. J., Spilde, M. N., *et al.* (2001). *Astrobiology* **1**(1), 25–55.
Bowring, S. A., and Williams, I. S. (1999). *Contrib. Mineral. Petrol.* **134**, 3–16.
Bowring, S. A., Williams, I. S., *et al.* (1989). *Geology* **17**, 971–975.
Brasier, M. Earth System Processes, Meeting of Geol. Soc. London and Geol. Soc. America, 24–28, June 2001.
Brocks, J. J., Logan, G. A., *et al.* (1999). *Science* **285**, 1033–1036.

Buick, R. (1990). *Palaios* **5**, 441–459.

Buick, R., and Dunlop, J. S. R. (1990). *Sedimentology* **37**, 247–277.

Buick, R., Dunlop, J. S. R., *et al.* (1981). *Alcheringa* **5**, 161–181.

Buick, R., Rasmussen, B., *et al.* (1998). *Am. Assoc. Petrol. Geol. Bull.* **82**(1), 50–69.

Cady, S. L. (1998). *Palaios* **13**(2), 95–97.

Cady, S. L. (2001). *In* "Signs of Life: Workshop on Life-Detection Techniques." National Research Council, National Academy Press, Washington, DC (in press).

Cady, S. L., and Farmer, J. D. (1996). *In* "Evolution of Hydrothermal Ecosystems on Earth (and Mars?)" (G. R. Bock and J. A. Goode, eds.), Vol. 202, pp. 150–173. John Wiley & Sons, Chichester, UK.

Cady, S. L., Farmer, J. D., *et al.* (1995). *Proc. Geol. Soc. Am.* **27**(6), 305.

Chang, S., Des Marais, D., *et al.* (1983). *In* "Earth's Earliest Biosphere" (J. W. Schopf, ed.), pp. 53–92. Princeton University Press, Princeton, NJ.

Chapman, D. J., and Schopf, J. W. (1983). *In* "Earth's Earliest Biosphere" (J. W. Schopf, ed.), pp. 302–320. Princeton University Press, Princeton, NJ.

Cloud, P., and Morrison, G. R. (1979). *Precambr. Res.* **9**, 81–91.

Conrad, P. G., and Nealson, K. H. (2001). *Astrobiology* **1**(1), 15–24.

Dawson, J. W. (1875). "The Dawn of Life." Hodder and Stoughton, London.

Des Marais, D. J. (1997). *In* "Geomicrobiology: Interactions Between Microbes and Minerals" (J. E. Banfield and K. H. Nealson, eds.), pp. 429–448. Mineralogical Society of America, Washington, DC.

Des Marais, D. J. (2001). *Science* **289**, 1703.

Doolittle, W. F. (2000). *Sci. Am.* 72–77.

Farmer, J. D., and DesMarais, D. J. (1999). *J. Geophys. Res.* **104**(E11), 26977–26995.

Galtier, N., and Tourasse, N., (1999). *Science* **283**, 220–221.

Gest, H., and Schopf, J. W. (1983). *In* "Earth's Earliest Biosphere" (J. W. Schopf, ed.), pp. 135–148. Princeton University Press, Princeton, NJ.

Grassineau, N. V., Nisbet, E. G., *et al.* (2001). *Proc. Roy. Soc. Lond. B* **268**, 113–119.

Grotzinger, J. P., and Rothman, D. R. (1996). *Nature* **383**, 423–425.

Groves, D. I., Dunlop, J. S. R., *et al.* (1981). *Sci. Am.* **245**, 64–73.

Gruner, J. W. (1923). *J. Geol.* **31**, 146.

Gruner, J. W. (1925). *J. Geol.* **33**, 151–152.

Hawley, J. E. (1926). *J. Geol.* **34**, 441–461.

Hayes, J. M. (1994). *In* "Early Life on Earth" (S. Bengtson, ed.), pp. 220–236. Columbia University Press, New York.

Hayes, J. M., Kaplan, I. R., *et al.* (1983). *In* "Earth's Earliest Biosphere" (J. W. Schopf, ed.), pp. 93–134. Princeton University Press, Princeton, NJ.

Hofmann, H. J. (1972). *Proc. 24th Int. Geol. Congr., Sect.* **1**, pp. 21–30.

Hofmann, H. J. (1973). *Earth Sci. Rev.* **9**, 339–373.

Hofmann, H. J. (1990). *Geology* **18**, 340–341.

Hofmann, H. J. (1992). *Episodes* **15**(2), 122–123.

Hofmann, H. J. (2000). *In* "Microbial Sediments" (R. E. Riding and S. M. Awramik, eds.), pp. 315–327. Springer-Verlag, Berlin Heidelberg.

Hofmann, H. J., Grey, K., *et al.* (1999). *Geol. Soc. Am. Bull.* **111**, 1256–1262.

Holland, H. D. (1984). "The Chemical Evolution of the Atmosphere and Oceans." Princeton University Press, Princeton, NJ.

Kasting, J. F., and Chang, S. (1992). *In* "The Proterozoic Biosphere" (J. W. Schopf and C. Klein, eds.), pp. 9–12. Cambridge University Press, New York.

Knoll, A. H. (1985). *Phil. Trans. R. Soc. Lond. B* **311**, 111–122.

Knoll, A. H. (1996). *In* "Palynology: Principles and Applications" (J. Jansonius and D. C. McGregor, eds.), Vol. 1 pp. 51–80. American Association of Stratigraphic Palynologists Foundation.

Knoll, A. H., and Golubic, S. (1979). *Precambr. Res.* **10,** 115–151.
Lawson, A. C. (1912). *Geol. Surv. Can. Mem.* **28,** 7–15.
Logan, B. W. (1961). *J. Geol.* **69,** 517–533.
Logan, G. A., Hinman, M. C., *et al.* (2001). *Proc. 2nd Gen. Meet. NASA Astrobiol. Inst.,* pp. 246.
Lowe, D. R. (1994). *Geology* **22,** 387–390.
Lowe, D. R. (1994). *In* "Early Life on Earth" (S. Bengston, ed.), Vol. 84, pp. 24–35. Columbia University Press, New York.
Lowe, D. R., and Byerly, G. R. (eds.) (1999). "Geologic Evolution of the Barberton Greenstone Belt, South Africa," Geological Society of America, Boulder, CO.
Lowe, D. R., and Ernst, W. G. (1992). *In* "The Proterozoic Biosphere" (J. W. Schopf and C. Klein, ed.), pp. 13–19. Cambridge University Press, New York.
Maher, K. A., and Stevenson, D. J. (1988). *Nature* **331,** 612–614.
McClendon, J. H. (1999). *Earth Sci. Rev.* **47,** 71–93.
McKay, D. S., Gibson, E. K. J., *et al.* (1996). *Science* **273,** 924–930.
Mojzsis, S. J., and Arrhenius, G. (1998). *J. Geophys. Res.* **103,** 28495–28511.
Mojzsis, S. J., and Harrison, T. M. (2000). *GSA Today* **10**(4), 1–6.
Mojzsis, S. J., Arrhenius, G., *et al.* (1996). *Nature* **384,** 55–59.
Mojzsis, S. J., Harrison, T. M., *et al.* (2001). *Nature* **409,** 178–181.
Muir, M. D. (1978). *Univ. West. Austral., Geol. Dept. Univ. Extens. Publ.* **2,** 11–21.
Myers, J. S., and Crowley, J. L. (2000). *Precambr. Res.* **103,** 101–124.
Nisbet, E. G. (1995). *In* "Early Precambrian Processes" (M. P. Coward and A. C. Ries, eds.), Vol. 95, pp. 27–52. Geological Society, London.
Nisbet, E. G., and Sleep, N. H. (2001). *Nature* **409,** 1083–1091.
Oehler, J. H. (1976). *Alcheringa* **1,** 314–349.
Oehler, J. H., and Logan, R. G. (1977). *Econ. Geol.* **72,** 1393–1409.
Pace, N. R. (1997). *Science* **276,** 734–740.
Papanastassiou, D. A., and Wasserburg, G. J. (1971). *Earth Planet. Sci. Lett.* **11,** 37–62.
Patterson, C. C. (1956). *J. Geophys. Res.* **82,** 803–827.
Rasmussen, B. (2000). *Nature* **405,** 676–679.
Rasmussen, B., and Buick, R. (2000). *Geology* **28**(8), 731–734.
Rosing, M. T. (1999). *Science* **283,** 674–676.
Schidlowski, M. (1988). *Nature* **333,** 313–318.
Schidlowski, M., and Aharon, P. (1992). *In* "Early Organic Evolution: Implications for Mineral and Energy Resources" (M. Schidlowski, ed.), pp. 133–146. Springer-Verlag, Berlin.
Schidlowski, M., Hayes, J. M., *et al.* (1983). *In* "Earth's Earliest Biosphere" (J. W. Schopf, ed.), pp. 149–186. Princeton University Press, Princeton, NJ.
Schopf, J. M. (1975). *Rev. Palaeobot. Palynol.* **20**(4), 27–53.
Schopf, J. W. (1983). *In* "Earth's Earliest Biosphere, Its Origin and Evolution." Princeton University Press, Princeton, NJ.
Schopf, J. W. (1992). *In* "The Proterozoic Biosphere" (J. W. Schopf, ed.), pp. 25–39. Cambridge University Press, New York.
Schopf, J. W. (1993). *Science* **260,** 640–646.
Schopf, J. W. (1999). "Cradle of Life, the Discovery of Earth's Earliest Fossils." Princeton University Press, Princeton, NJ.
Schopf, J. W., and Klein, C. (eds.) (1992). "The Proterozoic Biosphere." Cambridge University Press, New York.
Schopf, J. W., and Packer, B. M. (1987). *Science* **237,** 70–73.
Schopf, J. W., and Walter, M. R. (1983). *In* "Earth's Earliest Biosphere" (J. W. Schopf, ed.), pp. 214–239. Princeton University Press, Princeton, NJ.

Schopf, J. W., Hayes, J. M., *et al.* (eds.) (1983). "Evolution of Earth's Earliest Ecosystems: Recent Progress and Unsolved Problems." Princeton University Press, Princeton, NJ.

Semikhotov, M. A., Gebelein, C. D., *et al.* (1979). *Can. J. Earth Sci.* **19**, 992–1015.

Shen, Y., Buick, R., *et al.* (2001). *Nature* **410**, 77–81.

Shock, E. L., and Schulte, M. D. (1998). *In* "Origins of Life and Evolution of the Biosphere."

Sleep, N. H., and Zahnle, K. (1998). *J. Geophys. Res.* **103**, 28529–28544.

Sleep, N. H., Zahnle, K., *et al.* (2001). *Proc. Natl. Acad. Sci.* **98**(7), 3666–3672.

Summons, R. E., and Walter, M. R. (1990). *Am. J. Sci.* **290-A**, 212–244.

Summons, R. E., Jahnke, L. L., *et al.* (1999). *Nature* **400**, 554–557.

Turcotte, D. L. (1980). *Earth Planet. Sci. Lett.* **48**, 53–58.

Tyler, S. A., and Barghoorn, E. S. (1954). *Science* **119**, 606–608.

Walcott, C. D. (1912). *Geol. Surf. Can. Mem.* **28**, 16–23.

Walker, J. C. G., Klein, C., *et al.* (1983). *In* "Earth's Earliest Biosphere" (J. W. Schopf, ed.), pp. 260–290. Princeton University Press, Princeton, NJ.

Walsh, M. M. (1989). "Carbonaceous Cherts of the Swaziland Supergroup, Barberton Mountian Land, Southern Africa," Louisiana State University, Baton Rouge.

Walsh, M. M. (1992). *Precambr. Res.* **54**, 271–293.

Walsh, M. M., and Lowe, D. R. (1985). *Nature* **314**, 530–532.

Walter, M. R. (ed.) (1976). "Stromatolites," Developments in Sedimentology. Elsevier, Amsterdam.

Walter, M. R. (1983). *In* "Earth's Earliest Biosphere" (J. W. Schopf, ed.), pp. 187–213. Princeton University Press, Princeton, NJ.

Walter, M. R. (1994). *In* "Early Life on Earth" (S. Bengtson, ed.), pp. 270–285. Columbia University Press, New York.

Walter, M. R., and DesMarais, D. (1993). *Icarus* **101**, 129–143.

Walter, M. R., DesMarais, D. J., *et al.* (1996). *Palaios* **11**(6), 497–518.

Walter, M. R., McLoughlin, S., *et al.* (1998). *Alcheringa* **22**, 285–314.

Westall, F. (1999). *J. Geophys. Res.* **104**(E7), 16437–16451.

Wilde, S. A., and Valley, J. W., *et al.* (2001). *Nature* **409**, 175–178.

Xiong, J., Fischer, W. M., *et al.* (2000). *Science* **289**, 1724–1730.

Section II. Molecular Phylogeny and the Tree of Life

Whole-genome sequencing of archaeal chromosomes provides emphatic support for the notion that these organisms are fundamentally distinct from bacteria and eukaryotes. More importantly, the magnitude of their phylogenetic distinction is underscored by this information. The identification of archaeal-specific genes implicates the existence of novel solutions to basic subcellular processes. Some of these are explored in Section III. The existence of archaeal genes orthologous to those in bacteria may have occurred by horizontal transmission, perhaps between archaea and bacteria inhabiting common niches. Other orthologues presumably reflect more vertical processes of inheritance and therefore the existence of a common ancestor. Of particular relevance to this text are the archaeal genes which are orthologous to those in eukaryotes. The list of such genes continues to grow and, for now, remains focused on functions involved with information processing.

What can we look forward to as more archaeal genomes are sequenced? Several predictions are apparent. Organisms called korarchaeota by Norman Pace and co-workers constitute a cluster of archaea whose phylogenetic position places them closest to the last common ancestor at the "root" of the universal tree of life. It seems that their genes may constitute the earliest forms of modern-day proteins. So they too may be unique. In addition to the korarchaea, two other mains groups of archaea are recognized. These are the euryarchaea, including the methanogens and halophiles, and the crenarchaea, still largely hyperthermophiles. However, members of the crenarchaea have been described by several groups which are unlike their hyperthermophilic crenarchaeal relatives. Instead they are found in cold environments including oceans, sediments, and soils. 16S rRNA sequence comparisons place them at deeply branching positions, indicating that they are highly evolved. It has been proposed that they derive from hyperthermophilic progenitors and over time adapted to lower temperatures or mesophilic conditions. Perhaps this hypothesis can be tested by molecular investigations.

The second chapter in this book, by Liang and Riley, addresses the rapid impact the field of genomics is having on our understanding of

Archaea. New methods for analyzing protein phylogenies and evolutionary relatedness are emerging. The identification and comparison of a protein's functional essence, termed a module, are reviewed by Liang and Riley. This information constitutes a powerful new approach in a process analogous to triangulation to unravel the origin of a particular gene. Comparisons across large phylogenetic distances are particularly appropriate for this process and lead to compilations of gene inventories as a novel means of looking at genomic information.

A Comparative Genomics Approach for Studying Ancestral Proteins and Evolution

PING LIANG[*] AND MONICA RILEY

Josephine Bay Paul Center for Molecular Evolution and Comparative Biology
Marine Biological Laboratory
Woods Hole Massachusetts 02543

I. The Genome Era: A Great Opportunity for Evolutionary Biology

Large biological molecules in present-day life, mainly RNAs, DNAs, and proteins, contain traces of their ancestors in primary sequences as well as in structures. Among these molecules, ribosomal RNAs (rRNAs) have been the basis for the current version of the tree of life as is well described in the preceding article. However, rRNAs represent only a small fraction of cellular functions and do not reflect the vast variety and complexity of present-day cellular life. In this sense, the whole set of DNAs in a cell (the genome) contains all the information coding for an organism, while all proteins in an organism (proteome) represent all the units that carry out and reveal the functions of the genes.

Proteins preserve the sequence information over evolutionary time much better than DNA due to the existence of multiple alternative codons for many amino acids and the fact that amino acids have

[*] Current address: Roswell Park Cancer Institute, Buffalo, New York 14263.

ADVANCES IN APPLIED MICROBIOLOGY, VOLUME 50

distinctive structural and chemical properties. Thus substitutions of certain residues are allowed without changing the functional character-istics of the proteins. Therefore, while information from DNA sequences is very useful for studying recent evolution, proteins are more useful for studying distant homology and ancient proteins. For this reason, we deal mainly with protein sequences in the studies reported here.

Before complete genome sequences became available, evolution stud-ies at the protein level were focused mainly on using individual similar proteins from different organisms to build the phylogeny of species or to study ancestral proteins based on a limited number of protein fam-ilies (Gogarten *et al.*, 1989; Brown and Doolittle, 1997; Iwabe *et al.*, 1989). The availability of complete genome sequences for an ever-increasing number of organisms covering all three domains of life (a comprehensive list of the genomes and their sequences is available at http://www.ncbi.nlm.nih.gov/Genomes/) presents us with an excit-ing opportunity to study evolution at a brand new level. With the data from all the complete genomes, many fundamental questions for evolu-tion as well as for biology in general can now be addressed (Doolittle, 1998; Pennisi, 1999; Ouzounis *et al.*, 1996; Lander and Weinberg, 2000; Koonin *et al.*, 2000). Most excitingly, it is possible for us to start toward reconstructing the set of ancient genes/proteins that existed in the last universal common ancestor (LUCA), commonly found in most life on Earth (Koonin and Mushegian, 1996). It should be possible to reach even earlier in time to disclose evolutionary paths from a limited number of ancestral proteins that existed before the LUCA, leading to generation of the variety of proteins presumed to have been present in the LUCA.

II. Orthology and Paralogy: Evolution in Different Dimensions

A. BASIC CONCEPTS

Gene duplication and divergence are generally believed to have been the main mechanism for evolution of genome and proteome complexity in the biological world (Ohno, 1970; Li and Graur, 1991; Gogarten and Olendzenski, 1999; Dayhoff, 1976). A generally shared view of protein evolution is that a diversified set of proteins was present in the LUCA, many of the proteins whose descendants are present in the majority of known life forms today. Further elaboration and diversification since the LUCA to generate other nonuniversal proteins are believed to have en-tailed duplication and divergence. One can look back in the time before the LUCA, and although we have no firm view of the nature of protocells, we believe that the variety of proteins in LUCA must itself have been generated by some process akin to gene duplication and divergence.

Gene duplication generates gene families and protein families, within which members share common ancestors as indicated by detectable similarity in sequence and, in most cases, similarity in function as well. The relationship among family members is homology, and genes can share homology mainly in two ways according to the fundamental distinction initially made by Fitch (1970). Homologues found in different genomes are called orthologues, and their homology is specifically termed orthology. Orthologues are the descendants of common ancestral genes that were distributed into different organisms along with the speciation processes. Therefore, orthologues are useful for studying the phylogeny of organisms, a topic that has been the focus of most evolution studies. In addition to orthologues, homologues also exist as the descendants of a common ancestor due to gene duplication within the same genome, and these are called paralogues. The homology between paralogues is also termed paralogy. Let us say that a pair of paralogues was generated in a genome by one duplication event. When speciation next occurs, the two paralogues are distributed to both lines of descent, making two pairs of orthologues between the two species. The paralogues in each genome, although at first identical, are then free to evolve further separately as the independent species continue to evolve.

Paralogous groups from one organism are particularly useful in molecular evolution since, barring horizontal transfer from other sources, families of paralogues have descended together under a common, shared set of cellular circumstances. In contrast, when relating orthologous genes in different organisms, often physiologically quite different organisms, one must take into account a succession of noncomparable events in each organism's history that may have influenced the development of orthologous genes in ways we cannot correct for. Except for artifacts introduced by horizontal transfer of genes between organisms or intragenomic recombination, sequence-related paralogous groups within one genome constitute an uncomplicated set of descendants from earlier ancestors. Independent events following speciation, such as extinction and divergence along different lines, contribute to the difficulty of correctly identifying true orthologues. Collections of orthologues and paralogues have been assembled into units known as COGs (Tatusov *et al.*, 2000). They differ from purely paralogous groups from one organism in that they contain both orthologues and at least some paralogues.

B. METHODOLOGY FOR DETECTING HOMOLOGY ON A GENOME SCALE

To take advantage of the enormous amount of sequence data generated from the genome projects, it is necessary to use tools that allow systematic and yet sensitive detection of sequence similarity at the genome level. Among the existing methods, a locally installed BLAST (Basic Local Alignment Similarity Test) (Altschul *et al.*, 1997) provided by the National Center of Biological Information (NCBI; http://www.ncbi.nlm.nih.gov) offers the fastest all-against-all sequence similarity search. But based on our experience, the DARWIN (Data Analysis and Retrieval with Indexed Nucleotide/Peptide Sequences) package developed by the Computational Biochemistry Research Group at ETH, Zurich, Switzerland (Gonnet *et al.*, 1992, 2000; http://cbrg.inf.ethz.ch/), fits this need very well and is particularly valuable for the more distant protein relationships. DARWIN uses amino acid scoring matrices twice. It first uses the Needleman–Wunsch (1970) algorithms with PAM250 (Dayhoff *et al.*, 1978) to find the sequence match and approximate level of sequence similarity. It then employs the Smith–Waterman (1981) dynamic programming algorithm to optimize each local alignment, using an appropriately chosen score matrix consistent with the level of sequence similarity. Thus, unlike the Standard BLASTP, which uses a single BLOSUM matrix, DARWIN uses an entire set of PAM matrices appropriate to a related set of sequences (Gonnet *et al.*, 1992). This allows emphasis on codon relationships for amino acid substitutions made early in divergence between two highly similar proteins, but also takes into consideration the properties of amino acid residues in distantly related proteins. In the latter case, emphasis is more on the similarities in chemistry of the original amino acids and the substitutes, their relative steric properties, and influences on secondary structure of the polypeptide in calculating the sequence similarity. Outputs from DARWIN can be easily controlled to contain the list of all qualifying aligned pairs of peptides with the information related to the alignment, such as the start and end positions of alignment regions for both sequences, the PAM and the percentages of sequence identity and similarity. Thus the DARWIN program provides a quantitative measure for each sequence alignment in PAM distance over a wide range. To store and manage efficiently the huge amount of data generated from these genome analysis and, more importantly, to be able to relate or extract specific information in any fashion, the data are best dealt with in a relational database. We choose to use Visual FoxPro from Microsoft because it offers menu-driven functions to general computer users, but it also allows commands/scripting execution of tasks for advanced users.

The data from such exhaustive sequence pairwise comparisons among all the sequences from completed genomes can be used to address, at least as a start, the following issues: universal proteins that are present in all organisms, proteins that are characteristic of a species or a specific phylogenetic lineage, the characteristics of protein families for individual genomes, the amount and direction of horizontal gene transfer, and the evolution of genome structures. This article presents early work relevant to the first three issues.

C. Gene Duplication and Divergence: The Main Mechanism for Genome and Protein Evolution

As a part of our ongoing research projects, an all-against-all pairwise comparison for a data set covering all the predicted protein-coding open reading frames (ORFs) for 22 genomes (see Table I for the list of organisms) with representatives from each domain of the tree of life was performed using DARWIN. Data for both the paralogues within each genome and the orthologues with respect to each of the other genomes were collected. Table I lists the number of paralogues in each genome in relation to the genome size and the total number of protein-coding ORFs.

TABLE I

Genome Size, Number of ORFs, and Paralogues

Domain[a]	Genome	Size (Mbp)	ORFs[b]	Self matches[c]	No. of paralogues (%)[d] 200/100	250/83
B	Mycoplasma genitalium (mg)	0.58	479	472	174 (36)	318 (66)
B	Chlamydia trachomatis (ct)	1.05	894	498	222 (25)	468 (52)
B	Rickettsia prowazekii (rp)	1.10	834	651	269 (32)	542 (65)
B	Treponema pallidum (tp)	1.14	1,030	852	298 (29)	602 (58)
B	Borrelia burgdorferi (bb)	1.44	850	1,115	410 (48)	664 (78)
B	Aquifex aeolicus (aa)	1.50	1,522	2,353	755 (47)	1,223 (80)
B	Helicobacter pylori (hp)	1.64	1,553	2,670	620 (40)	1,086 (70)
A	Aeropyrum pernix (ap)	1.67	2,694	3,099	785 (29)	1,621 (60)
A	Methanococcus jannaschii (mj)	1.66	1,773	2,700	849 (48)	1,378 (78)
A	Methanobact. themoauto. (mt)	1.75	1,869	2,603	808 (43)	1,253 (67)
B	Thermotoga maritima (tm)	1.80	1,846	6,611	884 (48)	1,434 (77)
A	Pyrococcus horikoshii (ph)	1.80	2,064	3,508	930 (45)	1,451 (70)
B	Haemophilus influenzae (hi)	1.83	1,709	2,756	610 (36)	1,078 (63)
A	Archaeoglobus fulgidus (af)	2.18	2,409	4,158	999 (41)	1,744 (72)
A	Halobacterium sp. NRC1 (hs)	2.50	3,414	22,521	1,804 (53)	2,509 (73)
B	Deinococcus radiodurans (dr)	3.28	3,117	8,876	1,694 (54)	2,526 (81)
B	Synechocystis sp. (sy)	3.57	3,169	12,166	1,537 (49)	2,363 (74)
B	Bacillus subtilis (bs)	4.20	4,100	16,409	2,365 (58)	3,254 (83)
B	Mycobact. tuberculosis (my)	4.40	3,918	26,308	2,567 (66)	3,415 (87)
B	Escherichia coli (K12; ec)	4.60	4,290	16,706	2,416 (56)	3,364 (78)
E	Saccharomyces cerevisiae (sc)	13.1	6,351	150,108	4,258 (67)	5,453 (85)
E	Caenorhabditis elegans (ce)	100	19,105	348,934	12,439 (65)	18,405 (96)

[a]A, Archaea; B, Eubacteria; E, Eukarya.
[b]Limited to protein-coding ORFs.
[c]Number of sequence matches within each genome based on the 200/100 criteria.
[d]Number of paralogues and percentages of the total number of protein-coding ORFs using two criteria: 200/100, 200 as the maximal PAM value and 100 residues as the minimal alignment length; and 250/83, 250 as the maximal PAM value and 83 as the minimal number of residues for an alignment.

Although the actual number and the percentage of paralogues in rela-tion to the total number of protein-coding ORFs vary among genomes, there seems to be a trend: the paralogues, a result of gene duplication within the same genome, comprise a substantial portion of the proteome, and the bigger the genome, the higher is the percentage of paralogues in relation to the total number of proteins. For instance, over 50% of *Escherichia coli* proteins are paralogues, while a much smaller genome, *Chlamidia trachomatis,* has 25% of its proteins as paralogues, the lowest percentage among the 22 studied genomes. The biggest genome on the list, *Caenhorabditis elegans,* has the highest percentage of paralogues (65%). These numbers were obtained using relatively conservative cri-teria requiring a minimal alignment length of 100 residues and a max-imal PAM value of 200 for two proteins to be considered homologous. The number of paralogues was almost doubled when the stringency was lowered to 83 for the minimal number of residues aligned and 250 for the maximal PAM score, a criterion which is used by other researchers (Gonnet *et al.,* 2000; Cannarozzi *et al.,* 2000). In this case, almost all *C. elegans* proteins (95%) have at least one paralogue, and all other genomes have at least 50% of proteins as paralogues. However, one has to be cautious about the reliability of the homology at these marginal levels, and some false positives are surely present. Nevertheless, we can be confident in predicting that the percentage of paralogues in higher eukaryotes, especially in humans, may be much higher, to a point that it is possible that every gene has at least one counterpart, in addition to the fact that most eukaryotes are at least diploid. We can speculate that this store of similar genes could provide the organisms with great flexibility and resistance to accidental gene inactivation. These data strongly sug-gest that the number of ancestor proteins in the LUCA may have been limited, at least many fewer than the number of present-day proteins in larger genomes. Repeated gene duplication within genomes over evolu-tionary time would have generated more and more paralogous protein families which expanded the genomes.

D. Divergence of Genes within and between Genomes

After gene duplication, the duplicated gene copies may start to di-verge by changes in the sequences through a variety of mechanisms, such as random point mutations and small insertions and deletions (Li and Graur, 1991). The distribution patterns of homology for paralogues give some pictures of the process of divergence for duplicated genes within the same genome, while the distribution of sequence similar-ity among the orthologues from two genomes represents the divergence processes of genes that are duplicated as a consequence of speciation

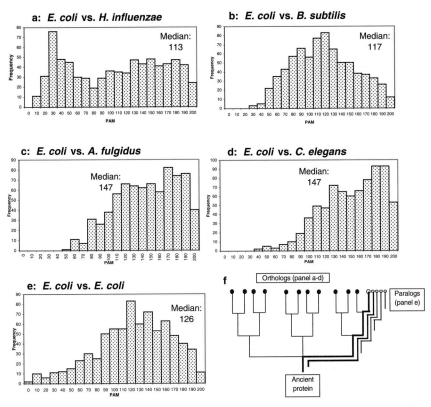

FIG. 1. Frequency of PAM scores for a selected set of *E. coli* paralogues and their ortho-logues from four genomes; 755 *E. coli* proteins that have both paralogues and orthologues in four other genomes were used in this analysis. For each of these proteins, the best PAM scores (lowest in value) for homologues in each genome were collected and the frequen-cies of PAM scores in each range of every 10 were plotted. The median value for each genome is also indicated.

events. Figure 1 shows the distribution patterns of homology measured by the PAM scores for the best paralogues of *E. coli* proteins and, also, their best orthologues from four other genomes: one genome from the relatively closely related *H. influenzae;* one from the distantly related bacterial genus, *Bacillus subtilis;* and one genome each from the other two domains of the tree of life—an archaeon, *Archaeoglobus fulgidus,* and a eukaryote, *C. elegans.* For this purpose, the 755 *E. coli* proteins which have homologues within their own genome (paralogues), as well as in each of the other four genomes (orthologues), were used. Since all three domains of the tree of life are represented, these proteins may also represent a partial list of ancestral and, likely also, essential proteins that existed in the LUCA. For each of these proteins, the best paralogues and best orthologues from each of the other genomes indicated by the lowest PAM score were identified and collected. The PAM scores were plotted against the frequency of each value range. A PAM score near 0 indicates a high level of sequence similarity, up to near-identity, while a PAM score of 200 represents a low level of sequence similarity, with the sequence identity mostly below 25%. So the distribution of PAM scores at the genome level reflects an overall pattern of homology (or divergence) at different levels.

One would probably argue, in this case, that such data would be neu-tralized by the behavior of a mixed group of proteins evolving at different speeds. However, since the comparison is made between the four ortho-logues and the one paralogue of the same gene, the results should reflect a meaningful pattern. As shown in Fig. 1, orthologues between *E. coli* and the four other genomes show distinctive patterns, as we would expect. The orthologue patterns, as well as the median PAM value for each genome, clearly reflect the evolutionary distances of these genomes from *E. coli*, in agreement with what we know based on the rRNA tree. A substantial portion of the orthologues from *H. influenzae*, the closest to *E. coli* among the four genomes, has PAM scores below 40 (identity above 85%) (Fig. 1b), a range that is almost absent for orthologues from the other three genomes (Figs. 1b–d). In contrast, the majority of PAM scores for orthologues from *A. fulgidus* and *C. elegans*, which are very evolutionarily distant from *E. coli*, are distributed above 100 (identity below 35%), with the majority ranging from 120 to 190 (Figs. 1c and d). The orthologues from *B. subtilis*, a distant bacterial relative, have PAM scores mostly distributed between 50 and 150 (Fig. 1b), while the PAM scores for orthologues from *Salmonella typhimurium*, a very close rela-tive to *E. coli*, are distributed almost exclusively within the 0 to 20 range (unpublished data).

The fact that there are almost-even numbers of the spectrum of PAM scores for orthologues of *E. coli* proteins in *H. influenzae* distributed from 50 to 190 (Fig. 1a) indicates that there are genes diverging at differ-ent speeds, some that have evolved very rapidly and some much more slowly. Different requirements for conservation of function and even small differences in the particular physiological function performed in the two bacteria can account for some of the observed differences. Also, some of the higher PAM scores may be due to the loss of the true ortho-logues in either or both genomes, more likely in *H. influenzae*, which has a reduced genome, leaving us with apparent orthologous relationships that are actually between paralogues, not the original two orthologues.

Compared with the PAM distributions for orthologues, the paralogues of *E. coli* (Fig. 1e) have a full range of PAM scores, from near-identity to the low limit we set. This indicates that gene duplication within the same genome has probably been occurring at various time points over the entire long history of the *E. coli* species and before, with the duplicated genes likely diverging at different speeds for different sets of genes as mentioned above. For example, TufA and TufB, two genes that are almost identical in sequence and function, are known to be the result of a very recent gene duplication event (Furano, 1977; Sharp, 1991; Johanson and Hughes, 1992; Miller, 1978), while some other paralogues, such as the P-type ATPase transport proteins, CopA and ZntA, in *E. coli* are due to a very early gene duplication event that occurred in the LUCA or even before (Palmgren and Axelsen, 1998; Axelsen and Palmgren, 1998). Paralogues and orthologues reflect the gene and genome evolution at different dimensions as illustrated in Fig. 1f.

III. Paralogous Protein Families: Windows on Ancestral Proteins and Protein Evolution

Paralogous proteins comprise sequence-related protein families within the same organism. Each of those families represents no more than one ancestral protein in the LUCA. Some families may have diverged from the same ancestor, hence the number could well be smaller than one. A complete assembly of paralogous protein families and unique proteins within an individual organism gives us the set of ancestral proteins represented by one species, and by expanding such an analysis to enough species to cover all the phylogenetic lineages from all three domains of life, we could in theory obtain the whole set of ancestral proteins that were necessary to generate all the existing proteins in the present-day organisms. By studying the function of the individual members and their shared primary sequence and structural properties in these families, information on the general physiological roles, as well as the sequence and structural characteristics, can be inferred for the ancestral proteins represented by the families. The earliest proteins are generally thought of as having had a broader specificity than the proteins of today. Upon repeated duplication and divergence, it seems likely that progeny proteins probably evolved with a narrower specificity and/or altered functions. Therefore, examination of the divergence of function under a general theme and the generation of new functions in exceptional cases among the members of paralogous families will allow us to trace the paths of protein evolution.

A. The Module, a Unit of Protein Function and Evolution

To assemble the paralogous protein families within a genome, proteins are grouped so that each member is related to others transitively and no protein is allowed to belong to more than one group. A transitive

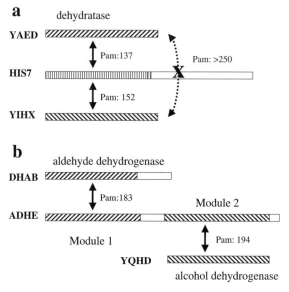

Fig. 2. True transitive homology relationship vs unrelated sequence relationship due to the existence of multimodularity. (a) A transitive homology relationship between protein YAED and protein YIHX was established by their sequence similarity with the same region of protein HIS7. (b) Proteins DHAB and YQHD do not share a transitive sequence relationship because their sequence similarities to ADHE were located in different regions of ADHE. In this case, each region of ADHE is called a module, with module 1 located in the N-terminal region corresponding to aldehyde dehydrogenase and module 2 located in the C-terminal region corresponding to alcohol dehydrogenase. Protein IDs were obtained from Swiss-Prot. The level of sequence similarity between two sequences is indicated by a solid arrow, with the PAM score shown.

relationship between the members of the group means that not every member has to be related to all other members directly. One relationship is sufficient for membership. As shown in Fig. 2a, protein HIS7 is related to proteins YAED and YIHX by sequence, but there is no detectable sequence similarity between YAED and YIHX; rather they share a transitive sequence relationship via a bridging protein, HIS7. This transitive manner extends beyond the sequence similarity detectable by simple pairwise comparison. However, in this case, to say that proteins YAED and YIHX belong to the same family together with HIS7, we need to make sure that HIS7 bridges the other two proteins using the same part of its sequence to exclude the situation shown in Fig. 2b. In the latter case, by using different parts, one protein, ADHE, aligns with two individual proteins, DHAB and YQHD, which are functionally and evolutionarily unrelated. Each region in ADHE representing an individual protein is termed a module. Thus, ADHE is a bimodular protein,

presumably as a result of gene fusion between an aldehyde dehydrogenase and an alcohol dehydrogenase. Therefore, unlike a motif or a domain, a module represents an independent individual functional protein of independent ancestry representing a unit of protein evolution
and cellular function (Riley and Labedan, 1997; Campbell and Baron,
1991; Baron *et al.*, 1991; Hegyi and Bork, 1997). A module can exist as
an individual protein, such as the *E. coli* proteins DHAB and YQHD, or
as a part of larger proteins, such as ADHE. The situations for corresponding modules in different organisms are not necessarily the same. Unless
proteins consisting of more than one module (multimodular) are identified and each module is treated as a separate entity, false connections
can be made. For instance, in the case of ADHE, without considering
its multimodularity, two proteins, aldehyde dehydrogenase and alcohol
dehydrogenase, would be incorrectly grouped into one family. Based on
our experience, if no modules are identified and treated as separate en-
tities from other modules in the same protein, over 70% of all *E. coli*
paralogous proteins can be grouped into one large group, which is certainly far from the truth.

Multimodularity of proteins has long been known. Many enzymes
known to have more than one enzyme activity in a single protein are due
mostly to the existence of multiple modules, such as ADHE, in Fig. 2b
(Kessler *et al.*, 1991). Another well-known example is FadB, which has
NADH-dependent dehydrogenase activity at its C-terminal region and
carries out three separate catalytic reactions—isomerization, hydration,
and epimerization—in its N-terminal region, seemingly by a single active site, at least for the latter two activities (Yang *et al.*, 1988; Muller-
Newen and Stoffel 1993; McCormack, Teichmann, and Riley, submitted
for publication). In addition to enzymes, multimodularity also exists for
other types of proteins. Interesting examples are the phosphotransferase
sytems (PTS), which both transport and phosphorylate carbohydrates.
The EII part of these systems consists of at least three components (EII
A/B/C), which can exist as individual proteins or fusion (multimodular)
proteins of any two or all three of the components (Postma *et al.*, 1996).

In recent work on the complete *E. coli* genomic sequence (Liang, Labedan, and Riley, submitted for publication), for the 2415 *E. coli* proteins
which are paralogues, 2749 modules were identified, indicating the existence of proteins that have more than one module, or are multimodular. In total, 286 proteins (11.8% of all paralogues) were identified as
multimodular, among which 226 have modules in more than one paralogous group and 86 have two or more modules in the same group,
representing the occurrence of gene fusion and internal duplication, respectively. A few proteins have a combination of internal duplication
and gene fusion. Table II lists examples for each of the three situations.

TABLE II

Selected Multimodular Proteins

SW ID	Module	Start[a]	End[a]	Group ID[b]	Function
MODF	b0760_1	1	273	32	Transporters of ABC superfamily (ATP-binding component)
	b0760_2	201	482	32	Transporters of ABC superfamily (ATP-binding component)
LPXD	b0179_1	37	142	40	UDP-3-O-(3-hydroxymyristoyl)-glucosamine n-acyltransferase
	b0179_2	106	306	40	UDP-3-O-(3-hydroxymyristoyl)-glucosamine n-acyltransferase
ILVB	b3671_1	15	344	39	Acetolactate synthase I, large subunit
	b3671_2	384	560	39	Acetolactate synthase I, large subunit
PTAA	b0679_1	1	311	33	Sugar-specific PTS: enzyme IIC, membrane component and sugar binding
	b0679_2	341	466	210	Sugar-specific PTS: enzyme IIB, phosphorylation of sugar
	b0679_3	475	648	211	Sugar-specific PTS: enzyme IIA, phosphotransferase EC 2.7.1.69
MALK	b4035_1	1	210	288	Maltose transport protein (ATP-binding component)
	b4035_2	179	291	289	Repressor of mal operon
MRCB	b0149_2	228	354	33	Glycosyl transferase
	b0149_3	656	768	210	Transpeptidase
ADHE	b1241_1	8	373	190	Aldehyde reductase (dehydrogenase)
	b1241_2	456	857	190	Aldohol dehydrogenase
FADB	b3846_1	8	206	15	NADH-dependent dehydrogenase
	b3846_2	314	643	323	Isomerase, hydratase, epimerase
DOGA	b3692_1	8	121	383	2-Oxo-3-deoxygalactonate 6-phosphate aldolase
	b3692_2	206	587	341	Galactonate dehydratase
HRSA	b0731_1	16	178	223	Sugar-specific PTS: enzyme IIA, phosphotransferase EC 2.7.1.69
	b0731_2	384	492	33	Sugar-specific PTS: enzyme IIC, membrane component and sugar binding
	b0731_3	468	628	33	Sugar-specific PTS: enzyme IIC, membrane component and sugar binding
YPIA	b3486_1	82	602	139	Conserved protein
	b3486_2	832	1170	139	Conserved protein
	b3486_3	1221	1363	196	Conserved protein, transmembrane domain
	b3486_4	1364	1476	139	Conserved protein

[a] The start and end position of the module.
[b] Arbitary group numbers for *E. coli* paralogous families.

B. PARALOGOUS PROTEIN FAMILIES IN *E. coli*

Paralogous protein families in *E. coli* had already been studied before the completion of the genome (Riley and Labedan, 1996, 1997; Labedan and Riley, 1995a,b, 1999). A recent analysis was performed with the complete *E. coli* (K-12) genomic sequence (Blattner *et al.*, 1997) and updated information on the functions of *E. coli* genes (Berlyn, 1998; Rudd, 2000; Riley and Serres, 2000; Liang, Labedan, and Riley, submitted for publication).

Among the 4289 protein-coding ORFs in *E. coli,* 2415 were identified as having at least one homologous partner in the same genome, and they are assembled into 611 sequence-related paralogous groups, while 1983 proteins do not match any other sequences in the genome at a level above the threshold we set and are, thus, called singles (nonparalogues). The major paralogous groups of proteins for each major type of function in *E. coli* are listed in Tables III–VI). Detailed information for the full list of paralogous groups and for the multimodular proteins, including the range of each module, is available at the GenProtEC web site (http://genprotec.mbl.edu).

We classified all genes of *E. coli* K-12 by the functions of their products, both the experimentally known and the predicted functions, as summarized in Table VII. The distribution of genetic resources in *E. coli* is in three large functional classes and several smaller ones. The largest number of genes code for enzymes (34%), followed by the genes for transport functions (11%) and then the genes for regulatory processes (9%). Genes in the three categories together account for more than 55% of all genes and for more than 75% of genes with known products in *E. coli* K-12. Moreover, almost all of the largest paralogous groups fall into these three categories (Tables III–VI).

Interestingly, proteins in each of the three categories show distinctive characteristics. Enzymes tend to be clustered into smaller paralogous groups as indicated by the smallest average group size and the highest percentage (32%) as singles (nonparalogues). Single proteins that have no partner in *E. coli* can be considered as the only surviving members of preexisting paralogous groups or as former members of groups that have diverged too far from their former partners to allow the similarities to be detected today. Transporters and regulators are in far fewer groups than enzymes but they aggregate into much larger groups (Fig. 3) and have lower percentages as nonparalogues (17 and 20%, respectively) compared with enzymes. While regulators have the highest proportion of groups larger than seven (Fig. 3), the three largest paralogous groups are all transporters and they account for over 60% of all transporter proteins (Table IV). In addition, transporters have the highest percentage

TABLE III

MAJOR PARALOGOUS ENZYME FAMILIES IN *E. coli*

Group ID[a]	Size[b]	Group function[c]
121	34	Mainly putative oxidative proteins
253	29	Fe–S subunit of oxidoreductase
168	18	NAD(P)-dependent oxidoreductases
128	18	NAD(P)-dependent alcohol dehydrogenase
27	17	ATP-dependent helicases
96	15	Aldehyde dehydrogenases
231	14	Methyltransferases
14	14	Acyl transferases
53	13	FAD/NAD(P) oxidoreductases
252	13	Formate dehydrogenase/DMSO reductase
87	12	Mainly P-type ATPases
308	11	Dehydratases/deaminase for cysteine, serine, threonine, tryptophan
368	11	Sugar kinases
170	11	Oxidoreductases; NAD(P)-binding subunit
16	10	Acetyl-CoA synthetases/ligases
543	9	IS tranposase
82	9	Oxidoreductases; all share an NAD(P)-linked oxidoreductase domain
473	9	Phage-related integrases/recombinases
39	8	Acetolactate synthase (large unit)
190	8	Dehydrogenases (iron-dependent)
91	7	Acyl-CoA *N*-acyltransferases
69	7	Aminotransferases
105	7	Endoglucanases/dihydrodipicolinate synthases
15	7	Dehydratases/racemases/CoA hydratase (involved in metabolism of fatty acids)
90	7	tRNA synthetases for asparagine and lysine
228	7	Dehydratases/epimerase for NDP-glucose/mannose
124	7	Acetyl-CoA synthetases
244	7	Glutathione *S*-transferases
29	7	Sugar kinases for ribulose, fuculose, xylulose, glycerol
28	6	Phosphate 4-epimerase/aldolase for L-ribulose, fuculose, rhamnulose
286	6	NADH/(P)-binding dehydrogenases
249	6	Reductases
324	5	Acyl-CoA transferases
77	5	Decarboxylases for lysine, ornithine, arginine
501	5	Phosphate epimerases/sythases for ribulose, thiamine, and hexose
305	5	Glucose-1-phosphate transferase (uridylyl, adenynyl, thymidylyl, pyrodylyl)
265	5	DNA-dependent helicases
242	5	Pyruvate formate lyase-activating enzymes
336	5	Oxidoreductases (for altronate, mannitol, mannonate; mostly N-terminal)
2	4	Transaldolases
434	4	*N*-Acetylmuramoyl-l-alanine amidase
372	4	Bisphosphate aldolases for tagatose, fructose

[a] Arbitary number for *E. coli* paralogous groups.
[b] Size excludes multimodules in the same group.
[c] General function based on the known functions of the group members.

TABLE IV

MAJOR PARALOGOUS TRANSPORTER FAMILIES IN *E. coli*[a]

Group ID	Size	Group function
521	94	Transport proteins (membrane component, mainly MFS superfamily)
32	83	Transport proteins of ABC superfamily (ATP-binding component)
33	68	Transport proteins of ABC superfamily (membrane component), also PTS
158	39	Mostly membrane component of transporter, regulators. Single hybrid motif
524	31	Transporter (mainly APC superfamily for amino acids)
278	12	Transport proteins (SSS, DAACS family)
68	12	Transport proteins (RND, GntP family)
223	12	PTS transport proteins, IIA component
530	11	Transport proteins of ABC superfamily (membrane component)
34	11	Transport proteins of ABC superfamily (peripliasmic-binding)
201	11	ABC transporters for amino acids (periplasmic-binding component)
532	10	Transport proteins (NCS2 and GntP families)
526	9	Transport proteins (GPH, MFS family)
467	7	Transport proteins (HAAAP family)
529	7	Transport proteins (N-terminal; membrane component; ABC, MFS family)
193	7	Part of transport proteins (mainly DASS family)
356	6	Putative transmembrane proteins
160	6	Putative transport proteins
531	6	Putative transport proteins
172	5	Putative transport proteins
523	5	MFS family of transport proteins
210	4	Sugar-specific PTS system (IIB component)
378	4	Sugar-specific PTS system (IIC component)
343	4	DMT family of transport proteins
405	4	NUP family of transport proteins
229	4	ABC superfamily (periplasmic-binding component for spermidine, molybdate, putrescine)
19	4	Transporter proteins (BCCT and MATE family)

[a] Headings the same as in Table III. (see footnotes *a–c*).

(18%) of multimodular proteins (both internal duplication and gene fusion), almost double the corresponding numbers for enzymes (8%) and regulators (8.6%), suggesting that transporter proteins are more often involved in gene fusion and internal gene duplication. Fused subunits of complex transporters are common.

TABLE V

MAJOR PARALOGOUS REGULATOR FAMILIES IN *E. coli*[a]

Group ID	Size	Group function
8	45	LysR-type transcription regulators
145	43	Mainly LuxR/UhpA-type two-component regulators
146	36	Kinase components of two-component regulators
51	19	Mainly GntR-type transcription regulators
30	27	AraC/XylS-type transcritional regulators
40	24	Mainly GalR/LacI-type transcription regulators, some periplasmic-binding component of ABC
130	15	Mainly EBP transcription regulators
245	12	DeoR-type transcription regulators
107	8	LclR-type transcription regulators
143	8	NagC/XylR-type transcription regulators
328	6	Methyl-accepting chemotaxis proteins (N-terminal of multimodular proteins)
165	5	MerR-type transcription regulators
83	4	LysR-type transcription regulators (mostly N-terminal)
156	3	AsnC-type transcription regulators
215	3	OmpR-type regulators
337	3	MarR-type transcription regulators

[a] Headings the same as in Table III. (see footnotes *a–c*).

TABLE VI

MAJOR PARALOGOUS FAMILIES IN *E. coli* FOR OTHER CATEGORIES[a]

Group ID	Size	Group function
59	30	Fimbrial proteins
139	15	Mainly conserved proteins, 2 flagellar-related
62	12	Periplasmic chaperones
61	11	Outer membrane proteins (N-terminal of group 60)
52	10	Mixed functions but mostly related to DNA
60	10	Outer membrane proteins (mostly C-terminal of group 61)
31	8	Unknown
155	6	Unknown
5	5	HSP70
472	5	Conserved proteins
238	5	Conserved proteins
218	5	Conserved proteins
283	5	Cytochrome
88	4	Conserved proteins
277	4	Conserved proteins

[a] Headings the same as in Table III. (see footnotes *a–c*).

TABLE VII

FUNCTION DISTRIBUTION OF ALL *E. coli* PROTEINS

Function	Paralogues	Singles	Total (%)[a]
Enzyme	1030	488	1518 (34.5)
Transporter	410	86	496 (11.2)
Regulator	312	84	396 (8.9)
Factor	62	69	131 (3.0)
Membrane	320	26	326 (7.4)
Structure	39	85	124 (2.8)
Carrier	8	21	29 (0.7)
RNA		114	114 (2.6)
Phage, IS	91	219	310 (7.0)
Leader		12	12 (0.3)
Cell process	11	14	25 (0.6)
Unknown	368	777	1145 (26.0)
Total[b]	2415	1983	4403

[a] Percentage in relation to ORFs.
[b] The number excludes the overlap situations between the function categories.

In addition, as illustrated in Fig. 4, enzymes, transporters, and regulators showed different distribution patterns of the sequence similarity level between paralogues as measured by PAM scores. The sequence similarity between enzyme paralogues ranges from the threshold we used (PAM, 200; residue identity, <25%) to near-identity (PAM, <1; residue identity, >98%), with a median PAM value of 149. Clearly enzymes range widely in their degree of sequence conservation. Since many enzymes are found in small paralogous groups or pairs, it appears that either there were many ancestors or, more likely, many enzymes have diverged far from each other in their specialization, maintaining closeness to only a few others of similar sequence and function. Regulator paralogues have the lowest median PAM (137), but with very few at a very high sequence similarity level, indicating that very few recent duplications have occurred for genes in this category. Since regulators tend to belong to large groups, it appears that a few kinds of regulation mechanisms have been widely used. Interestingly, transporters have the highest median PAM value (165) among the three types of proteins, and the majority of transporter paralogues have a PAM value over 180, indicating that they require the least conservation for their function as transporters or that many transporters are very old paralogues. Like regulators, there is not as much variety in types of transporters as in types of enzymes. These data suggest that different types of proteins, at least in the cases of enzymes, transporters, and regulators, have separate evolution histories. To account for the differences in the sequence similarity

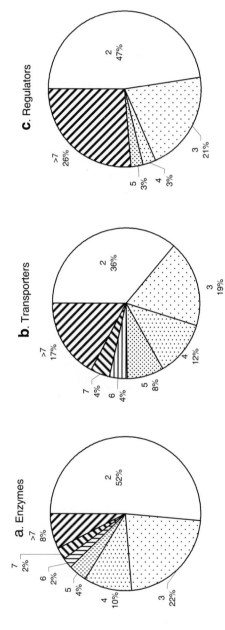

FIG. 3. The frequency of group sizes for enzymes, transporters, and regulators in *E. coli*. The sizes of paralogues groups for each type of proteins are divided into the following size categories: 2, 3, 4, 5, 6, 7, and >7. The percentage of groups in each size category in relation to the total number of groups in each function category is also indicated.

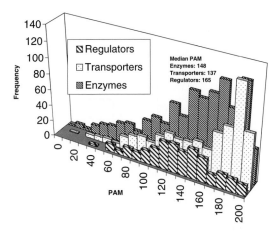

FIG. 4. Frequency of PAM scores for paralogous enzymes, transporters, and regulators. For each protein, only the best match as represented by the lowest PAM score was used. The frequency of PAM scores in each value range (every 10) is plotted for each of the three protein types.

level for the paralogues in the three types of proteins, it may be that transporters underwent gene duplication very early in evolution, while the gene duplication for most regulators happened at a later time and duplication for enzymes happened continuously over time. While many paralogous transporter proteins and enzymes are known to be very ancient and existed in the LUCA (Gogarten, 1994; Makarova *et al.*, 1999; Koonin and Bork, 1996; Mehta *et al.*, 1993; Tomii and Kanehisa, 1998), there have been a few reports suggesting a later emergence of transcription regulators. Aravind and Koonin reported that the core transcription regulators in archaea and eukaryotes (TFIIB/TFB, TFIIE-α, and MBF-1), and those in bacteria (the sigma factors) share no sequence similarity beyond the presence of a distinct HTH domain, and suggested that HTH domains might have been independently recruited for a role in trascription regulation in bacterial and archaeal/eukaryotic lineages (Aravind and Koonin, 1999). In another report, the derivation of eukaryotic transcription regulators from ancient enzymatic domains was proposed (Aravind and Koonin, 1998).

C. Relationships of Paralogous Protein Family Members

The sequence relationships among the members of one family may differ from those in other families. Some families have a tight and mutually connected sequence relationship, while others may have a smaller core set of members that are tightly related but the rest of the members

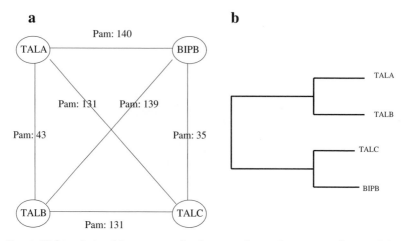

a

Pam: 140

TALA — BIPB

Pam: 131   Pam: 139

Pam: 43   Pam: 35

TALB — TALC
Pam: 131

b

TALA
TALB
TALC
BIPB

Fɪɢ. 5. Tight relationships among the four members of a group of transaldolases. (a) Sequence similarity exists between each member and all of the other members as indicated by the solid lines and the PAM scores along the lines. (b) A possible phylogenetic tree based on the PAM distance matrix. Proteins are labeled by the first section of the Swiss-Prot ID.

are connected to only one or a few of the core members. As shown in Fig. 5a, each of the four members in a group of transaldolases shares sequence similarity to the other three members, although the similarity levels are not the same. In contrast, in a family containing oxidases, the four members have a much looser relationship in sequence (Fig. 6a). The degrees of similarity between YDIJ and GLCF, YHCU and GLCD; and GLCF and GLCD are below the threshold we set. The different anatomy of protein families no doubt reflects the fact that each family of proteins has experienced different evolution paths and has different requirements of sequence conservation. Figures 5b and 6b represent the possible evolution paths for the genes corresponding to the proteins in two groups.

Members of each of the paralogous groups almost always exhibit a clear relationship in functions, supporting evidence for the homology shared among the members of a paralogous family. This is true even in groups formed transitively so that not all members are related by our criteria for sequence similarity to all other members. An example is the group of oxidases discussed above (Fig. 6), in which all four members have "oxidase" as their shared function (including predicted function) and "FAD-binding domain" as their shared reaction type and structural property, despite the loose sequence relationships among these four proteins.

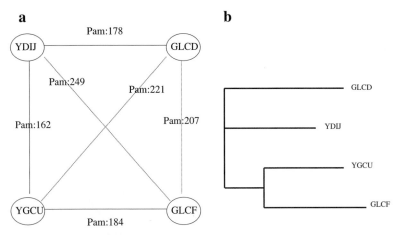

FIG. 6. Loose relationships among the four members of a group of oxidases. (a) Sequence similarity within the threshold exists only between three pairs of the members (indicated by solid lines). Dotted lines indicate that the sequence similarity is beyond the threshold. The PAM score for each pair of proteins is shown beside or on the line connecting the two proteins. (b) A phylogenetic tree based on the PAM distance matrix. Proteins are labeled by the first section of the Swiss-Prot ID.

For larger paralogous protein families, there is an observable relationship between the functions and deeply branching subfamilies. The diagram for the members in a group of ATP-binding subunits of the multimeric ABC transporters demonstrates the detectable conservation for different types of substrates transported, such as amino acids, oligopeptides, sugars, metals, and metal-containing proteins (Fig. 7). Evidence of functional divergence among the paralogous members in gaining specificity in substrate binding, cellular location of function, or even new types of function in some exceptional cases provides us with abundant materials for studying the modes of functional divergence of proteins originally descended from a common ancestor by sequence divergence (Kerr and Riley, submitted for publication; Nahum and Riley, 2001).

D. PROTEINS AND PROTEIN FAMILIES SHARED BETWEEN *E. coli* AND *A. fulgidus*

Proteins shared by *E. coli* and *A. fulgidus* can be thought of as representing the proteins shared by the eubacterial and archaeal domains except for those horizontally transferred between the specific lineages related to these two species. Therefore, groups of these proteins with similar sequences probably also represent ancestral proteins at nodes

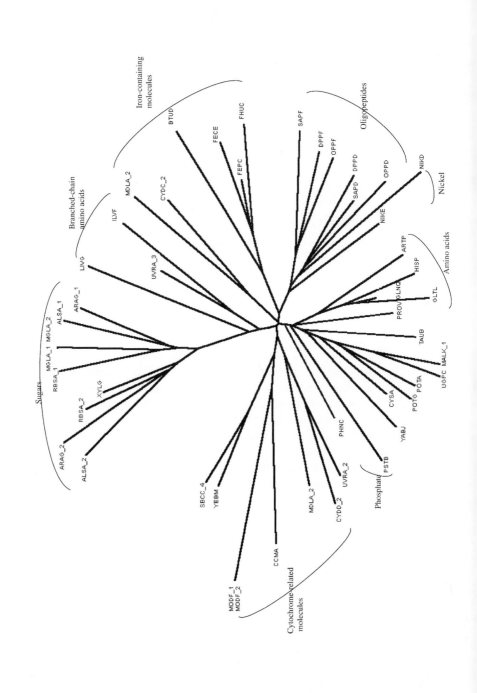

of evolutionary development prior to separation of the two domains. When conservative criteria (maximal PAM of 200 and minimal alignment length of 100 residues) are used to detect similarity in amino acid sequence, 1535 (36%) *E. coli* proteins are orthologous to 1146 (48%) *A. fulgidus* proteins. These numbers increase to 2581 (60%) and 1565 (65%) for *E. coli* and *A. fulgidus,* respectively, when the criteria are loosened to a maximal PAM value of 250 and a minimal alignment length of 83 residues. These data suggest that in these two very distantly related genomes, up to 50% of their proteins still share detectable homology today, while each species also has a substantial portion of genes whose counterparts either do not exist in the other genome or have diverged to an undetectable distance.

Among the 1535 *E. coli* proteins that have counterparts in *A. fulgidus,* 1200 (78%) are on the *E. coli* paralogue list, while that number is 815 (71%) for *A. fulgidus,* indicating that the majority of these "universal" and likely essential proteins have a history of duplication. Among the 610 paralogous protein families identified in *E. coli,* 324 contain at least one orthologue in *A. fulgidus,* while the rest of the groups contain proteins that have no *A. fulgidus* homologue. Table VIII shows a list of large paralogous groups shared by *E. coli* and *A. fulgidus,* with the group sizes in both genomes indicated and the groups ordered by the group size in *E. coli* in a descending manner. All large *E. coli* groups have counterparts in *A. fulgidus.* Both genomes have membrane components of transporters as the largest group and the ATP-binding component of ABC transporters as the second largest group, but the order of the other groups by size is not the same in the two genomes. For example, the next largest group in *A. fulgidus* is the two-component regulators, while it is the LysR type of transcriptional regulators in *E. coli.*

Notably, although its genome size is only about 50% of the genome size of *E. coli, A. fulgidus* seems to have unusually large groups of acetyl-CoA synthetases/ligases as shown by the much larger group size than for the corresponding *E. coli* group (group 16 in Table VIII) and as reported from the genome sequencing project (Klenk *et al.,* 1997). The same situation is true for two additional acetyl-CoA related groups (not listed). The phyiological implications of the multiplicity of fatty acid metabolism are not yet understood. In addition, *A. fulgidus* seems to have relatively

---

FIG. 7. An unrooted tree for an *E. coli* paralogous group representing the ATP-binding components of ABC transporters. Among the 94 members, only the 59 proteins with experimental data on the substrates were included in the tree construction based on the PAM distance matrix to show the relationship between phylogenetic positions and the substrate binding specificity. Proteins are labeled with their Swiss-Prot IDs. Proteins with suffixes are partial-length modules.

TABLE VIII

Major Types of Paralogous Families Shared by *E. coli* and *A. fulgidus* and Their Group Sizes

Group ID[a]	EC size[b]	AF size[c]	Group function
521	94	52	Transport proteins (membrane component; mainly MFS superfamily)
32	83	57	Transport proteins of ABC superfamily (ATP-binding component)
33	68	27	Transport proteins of ABC superfamily (membrane component; also PTS)
8	45	7	LysR transcription regulators; all have periplasmic-binding protein-like II domains
145	43	26	Mainly LuxR/UhpA-type two-component regulators; all have a CheY-like domain
158	39	17	Mostly membrane component of transporter, regulators
146	36	34	Kinase component of two-component regulators
121	34	18	Mainly conserved proteins; many have no SCOP assignments
524	31	16	Transport proteins (mostly APC superfamily for amino acids); no SCOP assignments
59	30	3	Fimbrial proteins, all have bacterial adhesins
51	27	12	Mostly GntR-type regulators
40	24	4	Mostly GalR/LcI-type transcription regulators/PBP; all have periplasmic-binding protein-like I
128	18	8	Oxidoreductases; all share NAD(P)-binding Rossmann-fold domains and GroES-like
168	18	4	Oxidoreductases; all share NAD(P)-binding Rossmann-fold domains
27	17	10	ATP-dependent helicases
103	17	11	GTP-binding proteins
96	15	1	Aldehyde dehydrogenases (NAD-dependent)
130	15	19	Mostly EBP-type transcription regulators with P-loop-containing nucleotide triphosphate hydrolases
14	14	1	Acyl transferases; all have trimeric LpxA-like domains
231	14	9	Methyltransferases; all have *S*-adenosyl-L-methionine- dependent methyltransferases
53	13	14	Oxidoreductases (mostly N-terminal); all have one or more FAD/NAD(P)-binding domains
251	13	4	Mixed functions (mostly membrane proteins related to cell division)
252	13	9	Dehydrogenases/reductases for DMSO, N_2, formate, biotin sulfoxide
68	12	12	Transport proteins (RND, GntP family)
87	12	12	P-type ATPase mostly; 6 have an SIS domain

(continues)

TABLE VIII—*Continued*

Group ID[a]	EC size[b]	AF size[c]	Group function
223	12	2	PTS IIA component; share the IIA domain of mannitol-specific and ntr phosphotranspherase EII
245	12	2	DeoR-type transcription regulators (mostly N-terminal)
278	12	14	Transport proteins (SSS, DAACS family)
7	11	5	Membrane protein (most for dehydrogenases)
34	11	1	Transport proteins of ABC superfamily (periplasmic-binding component)
170	11	4	Oxidoreductases; all have NAD(P)-binding Rossmann-fold domains
201	11	6	ABC transport proteins (periplasmic-binding component) with periplasmic-binding protein-like II domains
308	11	9	Dehydratases (cystine, serine, threonine, tryptophan)
368	11	4	Sugar kinases; all have ribokinase-like domains
530	11	14	Transport proteins of ABC superfamily (membrane component)
16	10	26	Acetyl-CoA synthetases/ligases; all have Firefly luciferase-like
52	10	8	Mixed; 3 dehydrogenases, 3 HSPs
60	10	1	Outer membrane proteins (mostly C-terminal of group 61)
532	10	7	Transport proteins

[a] Group number for *E. coli*.
[b] Group size for *E. coli*.
[c] The number of proteins that have orthologues to the *E. coli* proteins in the group.

large groups of two-component transcriptional regulators but smaller groups of certain other types of transcriptional regulators, such as LysR, GalR/LacI, and DeoR, and it has a very small number of fimbrial proteins. The different characteristics for *E. coli* and *A. fulgidus* in the kinds and relative sizes of protein families are believed to reflect the different cell physiology, metabolism, and biology of these two genomes. An extensive examination of these cases in the context of the physiology of these two genomes will provide useful information for improving our understanding of their biological natures.

E. PROTEINS UNIQUE TO *E. coli* AND *A. fulgidus*

Data from genome sequence projects so far have revealed that almost every genome has at least 20% of genes that do not match any other sequences in the current databases. These unmatched parts contribute a major proportion of unknown genes and impose a big challenge to genome annotation. However, these genes may be the key elements responsible for the distinctive characteristics of each type of organism. Therefore, it would be very crucial and interesting to understand the nature of these genes. Comparative genomics provides a powerful approach to identify and study these genes. To collect the genes that are unique to a genome, it seems more rational to use more aggressive (loose) criteria for detecting homologues to ensure the true uniqueness of the identified genes. When the criteria of maximum PAM scores of 250 and minimum alignment lengths of 83 were used for detecting all orthologues in the 21 genomes listed in Table I, 105 E. coli proteins were found not to have any homologues among those genomes (Table IX). Twelve of the 105 are prophage genes, not bacterial in origin. Of the remaining 83, 72 (86%) have no match to the 21 microbial genomes, but among them are 7 "putative" ORFs with orthologues in other genomes. The remaining 65 are proteins related to a variety of functions, such as regulators and redox-related proteins. Understanding the contribution of unique genes to cell physiology will provide molecular information on the properties that differ between organisms.

It needs to be pointed out that the label "unique" is relative only to the specific set of genomes used in the analysis. The number of unique proteins drops when closely related species are included. It drops to 64 when Salmonella typhimurium genes are included (unpublished data), and we can assume that the list will be even shorter when Shigella and other E. coli species are included in the comparison. However, although the exact numbers of unique genes in each genome might be difficult to predict, it can be certain that the number will not drop to zero, because each differentiated species and strain is characterized by observable differences. In a practical vein, the small differences have value. Differences exist between nonpathogen and pathogen strains. This kind of information should help us to identify the virulent factors in the pathogen strains.

A comparable analysis for A. fulgidus against the other 21 genomes listed in Table 1 revealed that 60 protein-coding ORFs in this genome do not have any orthologues in the genomes on the list, and none of them has any known or predicted function information. We can expect that this fact will remain true for every genome, especially those organisms with little biology known.

F. THE MINIMUM NUMBER OF ANCESTRAL PROTEINS

A matter of ongoing interest is to figure out a minimal set of ancestral proteins necessary for generating all present-day proteins (Riley and Serres, 2000; Mushegian and Koonin, 1996; Mushegian, 1999; Maniloff,

TABLE IX

E. coli Unique Proteins Relative to 21 Completed Genomes

Gene	Gene product
b1998	CP4-44 prophage; putative outer membrane protein
yfjM	CP4-57 prophage
b2362	CPS-53 prophage
b2363	CPS-53 prophage
yfdM	CPS-53 prophage; putative transferase
ybcO	DLP12 prophage
hipB	DNA binding regulatory protein; interacts with HipA
ydfR	Qin prophage
b1556	Qin prophage
rem	Qin prophage
rpoZ	RNA polymerase, ω subunit
ynaK	Rac prophage
racC	Rac prophage; contains recE and oriJ
cedA	Cell division modulator; affects inhibition after overreplication of chromosome in dnaAcos mutants
minE	Cell division topological specificity factor; reverses MinC inhibition of FtsZ ring formation
dinJ	Damage-inducible protein J
ymfJ	E14 prophage
grxC	Glutaredoxin 3
grxA	Glutaredoxin1 (redox coenzyme for glutathione-dependent ribonucleotide reductase)
hfq	Host factor I for bacteriophage Q β replication, a growth-related protein
psiF	Induced by phosphate starvation
rof	Modulator of Rho-dependent transcription termination
chpS	Part of proteic killer gene system, suppressor of inhibitory function of ChpB
tatA	Part of sec-independent protein export, membrane protein
napD	Periplasmic nitrate reductase
csgC	Putative curli production protein
yfgJ	Putative cytochrome
paaB	Putative phenylacetic acid degradation protein; possibly part of multicomponent oxygenase
ykgM	Putative ribosomal protein
yiaG	Putative transcription regulator
flhC	Regulator of flagellar biosynthesis; acts on class 2 operons
nlp	Regulatory factor of maltose metabolism; similar to Ner repressor protein of phage μ
araH-2	Split gene, ABC superfamily (membrane) high-affinity L-arabinose transport protein, fragment 2
72 unknown ORFs	b0235, b0302, b0370, b0380, b0392, b0609, b0725, b0816, b1228, b1481, b1760, b1825, b1836, b1839, b1903, b1936, b2084, b2658, b3100, b3776, b3836, smf_1, yahM, yaiB, yajD, ybbV, ybfN, ybiI, ybjC, ybjH, yceP, ycgW, ychH, yciN, ydbL, ydiH, yebW, YedN, YefM, yehE, yfaE, yfcL, yfiM, ygaC, ygeP, ygfY, yghW, ygiA, ygjN, yhbQ, yhcR, yhhH, yifN, yjbD, yjbL, yjeN, yjfA, yjgG, yjgJ, yjgZ, ykfE, ymdA, ymgA, ymgB, ynaJ, yneG, ynfB, yoaF, yohH, ypdI, yrdB, ytfA

1996). Assembly of paralogous protein families for a large and sufficiently varied number of genomes provides a possibility to estimate the minimal number of ancestral genes. If molecular evolution were completely transparent on the basis of sequence similarity, then the minimal number of ancestral proteins for *E. coli* would be the sum of the number of paralogous groups and the number of singles (nonparalogues). However, the actual number of ancestral proteins represented by *E. coli* is probably much smaller and the exact number might not be possible to know on the basis of the primary sequence alone. Our current methods for detecting sequence similarity may not be sensitive enough to bring all the true homologous members into groups. After many rounds of duplication and divergence, some genes and groups of genes have diverged to a point beyond the capability of the detection method we use. This means that it should be possible to cluster some of the *E. coli* paralogous groups into larger supergroups and some groups may be extended to include some proteins that are currently identified as singles. The number of ancestors required for *E. coli* proteins will be reduced (1) when bridges between sequence-similar groups are found by similarity of motifs or by finding bridging sequences in other organisms and (2) when superfamilies of protein structure are used to extend the protein families by their similarity in structural properties.

G. THE LENGTH OF PRESENT-DAY PROTEINS AND MODULES: A HINT FOR THE SIZES OF ANCESTRAL PROTEINS

To examine the patterns for evolution by gene fusion and duplication, we examined the distribution patterns of protein length according to the number of residues for all proteins and modules in each of the three domains based on their representatives listed in Table 1. At least two conclusions may be drawn from the data in Fig. 8. First, most ancestral proteins were about 100 residues in length. Proteins in all three domains seem to have selective ranges for their protein lengths, but all three domains share one range, from 100 to 130 residues, although eubacteria and eukaryotes have additional preferred ranges. In contrast, modules in all three domains demonstrate very uniform preferences of about 100 residues, suggesting that this is the optimal length range for unimodular proteins, which are probably the case for most ancestral proteins. Second, the total protein length distribution patterns are not the same for the three domains of life. On average, eukaryotes (data limited to *S. cerevisiae* and *C. elegans*) have the longest protein length (443 amino acids), followed by eubacteria (320 amino acids), with archaea having the shortest protein length on average (288 amino acids). Eukaryotic proteins seem to show a trimodal distribution pattern

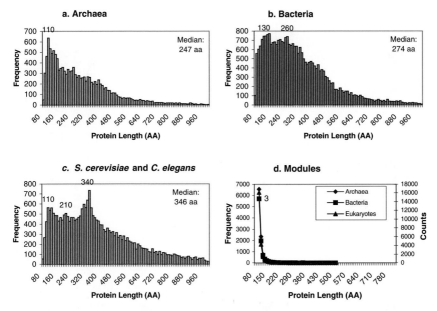

FIG. 8. Length distribution for proteins and modules. All proteins that are longer than 80 residues and their detected modules from the 22 genomes divided into the three domains of life. (a, b, c) The length frequency of proteins in Archaea, Eubacteria, and Eukarya, respectively. (d) The length frequency for modules in the three domains. Only the shortest lengths were used for modules which have alignment with more than one protein using slightly variable regions. The median values for proteins in each domain are indicated in the upper-right corners.

corresponding to unimodular, bimodular, and trimodular proteins at 110, 230, and 350 residues, respectively, while eubacteria have only the first two and archaea have only the first one. We believe that the average protein length in a genome is an indicator of the complexity of the genome and protein functions, and this number can be expected to be much higher for higher eukaryotes due to more frequent and complex gene fusions. In this sense, the archaeal domain seems to be the simplest in form among the three domains.

IV. Conclusions and Perspectives

The existence of paralogous groups of proteins of similar sequence and related function is entirely consistent with the proposal that processes of duplication followed by divergence have played a major role in the molecular evolution of proteins and genomes as shown in Fig. 9. It seems quite clear that limited kinds of proteins existed in the cells of

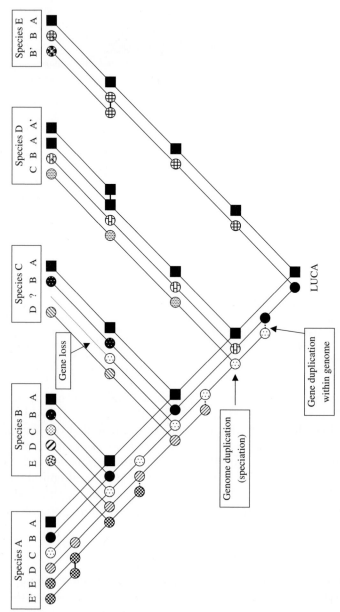

FIG. 9. A schematic illustrating simplified patterns of protein evolution. Two ancestral proteins (represented by the different shapes) from the last common ancestor (LUCA) have evolved differently, generating the set of proteins in the present-day species A, B, C, D, and E. One is very conservative (squares), with all of its descendants sharing detectable sequence similarity to each other, while the other (circles) has been actively involved in duplication within genomes and divergence at various points over the whole evolution path, generating descendant proteins that may not share detectable sequence similarity among the paralogues and orthologues (represented by different shading patterns of circles). Gene loss occurred in some lineages.

the LUCA, represented mainly by enzymes, transporters, and factors which were mostly unimodular and about 100 residues long. These proteins represent the core set of proteins that are involved in essential cellular functions, and they were used as the basic building blocks for generating all present-day large and complex multimodular proteins through internal gene duplication and gene fusion. Gene duplication within a genome has given rise to all paralogous families in present-day genomes and has led to the increase in the size and complexity of genomes, while duplication of genomes in the course of speciation has resulted in orthologous gene families and an increased number of species. Enzymes, transporters, and regulators seem to have experienced separate and distinctive evolution paths, with enzymes and transporters being relatively older than regulators.

Several biological processes complicate the analysis of the evolution process. Duplication of genes has continued over time. Through speciation and by duplication and divergence within individual genomes, proteins of any age have become ancestors and have continued to produce variants. Some variants, continually evolving, finally, over time, have lost recognizable similarities, leaving these genes outside of any paralogous families. The same process also has occurred among genes entering different genomes due to the speciation process, in which case orthologues may have followed different evolutionary paths and may have gained different functions because of the different physiological environments of separated species. In other cases, an entire gene lineage may have been lost in certain species, making attempts to track ancestral relationships more challenging. Horizontal gene transfers among different genomes have provided sources of new function and, at the same time, have complicated the analysis of evolution processes (Doolittle, 1999; Hilario and Gogarten, 1993). Alien genes have been introduced into bacteria, mainly from phages, from transposons, and by chromosomal recombination with plasmids. Extended sequence divergence and alien genes seem likely to be the main factors responsible for the existence of genes that are unique to certain phylogenetic lineages.

With the improving sequencing capability at industrial scales, we can expect that the number of completed genomes will increase very quickly. Experts are anticipating this number to be over 1000 genomes by the year 2005. As the number of sequenced genomes increases, the amount of comparative sequence data increases exponentially. To be able to manage and make full use of the astronomic amount of data, we are relying on more powerful computers with more memory and computation speed as well as more intelligent software (Kanehisa, 1998). One of the main tasks is to decipher the function of all genes based on information from traditional experimental biology, bioinformatics predictions using homology, and non-homology-based approaches including protein structure and microarray expression data. Along with the genome sequence projects, worldwide efforts are being made toward determine the structural properties for all existing proteins. This information not only will help us to understand better the function of individual proteins, but also will help to connect and expand our protein families and thus trace back farther the ancient connections to ancestors shared by present-day proteins.

Continued analysis will bring greater understanding of the fundamental events in evolution, the separation of the basic three domains of life: Archaea, Eubacteria, and Eukarya. The functional elucidation of lineage-specific unique genes will be a big challenge but it is important information for understanding the genetic basis of the characteristics of each domain and its constituent subdivisions. But what will be even more challenging and meaningful is to integrate all functions of individual genes within organisms in the context of gene regulation and metabolism, intercellular communication, and interaction with the physiological environment. Nevertheless, more complete genome data covering more phylogenetic lineages as well as closely related species will provide us answers about the nature of species and a more complete picture of ancient genomes.

ACKNOWLEDGMENTS

We thank Dr. Mitch Sogin for helpful discussions and suggestions on the manuscript and Drs. Margrethe Serres and Sarah Teichmann for the provision of some of their unpublished data. This work was supported by NASA Astrobiology Institute Grant NCC2-1054.

REFERENCES

Altschul, S. F., Madden, T. L., Schaffer, A. A., Zhang, J., Zhang, Z., Miller, W., and Lipman, D. J. (1997). *Nucleic Acids Res.* **25**, 3389–3402.
Aravind, L., and Koonin, E. V. (1998). *Curr. Biol.* **8**, R111–R113.
Aravind, L., and Koonin, E. V. (1999). *Nucleic Acids Res.* **27**, 4658–4670.
Axelsen, K. B., and Palmgren, M. G. (1998). *J. Mol. Evol.* **46**, 84–101.
Baron, M., Norman, D. G., and Campbell, I. D. (1991). *Trends Biochem. Sci.* **16**, 13–17.
Berlyn, M. K. B. (1998). *Microbiol. Mol. Biol. Rev.* **62**, 814–984.
Brown, J. R., and Doolittle, W. F. (1997). *Microbiol. Mol. Biol. Rev.* **61**, 456–502.
Campbell, I. D., and Baron, M. (1991). *Philos. Trans. R. Soc. Lond. B Biol. Sci.* **332**, 165–170.
Cannarozzi, G., Hallett, M. T., Norberg, J., and Zhou, X. (2000). *Bioinformatics* **16**, 654–655.
Dayhoff, M. O. (1976). *Fed. Proc.* **35**, 2132–2138.
Dayhoff, M. O., Schwartwz, R. M., and Orcutt, B. C. (1978). *In "Atlas of Protein Sequence and Structure,"* (M. O. Dayhoff, ed.), pp. 345–352. National Biomedical Research Foundation, Washington, DC.

Doolittle, R. F. (1998). *Nature* **392,** 339–342.
Doolittle, W. F. (1999). *Trends Cell Biol.* **9,** M5–M8.
Feng, D. F., Cho, G., and Doolittle, R. F. (1997). *Proc. Natl. Acad. Sci. USA* **94,** 13028–13033.
Fitch, W. M. (1970). *Syst. Zool.* **19,** 99–113.
Furano, A.V. (1977). *J. Biol. Chem.* **252,** 2154–2157.
Gogarten, J. P. (1994). *J. Mol. Evol.* **39,** 541–543.
Gogarten, J. P., and Olendzenski, L. (1999). *Curr. Opin. Genet. Dev.* **9,** 630–636.
Gogarten, J. P., Kibak, H., Dittrich, P., Taiz, L., Bowman, E. J., Bowman, B. J., Manolson, M. F., Poole, R. J., Date, T., and Oshima, T. (1989). *Proc. Natl. Acad. Sci. USA* **86,** 6661–6665.
Gonnet, G. H., Cohen, M. A., and Benner, S. A. (1992). *Science* **256,** 1443–1445.
Gonnet, G. H., Hallett, M. T., Korostensky, C., and Bernardin, L. (2000). *Bioinformatics* **16,** 101–103.
Hegyi, H., and Bork, P. (1997). *J. Protein Chem.* **16,** 545–551.
Hilario, E., and Gogarten, J. P. (1993). *Biosystems* **31,** 111–119.
Johanson, U., and Hughes, D. (1992). *Gene* **120,** 93–98.
Kanehisa, M. (1998). *Bioinformatics* **14,** 309.
Iwabe, N., Kuma, K., Hasegawa, M., Osawa, S., and Miyata, T. (1989). *Proc. Natl. Acad. Sci. USA* **86,** 9355–9359.
Kessler, D., Leibrecht, I., and Knappe, J. (1991). *FEBS Lett.* **281,** 59–63.
Klenk, H. P., Clayton, R. A., and Tomb, J. F., *et al.* (1997). *Nature* **390,** 364–370.
Koonin, E. V., and Bork, P. (1996). *Trends Biochem. Sci.* **21,** 128–129.
Koonin, E. V., and Mushegian, A. R. (1996). *Curr. Opin. Genet. Dev.* **6,** 757–762.
Koonin, E. V., Aravind, L., and Kondrashov, A. S. (2000). *Cell* **101,** 573–576.
Labedan, B., and Riley, M. (1995a). *Mol. Biol. Evol.* **12,** 980–987.
Labedan, B., and Riley, M. (1995b). *J. Bacteriol.* **177,** 1585–1588.
Labedan, B., and Riley, M. (1999). In *"Organization of the Prokaryotic Genome,"* (R. L. Charlebois, ed.), pp. 311–329. ASM Press, Washington, DC.
Lander, E. S., and Weinberg, R. A. (2000). *Science* **287,** 1777–1782.
Li, W.-H., and Graur, D. (1991). "Fundamentals of Molecular Evolution." Sinauer Associates Sunderland, MA.
Makarova, K.S., Aravind, L., Galperin, M. Y., Grishin, N. V., Tatusov, R. L., Wolf, Y. I., and Koonin, E. V. (1999). *Genome Res.* **9,** 608–628.
Maniloff, J. (1996). *Proc. Natl. Acad. Sci. USA* **93,** 10004–10006.
Mehta, P. K., Hale, T. I., and Christen, P. (1993). *Eur. J. Biochem.* **214,** 549–561.
Miller, D. L. (1978). *Mol. Gen. Genet.* **159,** 57–62.
Muller-Newen, G., and Stoffel, W. (1993). *Biochemistry* **32,** 11405–11412.
Mushegian, A. (1999). *Curr. Opin. Genet. Dev.* **9,** 709–714.
Mushegian, A. R., and Koonin, E. V. (1996). *Proc. Natl. Acad. Sci. USA* **93,** 10268–10273.
Nahum, L. A., and Riley, M. (2001). Divergence of Function in Sequence-Related Groups of *Escherichia coli* Proteins. Genome Research. (in press).
Ohno, S. (1970). "Evolution by Gene Duplication." Springer-Verlag, Berlin.
Ouzounis, C., Casari, G., Sander, C., Tamames, J., and Valencia, A. (1996). *Trends Biotechnol.* **14,** 280–285.
Palmgren, M. G., and Axelsen, K. B. (1998). *Biochim. Biophys. Acta* **1365,** 37–45.
Pennisi, E. (1999). *Science* **284,** 1305–1307.
Postma, P. W., Lengeler, J. W., and Jacobson, G. R. (1996). In *"Escherichia coli and Salmonella"* (F. Neidhardt, I. R. Curtiss, J. Ingraham, *et al.,* eds.), pp. 1149–1174. American Society for Microbiology, Washington, DC.
Ribeiro, S., and Golding, G. B. (1998). *Mol. Biol. Evol.* **15,** 779–788.

Riley, M., and Labedan, B. (1996). *In* "*Escherichia coli and Salmonella,*" (F. Neidhardt, I. R. Curtiss, J. Ingraham, *et al.,* eds.). American Society for Microbiology, Washington, DC.

Riley, M., and Labedan, B. (1997). *J. Mol. Biol.* **268,** 857–868.

Riley, M., and Serres, M. H. (2000). *Annu. Rev. Microbiol.* **54,** 341–411.

Rivera, M. C., Jain, R., Moore, J. E., and Lake, J. A. (1998). *Proc. Natl. Acad. Sci. USA* **95,** 6239–6244.

Sharp, P. M. (1991). *J. Mol. Evol.* **33,** 23–33.

Smith, T. F., and Waterman, M. S. (1981). *J. Mol. Biol.* **147,** 195–197.

Tomii, K., and Kanehisa, M. (1998). *Genome Res.* **8,** 1048–1059.

Yang, S. Y., Li, J. M., He, X. Y., Cosloy, S. D., and Schulz, H. (1988). *J. Bacteriol.* **170,** 2543–2548.

Section III. Archaea as Models for Eukaryotic Processes

We can see by examination of the universal phylogenetic tree that archaea emanate from a main branch ultimately leading to the eukaryotes. Bacteria, in contrast, diverged from the archaea at an earlier point in time. An essential feature distinguishing eukaryotes from other organisms is their nuclei. Accordingly, scientists often refer to eukaryotic evolution as the nuclear line of descent. The positioning of archaea on the eukaryotic branch predicts that they harbor features better known as components of the modern-day nucleus.

The final and largest section of this volume focuses on the existence of eukaryotic-like archaeal subcellular mechanisms. Close examination, however, reveals both similarities and fundamental distinctions from their eukaryotic cousins. Perhaps such distinctions provide a different view of the last common ancestor. The list of these conserved systems includes many examples involving the preservation and propagation of genetic information and its protein products. Kathleen Sandman and John Reeve describe the existence and roles of archaeal histones in chromosome packaging. Erica Seitz and co-workers review key aspects of DNA recombination and repair in archaea. Jörg Soppa reviews the important features of the eukaryotic-like basal transcription system of archaea. Protein folding and refolding enzymes, termed chaperones, constitute yet another eukaryotic-like feature of the archaea. Michel Leroux presents a comparative analysis of these enzymes from all three domains. Ultimately proteins are either degraded by the cell or diluted by cell division. Julie Maupin-Furlow and co-workers provide an in-depth comparison of the proteosome and other proteolytic systems found in members of the three domains. In the final article, Elisabetta Bini and Paul Blum examine the process of catabolite repression as currently understood from studies on bacteria and eukaryotes as a point of comparison for the emerging catabolite repression system discovered in the archaeon *Sulfolobus solfataricus.*

The list of eukaryotic-like archaeal systems is growing quickly. Additional systems not included in this text include mechanisms for RNA methylation, DNA synthesis, tRNA charging, protein phosphorylation, and protein secretion. Perhaps more will become apparent as studies on archaea continue.

Chromosome Packaging by Archaeal Histones

KATHLEEN SANDMAN AND JOHN N. REEVE

Department of Microbiology
Ohio State University
Columbus, Ohio 43210

I. Introduction

Chromosomes are condensed *in vivo*, that is, they occupy significantly less volume than an equivalent amount of DNA free in solution. Physical and chemical forces promote the condensed state, but for biological activity, condensed chromosomes must be structured so that genetic information can be accessed. Small, abundant proteins with positively charged surface residues positioned to interact with DNA are responsible for the organization of the condensed chromosome, but despite having a common function and similar biochemical features, there are several structurally distinct families of architectural chromosomal proteins. Many Bacteria employ members of the HU family, and most Eukarya have histones. In the Archaea, the distribution of chromosomal proteins parallels the primary phylogenetic subdivisions, with most Euryarchaeota having histones in common with the Eukarya, whereas several proteins, including Sac7 and homologues, are present in the Crenarchaeota. HU, histones, and Sac7 families apparently represent three solutions to the DNA compaction problem, and the existence of all three in prokaryotes suggests that all three have ancient origins.

ADVANCES IN APPLIED MICROBIOLOGY, VOLUME 50
Copyright © 2001 by Academic Press
All rights of reproduction in any form reserved.
0065-2164/01 $35.00

This article focuses on histones and the complexes they form with DNA, known as nucleosomes. Prokaryotes and eukaryotes differ vastly in the size and complexity of their genomes, but both contain nucleosomes. Here we discuss the versatility of these structures and emphasize the comparative simplicity of the archaeal system, which facilitates experimental molecular structure–function–stability dissections.

II. The DNA Packaging Problem

As a consequence of the linear nature of DNA, without compaction the physical dimensions of the genetic information required to encode a living organism would vastly exceed the dimensions of a cell and, certainly, of the subcellular region (nucleoid or nucleus), where the DNA is stored. Historically DNA compaction has been described in terms of linear reduction, but recently it has been argued that it is more appropriately treated as a reduction in volume (Bloomfield, 1997; Holmes and Cozzarelli, 2000).

Bacterial and archaeal prokaryote cells have similar dimensions, rods and cocci with lengths of $\sim$1–2 μm, and volumes that can be estimated by modeling them as spheres (cocci) or rectangular prisms (bacilli), e.g., the volume of a coccus with 1-μm diameter is $\sim$0.5 μm^3 and an *Escherichia coli*-sized bacillus of $1 \times 1 \times 2$ μm has a volume of $\sim$2 μm^3. Estimating the volume of a chromosome is significantly more complicated. The common image of a DNA molecule as a rigid cylinder is invalid for molecules longer than $\sim$140 bp (50 nm). As the length of a DNA molecule increases, so does its flexibility and the number of alternative spatial configurations. Chromosome-sized molecules are typically modeled as random coils, and Trun and Marko (1998) calculated that the 4.6-Mbp (mega-base pair) *E. coli* chromosome should adopt a random coil volume of $\sim$10 μm^3, much larger than the volume of the *E. coli* nucleoid. This example emphasizes clearly that DNA molecules must be condensed for accommodation within the confines of a nucleus or nucleoid.

Bloomfield (1996) has reviewed the forces that facilitate DNA condensation *in vitro,* and many of these also apply *in vivo.* Multivalent cations, such as polyamines, neutralize otherwise repulsive negative charges along a DNA molecule, and the crowding effects of high concentrations of proteins and nucleic acids disrupt DNA–solvent interactions. Under these conditions, there are no chemical or physical impediments to DNA aggregation into densely packed structures with a volume easily accommodated intracellularly. The DNA packaging problem is, in fact, an issue not of how to confine a "too-large" chromosome but, rather, of how to preserve the conformational flexibility of this DNA molecule after its packaging into an environment dominated by condensing forces.

Architectural chromosomal proteins solve the DNA packaging problem by altering the linear path of DNA and preventing the spontaneous
aggregation of the chromosome. Members of the bacterial HU family, as
exemplified by the *E. coli* IHF protein, bend DNA and maintain the curve
by binding to the resulting concave surface (Rice *et al.,* 1996). In contrast, the archaeal Sac7 bends and binds to the resulting convex surface
(Robinson *et al.,* 1998; Agback *et al.,* 1998). Histones also bend and bind
to the concave surface but are completely encircled by the DNA, which
is held in a superhelix around a histone core in a nucleosome (Luger
et al., 1997). It appears that the histone-based system of DNA packaging
is most efficient (Sandman *et al.,* 1998a). Histones block DNA aggregation (Minsky *et al.,* 1997) and, in so doing, organize the greatest length of
DNA per unit mass of protein. This is supported by the observations of Li
et al. (1999) that chromosome preparations from nucleosome-containing
organisms, both from Archaea and Eukarya, are more resistant to aggregation *in vitro* than those from organisms lacking nucleosomes.

III. Nucleosomes

First, a brief description of the eukaryal nucleosome is provided, followed by a detailed description of the archaeal nucleosome, noting the
similarities and differences between the two in the discussion of each
feature of the DNA or histones.

A. The Eukaryal Nucleosome

X-ray analyses of crystals of the histone protein core (Arents and
Moudrianakis, 1991) and of the entire histone–DNA complex (Luger
et al., 1997) have provided detailed views of the eukaryal nucleosome
core particle. The DNA molecule is wrapped 1.65 times around a histone
core in a left-handed superhelix (Luger *et al.,* 1997). The length of DNA
in a single nucleosome, 145 bp (base pairs), is close to the persistence
length of DNA (140 bp), the minimal length that can be circularized by
the inherent flexibility of a DNA molecule. That the DNA is wrapped
1.65 times (88 bp per circumference) around the histone core emphasizes the extent of distortion imposed on the DNA by histone binding
in a nucleosome.

Eight histone polypeptides form the core of the nucleosome (Fig. 1),
and these are organized as heterodimers, H2A·H2B and H3·H4, with
the central domain of each polypeptide ($\sim$65 amino acid residues)
assuming the common histone fold (Arents and Moudrianakis, 1995)
(Fig. 2). These four dimers form a protein superhelix, with positively
charged residues on the surface of each dimer dictating the path of the

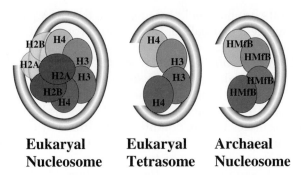

Eukaryal   Eukaryal   Archaeal
Nucleosome   Tetrasome   Nucleosome

FIG. 1. Arrangement of the individual polypeptides (ovals) and DNA molecule in the eukaryal nucleosome core particle, the eukaryal tetrasome, and the archaeal nucleosome.

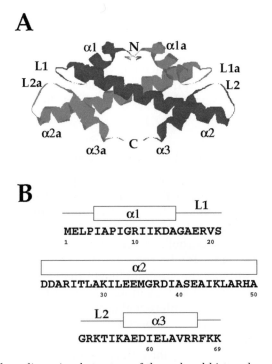

FIG. 2. (A) Three-dimensional structure of the archaeal histone homodimer (HMfB)$_2$ as determined by X-ray crystallography (Decanniere *et al.*, 2000), with the two monomers in different shades of gray. N indicates the paired N-termini and C indicates the paired C-termini. Alpha-helices and loop regions are indicated by α and L, respectively, and numbered as in the diagram in B. (B) Amino acid sequence of HMfB, with the secondary structural elements of the polypeptide indicated schematically above the sequence.

DNA in the nucleosome. As illustrated in Fig. 1, two H3·H4 dimers form the central tetramer by association of residues in each copy of H3, and one H2A·H2B dimer is attached on each side of this tetramer, by interactions between H4 and H2B, to complete the histone octamer.

B. The Archaeal Nucleosome

Electron microscopy (EM) first hinted at the presence of a nucleosomal organization in prokaryotic Archaea. Chromosomes from *Halobacterium salinarium* cells displayed the beads-on-a-string structure typical of eukaryal chromatin (Shioda *et al.*, 1989), and proteinase and nuclease treatments subsequently revealed that these "beads" contained protein and could be released as individual particles by limited nuclease exposure (Takayanagi *et al.*, 1992). EM visualization of complexes reconstituted *in vitro* using purified methanogen histones demonstrated the presence of similar structures and removal of the protein revealed that the DNA was constrained in loops, in a manner very reminiscent of eukaryal chromatin (Sandman *et al.*, 1990). Subsequent biochemical studies have defined the structure and properties of the archaeal nucleosome. These structures contain a histone tetramer (a dimer of dimers) circumscribed by ~80 bp of DNA (Grayling *et al.*, 1996; Pereira *et al.*, 1997; Bailey *et al.*, 1999) (Fig. 1). This structure is smaller than the eukaryal nucleosome, but the near-identity of the protein folds (Starich *et al.*, 1996; Decanniere *et al.*, 2000) and the mechanisms of DNA binding and wrapping justify the designation archaeal nucleosomes rather than "nucleosome-like" structures. However, the most meaningful comparison is not between the archaeal and the eukaryal nucleosomes but, rather, between archaeal nucleosomes and eukaryal tetrasomes (Pereira and Reeve, 1998; Sandman and Reeve, 2000), the structures formed by the eukaryal (H3·H4)$_2$ tetramers and DNA (Alilat *et al.*, 1999) (Fig. 1). These have the same histone:DNA stoichiometry and share DNA topology, DNA recognition, and nucleosome positioning features. Recently the tetrasome has been shown to play a biologically relevant role in transcription (see Section V).

IV. Nucleosomal DNA

A. Length

The length of DNA in a nucleosome is determined by the number of histone dimers directing its path. The length of DNA directly in contact with an archaeal histone tetramer is 60–70 bp based on micrococcal nuclease (MN) protection assays, performed on archaeal nucleosomes assembled *in vivo* (Pereira *et al.*, 1997) and *in vitro* (Grayling *et al.*, 1997;

Sandman and Reeve, 1999). Similarly, 60–70 bp has been estimated as the length of DNA in a tetrasome based on EM measurements (Prunell, 1998) and MN protection experiments (Dong and van Holde, 1991). This is consistent with predictions based on the nucleosome core particle crystal structure, in which 27–28 bp is associated with each histone dimer, and 4 bp spans the space between the two dimers (Luger *et al.,* 1997). In ligase-catalyzed circularization assays that used archaeal histones to juxtapose the ends of short DNA molecules, a molecule of at least 88 bp was required for efficient circularization (Bailey *et al.,* 1999), the length of DNA calculated to encircle the eukaryal nucleosome (see Section III.A). The difference between the two values, 60–70 bp from MN protection assays and 88 bp from circularization assays, can be explained by proposing that the DNA protected from MN digestion is the DNA in direct contact with the histone tetramer but that this structure is horseshoe-shaped and the additional base pairs are needed to span the open end of the horseshoe (Pereira and Reeve, 1999).

The value of 146 bp that is commonly cited as the length of DNA in a eukaryal nucleosome core particle is also based on MN protection studies. However, van Holde and Zlatanova (1999) have recently reviewed studies that reveal a variety of DNA lengths in the nucleosome core particle, and they propose that the length of DNA present is dynamic and changes with chromosome activity. For example, Protacio *et al.* (1997) demonstrated that DNA at either end of the nucleosome transiently unwraps, resulting in less than 146 bp organized by the histone octamer. Nucleosomes exist in arrays *in vivo,* not as isolated particles, and the length of DNA organized by each nucleosome most likely varies with the properties of the local sequence, the transcriptional state of the DNA, and the association of linker histones and other chromosomal proteins.

B. Topology

DNA is wrapped around the eukaryal histone octamer core in a left-handed supercoil, whereas the DNA wrapped around a histone tetramer, either archaeal (Musgrave *et al.,* 1991; Sandman *et al.,* 1994a) or eukaryal (Hamiche *et al.,* 1996), can be constrained in either a left-handed (negative) or a right-handed (positive) supercoil. The ability of histone tetramers to wrap DNA in either direction is explained by a reorientation of the interface between the two dimers (Hamiche and Richard-Foy, 1998). The horseshoe-shaped tetramer is not planar, but twisted, and to change the direction of DNA wrapping, the two halves of the horseshoe (each dimer) need only to move relative to each other, a movement made possible by swiveling around the axis of symmetry. Prunell (1998) estimated that transition between the two forms requires very little energy and could be triggered by DNA thermal fluctuations. The eukaryal nucleosome core particle seems to be limited to negative DNA super-coiling, apparently because the addition of the two H2A·H2B dimers to the surface of the (H3·H4)$_2$ tetramer locks the structure in a left-handed supercoiling configuration and prevents interconversion between the two histone tetramer topologies (Hamiche *et al.*, 1996).

The DNA wrapped around a nucleosome is overwound with a helical repeat of ~10.1 bp, relative to B-form DNA in solution, which has a helical repeat of 10.4 bp. This value, and the potential energy conserved by the DNA constraint, is a feature of both the archaeal and the eukaryal nucleosomes (Pereira *et al.*, 1997).

C. NUCLEOSOME POSITIONING

As a general chromosome packaging mechanism, nucleosomes must form regardless of the DNA sequence, but nevertheless, some sequences are preferentially incorporated into nucleosomes. Sequence features that result in preferential packaging in nucleosomes have been determined using SELEX technology (Tuerk and Gold, 1990). Experiments using archaeal histones and either a random DNA population or mixed genomic sequences produced consistent results (Bailey *et al.*, 2000). Sequences that conform to the general consensus of 5'-(A/T)$_3$NN(G/C)$_3$NN-3' were most readily incorporated into archaeal nucleosomes, and these are the sequences that readily accept the alternating major and minor groove compressions imposed on the DNA by nucleosome assembly. In light of the fact that the constraints of DNA wrapping are conserved in archaeal and eukaryal nucleosomes (10.1 bp per turn, 88 bp per wrap), it is not surprising that essentially the same result was obtained previously when SELEX was used with eukaryal histone octamers to identify sequences preferentially incorporated into eukaryal nucleosomes (Widlund *et al.*, 1997; Lowary and Widom, 1998).

Evidence of the influence of nucleosomal packaging on DNA sequence selection *in vivo* is present in genome sequences. Fourier transform analyses of the histone-selected SELEX fragments and some entire genome sequences have revealed the presence of AA (=TT), TA, and GC dinucleotides occurring nonrandomly at ~10-bp intervals (Thaström *et al.*, 1999; Bailey *et al.*, 2000). Such dinucleotide repeats facilitate nucleosome assembly, and their overabundance is observed only in genomes from organisms with a nucleosomal DNA organization, namely, Eukarya and Euryarchaeota.

V. Transcription on Nucleosomal Templates

DNA packaged in a nucleus or nucleoid must be accessible to enzymes that use it as a substrate. Struhl (1999) described a significant difference in promoter accessibility between Bacteria and Eukarya, attributable largely to the packaging of eukaryotic DNA in chromatin. In Bacteria there is a "nonrestrictive" ground state in which promoters are generally available to RNA polymerase, whereas in eukaryotes there is a "restrictive" ground state in which RNA polymerase cannot interact with promoters in the absence of activators and chromatin remodeling complexes that counteract the repressive effects of nucleosomal packaging. Not considered by Struhl is the nucleosomal organization in Archaea, with transcription regulated by repression, as in Bacteria (Soppa et al., 1999). Apparently the archaeal nucleosome-based system of chromosome organization is not repressive per se, and the difference between the archaeal and the eukaryal nucleosomal organization presumably reflects the use of histone tetramers versus histone octamers. The concept emerging is that the octamer-based nucleosome core particle is a DNA packaging and storage structure, whereas the tetrasome (or the archaeal nucleosome) is the compacting structure in transcriptionally active chromatin. Briefly, the arguments supporting the presence of tetrasomes in transcribed regions of eukaryal chromosomes are as follows: (i) actively transcribed chromatin has less than stoichiometric amounts of histones H2A and H2B relative to histones H3 and H4 (Baer and Rhodes, 1983); (ii) H2A·H2B dimers exchange rapidly between the chromosome and the free histone pool during transcription, suggesting a loose association with the tetrasome (Louters and Chalkley, 1985; Jackson, 1990); (iii) defects in the SWI/SNF chromatin remodeling complex that is needed for gene activation and transcription are suppressed by H2A·H2B depletion (Hirschhorn et al., 1992); (iv) the transcription elongation factor FACT binds nucleosomes and removes H2A·H2B dimers to facilitate RNA polymerase passage (Orphanides et al., 1999); (v) transcription factor binding sites are accessed in vitro more readily in tetrasomes than in nucleosome core particles (Spangenberg et al., 1998); (vi) DNA organized into tetrasomes is far less compact and rigid than DNA in nucleosome core particles (Hansen and Wolffe, 1994; Tse et al., 1998); and (vii) in vitro transcription from $(H3·H4)_2$ templates is as efficient as transcription on naked DNA templates (Chirinos et al., 1999).

Gene expression is apparently correlated with loss of H2A·H2B dimers from the nucleosome core, leaving a tetrasome as the resulting template structure for transcription. The topological flexibility of the tetrasome then allows it to accommodate the supercoiling inherently associated with RNA polymerase passage (Liu and Wang, 1987) without dissociation from the DNA template. In higher eukaryotes, where transcriptional units may extend for tens of kilobase pairs, it is presumably very important that the tetrasome preserves chromosome packaging and has the flexibility needed to accommodate the mechanical events involved in transcription.

VI. Histones

Histones, the protein components of archaeal and eukaryal nucleo-somes, are defined by a common three-dimensional structure, the histone fold (Arents and Moudrianakis, 1995) (Fig. 2). In this fold, ~65 residues form three α-helices (α1, α2, α3) separated by short β-strand loops (L1, L2). Histone-fold monomers dimerize spontaneously, with the hydrophobic faces of the six helices interacting to form the core of the dimer structure. The β strands pair to form short loops at each end of the dimer. Basic residues extend from the dimer surface to contact DNA at three sites, at each loop region and near the paired N-termini.

Archaeal histones are essentially only histone folds, whereas eukaryal histones have N- and C-terminal sequences that flank the histone fold. The N-terminal domains ("tails") of the eukaryal histones range from 25 to 60 amino acid residues and protrude outside the nucleosome through the DNA minor groove (Luger et al., 1997). They provide targets for enzymes that add or remove acetyl, methyl, or phosphate groups, and it has been proposed (Strahl and Allis, 2000) that these covalent modifi-cations, which can exist in a very large number of combinations, act as a complex code that signals the state of chromosome condensation. Direct interactions between the histone tails extending from different nucleo-somes compact nucleosome arrays locally, and by modifying these tails, nucleosome arrays may be decondensed to facilitate transcription factor access.

A. ARCHAEAL HISTONES

The sequences of ~30 archaeal histones are now known (http://www. biosci.ohio-state.edu/~microbio/Archaealhistones/index.html) and they constitute a conserved family of proteins with 65–69 amino acid residues in sequences that exhibit limited but clear homology with the eukaryal nucleosome core histones. Archaeal histones and histone-encoding genes have been isolated from several members of the Euryar-chaeota, and genome sequences have revealed that archaeal histones are present in species throughout the Euryarchaeota except for *Thermo-plasma acidophilum* (Ruepp et al., 2000). To date, however, there is no evidence for histones in Crenarchaeota. Four high-resolution archaeal histone structures have been determined by NMR spectroscopy and by X-ray crystallography (Starich et al., 1996; Zhu et al., 1998; Decanniere et al., 2000). These four structures are very similar, and given the high level of sequence conservation in this protein family (Fig. 3), these structures should be reliable models for all archaeal histone structures.

```
                          |---αI---|          |------------αII------------|          |--αIII---|
HMfB     MELPIAPIGRIIKDA--GAERVSDDARITLAKILEEMGRDIASEAIKLARHAGRKTIKAEDIELAVRRFKK*
HMfA     MGELPIAPIGRIIKNA--GAERVSDDARIALAKVLEEMGEEIASEAVKLAKHAGRKTIKAEDIELARKMFK*
HFoB     MELPIAPIGRIIKNA--GAERVSDDARIALAKVLEEMGEEIASEAVKLAKHAGRKTVKASDVELAVKRL*
HTzA1    MAELPIAPIDRLIRKA--GAERVSEDAAKALAEYLEEYAIEVGKKATEFARHAGRKTVKAEDVRLAVKA*
HPyA1    MGELPIAPVDRLIRKA--GAERVSEEAAKILAEYLEEYAIEVSKKAVEFARHAGRKTVKAEDIKLLAIKS*
MJ0168   MAELPVAPFERIIKKA--GAERVSEAAAEYLAEAVEEIALEIAKEAVELAKHAKRKTVKVEDIKLALKK*
MkaN     MAVELPKAAIERIFPQGI-GERRLSQDAKDTIYDFVPTMAEYVANAAKSVLDASGKKTLMEEHLKALADVLMVEGVED
MkaC     DGELFGRATVRRILKRA--GIERASSDAVDLYNKLICRATEELGEKAAEYADEDGRKTVQGEDVEKAITYSMPKGGEL*
HHbN     MSVELPFAPVDSLIRGHA-GDLRVSAGAAEELARRIQRHGAILAVDAAAAREDGRKTLMASDFEGIVGPRDDTAPDRR
HHbC     GDLALPVAPVDRIARLEIDDRFRVSEDARVALAGVLEAYAADIADGAAVLAEHAGRRTVQAEDIQTYVTLVE*
MJ1647   MLPKATVKRIMKQH--TDFNISAEAVDELCNMLEEIIKITTEVAEQNARKEGRKTIKARDIKQCDDERLKRKIMELSERTDKMPILIKEMLNVITSEL*
```

FIG. 3. Alignment of archaeal histone sequences. The positions of the α-helices are indicated above the alignment. An asterisk indicates the presence of a stop codon in the DNA. HMfB and HMfA—histones from *Methanothermus fervidus*; HFoB—histone from *Methanobacterium formicicum*; HTzA1—histone from *Thermococcus zilligii*; HPyA1—histone from *Pyrococcus* strain GB-3a; MJ0168 and MJ1647—histones from *Methanococcus jannaschii*; MkaN, MkaC, and HHbN, HHbC—histone-fold sequences of the N- and C-terminal domains of the histones from *Methanopyrus kandleri* and *Halobacterium* strain NRC1, respectively.

Unlike the eukaryal histones, which form only H2A·H2B and H3·H4 heterodimers, archaeal histones form dimers apparently without partner restriction. Expression of a single archaeal histone gene in *E. coli* results in homodimer formation (Sandman *et al.*, 1998) and histone preparations isolated from Archaea contain both homodimers and heterodimers at ratios that reflect an unbiased association of the monomer polypeptides. *Methanothermus fervidus* contains two histones, HMfA and HMfB, and their abundances, and therefore the proportions of (rHMfA)$_2$ and (rHMfB)$_2$ homodimers and (rHMfA-rHMfB) heterodimers are regulated by growth conditions (Sandman *et al.*, 1994a). (HMfB)$_2$ homodimers that predominate in the stationary phase are capable of introducing more supercoils into DNA than (HMfA)$_2$ homodimers, consistent with decreased transcriptional activity of the genome in the stationary phase.

1. *Details of Archaeal Histone Structure*

Figure 2A shows the (HMfB)$_2$ homodimer structure generated by X-ray crystallography analysis (Decanniere *et al.*, 2000), and Fig. 2B presents the amino acid sequence of HMfB, with the sequences of the secondary structural elements indicated. In the homodimer, the two N-termini are in close proximity near the center of the dimer and helices α1 and α1a (suffix "a" indicates the second monomer in a dimer), then diverge, placing the two loop1 structures (L1 and L1a) at opposite ends of the dimer. The long central helices α2 and α2a are antiparallel aligned through the length of the dimer and cross near residues 35 and 35a. The two L2 loops pair in parallel orientation with the L1 loops and the C-terminal helices α3 and α3a extend toward each other and away from the N-termini. The native fold of the dimer is determined primarily by hydrophobic interactions among the six α-helices together with intramonomer salt bridges between D22 and R25 and between R52 and D59 and intermonomer salt bridges between E18 and K53a and between E18a and K53.

2. *Advantages of Archaeal Histones and Nucleosomes as Experimental Models*

Archaeal histones and nucleosomes are considerably simpler than their eukaryal counterparts and therefore provide attractive and tractable model systems for the experimental dissection of the histone fold and histone fold–DNA interactions. For example, archaeal histones assemble spontaneously with DNA to form archaeal nucleosomes *in vitro,* whereas eukaryal nucleosome assembly *in vitro* requires either multiple dialysis steps from high to low salt to deposit, first, the (H3·H4)$_2$ tetramer and, then, H2A·H2B dimers on a DNA molecule or nucleosome transfer from previously assembled chromatin. In addition, archaeal nucleosomes can be generated using a single recombinant archaeal histone preparation purified from *E. coli*. The addition of such recombinant histone preparations to DNA results in the formation of homotetramer-containing archaeal nucleosomes (Bailey *et al.*, 1999), and site-directed mutagenesis of the cloned histone gene with subsequent expression in *E. coli* results in the availability of single histone variants that can be assayed for histone-fold stability and archaeal nucleosome formation. Many of the archaeal histones are from hyperthermophilic archaea, and purification of these stable recombinant proteins from *E. coli* is often as simple as boiling the crude extract to denature the host proteins and then recovering the histone of interest in a single heparin affinity step (Sandman *et al.*, 2001). Crystallization and data collection by NMR over periods of several days are also greatly facilitated by the availability of such stable proteins.

3. *Variants of rHMfB*

Given the experimental advantages, we have undertaken a detailed investigation of archaeal histones and the archaeal nucleosomes assembled from homotetramers of rHMfB, the recombinant (r) version of HMfB from *M. fervidus,* and from rHMfB variants generated by site-directed mutagenesis. An alignment of the amino acid sequences of archaeal histone family members (Fig. 3) identifies residues that are highly conserved and regions that accept residue variability. Differences among the archaeal histones and between archaeal and eukaryal histones have been investigated by replacing rHMfB residues with residues present in other histones. Alanine residues have also been introduced to determine the consequences of the presence of this relatively unreactive side chain and oppositely charged or similarly charged residues have been substituted to probe electrostatic interactions (Table I).

After genetic construction, the first assay performed on every histone variant is to confirm its synthesis in *E. coli*. Most, but not all, of the rHMfB variants do accumulate in *E. coli,* and those that do not are presumed to be grossly misfolded and therefore subject to rapid degradation. DNA binding by rHMfB variants has been studied by an electrophoretic mobility shift assay (EMSA) performed using an agarose gel and a long DNA molecule [4.5 kb (kilo-base pairs)] that assays for the assembly of complexes that migrate faster than histone-free DNA (Sandman *et al.*, 1990) and by a traditional EMSA in which the mobility of a shorter DNA fragment is retarded through a polyacrylamide gel by histone binding (Grayling *et al.*, 1997). Ligase-mediated circularization of short DNA molecules (Bailey *et al.*, 1999) and changes in the circular dichroism signal of DNA at 275 nm (Soares *et al.*, 2000) have also been used to assay the DNA binding of HMfB variants. Measurements

TABLE I

LIST AND PROPERTIES OF rHMfB VARIANTS GENERATED BY SITE-DIRECTED MUTAGENESIS

Variant(s)	Category[a]	Nucleosome formation[b]	Additional phenotypes/notes
M1G	AV	+	N-Terminal structural variability
E2D	CS	±	
E2K	CS	+	
L3I	CS	+	
L3C	CS	−	
P4C	CS	±	
P4S	CS	+	
I5V	CS	+	
A6P	CS		Recombinant protein not detected
P7C	CS	−	
I8V	CS	+	
R10G	CS		Recombinant protein not detected
R10S, R10K	CS	±	
I11L	AV	+	
K13R	AV	+	
K13E	AV	−	
K13T	AV	±	
D14E	AV	+	
D14K	AV	±	
D14K/R37E	AV	±	Higher salt required for stability
D14N	AV	+	Higher salt required for stability
D14N/R37E	AV	+	
A15P	CS		Recombinant protein not detected
A15S	CS	+	
G16K	CS		Recombinant protein not detected
A17I, A17V	CS	−	
E18D, E18P	AV	+	
E18K	EV	+	Greater affinity for DNA than wild-type histone
E18K/G51K	EV	+	Greater affinity for DNA than wild-type histone
R19S, R19I, R19Q	CS	−	
R19G	CS, EV	±	
V20C	CS	−	
V20D	CS		Recombinant protein not detected
V20I	AV	+	
S21C	CS	+	
S21A, S21T	CS	±	
R25K	AV	+	
I26E	AV, ST	±	
T27A	AV	+	
I31V	AV, ST	+	
I31A	AV, ST	+	5°C decrease in $T°$

(continues)

TABLE I—*Continued*

Variant(s)	Category[a]	Nucleosome formation[b]	Additional phenotypes/notes
I31Y/M35Y	AV	+	Introduction of aromatic residues
M35C		+	Monomers can be cross-linked by oxidation of these cysteine residues
M35K	AV, ST	+	19°C decrease in $T°$
M35Y	AV, ST	+	
G36A	AV, ST	+	5°C increase in $T°$
R37E	AV, ST	−	6°C decrease in $T°$
R37K		+	
R37L	AV	±	
R37Q		±	
D38E	AV	+	8°C increase in $T°$
D38T	AV	+	6°C increase in $T°$
I39A			Recombinant protein not detected
I39V	AV	−	
A43I	CS		Recombinant protein not detected
I44V	AV, ST	+	
L46A, L46I, L46Q, L46C, L46V	CS	±	
L46F	CS	+	
L46S	CS		Recombinant protein not detected
R48K	AV	+	
H49A	CS	±	
H49D	EV	±	
H49E, H49K		±	
G51A	CS		Recombinant protein not detected
G51K	AV, EV	+	Greater affinity for DNA than wild-type histone
R52K	CS	−	
R52H, R52Q, R52A	CS		Recombinant protein not detected
K53R, K53T	CS	±	
K53E	CS	−	
T54K, T54I, T54R, T54C, T54S, T54V, T54Y	CS	−	
I55L, I55T	CS	±	
I55M	CS	−	
I55C	CS		Recombinant protein not detected
I55V	AV, CS	+	
K56R	CS	+	
K56E	CS	−	
K56Q, K56I	CS		Recombinant protein not detected
E58S	AV, ST	+	Greater affinity for DNA than wild-type histone; 5°C increase in $T°$
D59A, D59E, D59N	CS		Recombinant protein not detected
I60A, I60G	CS		Recombinant protein not detected

TABLE I—*Continued*

Variant(s)	Category[a]	Nucleosome formation[b]	Additional phenotypes/notes
I60V	AV, CS, ST	+	
E61A	CS	+	Higher salt required for stability
E61K	CS	±	DNA bands are retarded in agarose EMSA
E61V, E61Q	CS	+	
E61R	CS	±	
L62I, L62V, L62A, L62W, L62R, L62C	CS	±	
L62M	CS		Recombinant protein not detected
L62Y	CS	+	
L62Q	CS	±	DNA bands are retarded in agarose EMSA
V64R	AV	±	
R66M	AV	+	
F67A	ST		5°C decrease in $T°$
K68*	AV		Recombinant protein not detected
K68E		±	
K68E/K69*	AV	±	
K69*	AV	+	8°C decrease in $T°$
K69E	AV	±	

[a] AV, mimics natural variation within the archaeal histone family; EV, mimics natural variation between archaeal and eukaryal histones; ST, constructed to investigate histone-fold stability; CS, constructed to investigate conservation of secondary or tertiary structure.

[b] +, indistinguishable from wild type; ±, nucleosome formation but somewhat impaired; −, no nucleosome formation.

of the number of supercoils introduced and the direction of the supercoiling that results from nucleosome formation (Musgrave *et al.,* 1991) by rHMfB variants on a closed circular DNA molecule have also been made. Both circular dichroism and differential scanning calorimetry have been used to quantitate the effects of amino acid residue substitution on the stability of the native fold of (rHMfB)$_2$ dimers (Li *et al.,* 1998).

Substitutions in the Hydrophobic Core of the (HMfB)$_2$ Histone Dimer. Histone-fold monomers are held together to form a dimer primarily by hydrophobic interactions between residues on the hydrophobic face of each amphipathic helix. As illustrated in the sequence alignment (Fig. 3), a hydrophobic side chain is present every three or four residues in each helix, although there is no strict residue conservation. For example, in α1, positions 5, 8, 11, and 12 are occupied predominantly by isoleucine, valine, or leucine, and changing these residues individually to other hydrophobic residues results in histone variants that are indistinguishable from the wild type (Table I) (Soares *et al.*, 2000).

In contrast, residue 55, located in L2 adjacent to DNA contacting residues, is most often a valine and, in a few cases, isoleucine or leucine. Substitutions in HMfB at I55 illustrate the importance of the hydrophobic packing in this region of the molecule. Only the I55V change is tolerated in the rHMfB context; all others result in histone variants that exhibit reduced DNA binding (Table I) (Soares *et al.*, 2000).

rHMfB has only one aromatic residue, phenylalanine 67, and this has precluded the use of spectroscopic techniques to quantitate this protein. The related archaeal histones HPyA1 (Sandman *et al.*, 1994b) and HTzA1 (Ronimus and Musgrave, 1996), from *Pyrococcus* strain GB-3a and *Thermococcus zilligii*, respectively, have tyrosine residues at positions 31 and 35 within the hydrophobic core (Fig. 3), and substituting tyrosine into rHMfB at positions 31 and 35 resulted in stable homodimers with spectral properties expected for a protein with four aromatic residues.

Structural Stability of Archaeal Histone Dimers. A comparison of rHMfB from *M. fervidus* [optimal temperature (T_{opt}) for growth of 83°C] with rHFoB from *M. formicicum* (T_{opt} for growth of 45°C) (Darcy *et al.*, 1995) revealed that, although the two histones share 78% sequence identity (Fig. 3), they unfold at temperatures that differ by more than 30°C (Li *et al.*, 1998). The 13 residue differences between HFoB and HMfB were introduced individually into HMfB, and the stability of the resulting HMfB variants was determined by following melting transitions using circular dichroism spectroscopy (Li *et al.*, 2000b). The largest decrease in stability resulted from M35K substitution, and in reciprocal experiments, replacing residue K35 in rHFoB with methionine, tyrosine, or phenylalanine increased the stability. A comparison of the high-resolution structures of HMfB and HFoB dimers reveals the presence of two solvent-accessible cavities in rHFoB dimers that are filled in rHMfB dimers (Zhu *et al.*, 1998). These cavities are adjacent to residues 36 and 36a, near the center of the histone dimer, where $\alpha2$ and $\alpha2a$ cross. Also, surrounding these cavities and contributing to the differences in thermal stability is the residue at position 31 (I in HMfB but A in HFoB). Overall it is clear that bulky hydrophobic residues that fill cavities more completely within the hydrophobic core make a substantial contribution to the higher stability of archaeal histone dimers from hyperthermophiles, although some additional stability is conferred by an increased number of surface-located electrostatic interactions (see DNA-Contacting Residues, below).

DNA-Contacting Residues. Structural studies (Starich *et al.,* 1996; Decanniere *et al.,* 2000) have identified surface-located basic residues that are positioned appropriately to interact with DNA. Near the N-terminus a hydrogen-bonded proline tetrad motif (P4/P4a and P7/P7a) unique to the archaeal histones orients the R10 side chain appropriately so that imino groups can make an electrostatic interaction with phosphates on each strand of the bound DNA. DNA contacts are also made in both loop regions, by residues R19, K53a, and T54a and R19a, K53, and T54, supported by the E18–K53a and E18a–K53 salt bridges and anchored to the hydrophobic core by V20/V20a and I55/I55a.

Not surprisingly, these residues, R10, R19, K53, and T54, are absolutely conserved in all archaeal histones (Fig. 3) and have clear homologues in the eukaryal histones. Changing these residues results in rHMfB variants with significantly impaired abilities to form nucleosomes based on gel shift assays (Table I) (Soares *et al.,* 2000). As these EMSAs require nucleosome persistence throughout extended periods of electrophoresis, the ligase-mediated circularization assay was used to determine if these histone variants might transiently form nucleosomes. Several rHMfB variants that gave a negative EMSA result were still able to facilitate the circularization of an 88-bp DNA fragment, demonstrating that they did still bind and bend DNA in solution (Soares *et al.,* 2000). In fact, construction of an HMfB variant that was totally incapable of DNA binding required introducing residue substitutions at all three DNA binding sites, namely, in α1, L1, and L2. These results underscore the cooperative nature of DNA binding within a nucleosome and that loss of DNA binding at one histone–DNA contact point, even repeated four times around a homotetramer, does not necessarily prevent nucleosome formation.

A rHMfB-R37E variant was generated to address the unique occurrence of an arginine at position 37 in HMfB (Fig. 3). Based on the agarose EMSA, this variant did not form nucleosomes (Decanniere *et al.,* 2000), but as residue 37 is located on the surface of α2, and distant from the three DNA contacting sites, this suggested that additional factors were influencing nucleosome formation. Based on the crystal structure of HMfB, R37 forms surface-located salt bridges with E33 and D14a, and these influence the orientation of E2a, which in turn positions R10a correctly for DNA binding (Decanniere *et al.,* 2000). Disruption of this network of ionic interactions by substituting glutamic acid for arginine at position 37 apparently therefore has a domino effect that ultimately interferes with DNA binding.

The surface-localized electrostatic interactions also contribute to stability in rHMfB dimers. The R37E change destabilized rHMfB and a reciprocal E37R change stabilizes rHFoB (Li *et al.,* 2000b). Although the magnitude of the increase in stability conferred by these charged residue substitutions was less than observed with cavity-filling hydrophobic changes, they nevertheless confirmed the importance of surface ionic interactions for fold stability.

Conserved Structures.   Many of the invariant residues in the archaeal histone family play important structural roles. They are needed to maintain the histone fold and do not tolerate substitutions. For example, alanine–glycine pairs are present at the C-termini of α1 and α2 that terminate these helices (Aurora *et al.*, 1994), and most substitutions at these positions result in variants that do not accumulate in *E. coli* (Table I). A notable exception is the rHMfB-G51K variant that was designed to mimic the eukaryal histones and the archaeal histones from *Methanococcus jannaschii*. This HMfB variant has an increased affinity for DNA (Soares *et al.*, 2000).

A salt bridge is formed between R52 and D59 in all archaeal histones and between the structurally homologous residues in all eukaryal histones. Substitutions of alanine, glutamic acid, or glutamine at position D59 resulted in variants that did not accumulate in *E. coli,* and of the four substitutions attempted at R52, only the conservative rHMfB-R52K variant accumulated in *E. coli.* Still, the variant did not form nucleosomes as judged by the agarose EMSA (Table I).

The N-terminal sequence ELPhAP (h indicates a hydrophobic residue) is conserved in all the archaeal histones. These residues position R10 to contact DNA, by forming the proline tetrad structure and by electrostatic interactions mediated through E2, although residue substitutions surprisingly revealed that the presence of a proline at positions 4 and 7 is not absolutely required for DNA binding (Decanniere *et al.*, 2000). This proline-rich sequence is not found in the eukaryal histones, which have much longer N-terminal tail domains and which must, therefore, use a different structural mechanism to orient the side chains of arginines in α1 toward the DNA.

Dimer–dimer Interactions.   The similarity between the archaeal and the eukaryal histone dimers predicts that the molecular basis for tetramer formation is also similar. As illustrated in the crystal structure of the eukaryal nucleosome core (Luger *et al.*, 1997) the two copies of H3 in H3·H4 dimers interact (Fig. 1) to assemble the (H3·H4)$_2$ tetramer by forming a four-helix bundle that contains the C-terminal regions of two α2's and the N-terminal regions of two α3's. A histidine residue in each copy of H3 makes a buried hydrogen bond to an aspartic acid residue, and additional stability is conferred by hydrophobic interactions.

Homologues of all residues responsible for these interactions are conserved in HMfB, namely, H49, D52, L46, and L62. Of these, H49 and D52 are present in all archaeal histones and L46 and L62 are highly conserved but occasionally are replaced by other large hydrophobic residues. As noted above (see Conserved Structures), D52 cannot be substituted in rHMfB, but surprisingly, substitutions are acceptable at positions 46, 49, and 62 (Table I), resulting in HMfB variants that are capable of nucleosome formation (Soares *et al.*, 2000).

B. Natural Archaeal Histone Variants

The genome of *M. jannaschii* encodes six histone-fold proteins (Bult *et al.*, 1996), five predicted to have typical archaeal histone features and one, encoded by MJ1647, that should have an additional C-terminal 30-residue domain (Fig. 4). The histone fold of MJ1647 has many of the invariant features described above, but conspicuously absent are homologues of P7, R19, and H49. To evaluate these variations, both the full-length MJ1647 and a deletion encoding only the histone-fold region (MJ1647Δ) were expressed in *E. coli* (Li *et al.*, 2000a). Both resulting recombinant proteins were found to be capable of nucleosome formation, although archaeal nucleosomes formed by the full-length MJ1647 were not as stable as nucleosomes formed by MJ1647Δ during prolonged electrophoresis. *M. jannaschii* is a hyperthermophile (T_{opt} for growth of 85°C), and thermal unfolding measurements were therefore made on both recombinant proteins. The presence of the C-terminal domain clearly increased the overall stability of the protein and the standard-state (1M) midpoint unfolding temperature ($T°$) calculated for MJ1647 was above 120°C, the highest value observed for any archaeal histone to date. The somewhat paradoxical instability of the MJ1647-containing nucleosome compared with the extreme stability of the MJ1647 dimers

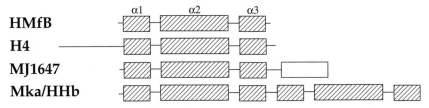

FIG. 4. The organization of the histone-fold domains in archaeal and eukaryal histones. Hatched boxes represent the three α-helices in a histone-fold domain. The open box represents an amino acid sequence in MJ1647 predicted to form an α-helix with no recognizable homology in other histone fold-containing proteins.

suggests that this polypeptide may not normally form homodimers but, rather, functions as a partner in a heterodimeric complex with one of the other five histones in *M. jannaschii*.

Methanopyrus kandleri contains a histone that is approximately twice the size of the typical archaeal histone (Slesarev *et al.*, 1998), and *Halobacterium* sp. NRC1 is predicted to encode such a histone (Fig. 4) (Ng *et al.*, 2000). These appear to have resulted from gene fusions and actually contain two tandem histone-fold motifs. Based on the 17-kDa molecular weight determined for the histone purified from *M. kandleri,* this "dimeric" archaeal histone is not proteolytically cleaved *in vivo* but must fold to form a structure that is effectively a histone-fold dimer stabilized by intramolecular interactions.

C. Nonhistone Histone Fold-Containing Proteins

The histone fold is present as a structural element in several eukaryal proteins in addition to the nucleosome core histones (Table II). These proteins have a variety of functions, ranging from atypical core histones found in a subset of nucleosomes to subunits of transcription factors (TATA-binding protein associated factors; TAFs) and DNA polymerase. In all cases, the histone folds in these proteins exist in dimer configurations, consistent with the requirement of dimer formation for histone-fold stability. The most conserved feature is that all these proteins have three α-helices with hydrophobic surfaces which mediate dimerization. Mutagenesis studies of the hydrophobic residues of $hTAF_{II}20$ have confirmed their importance for dimer formation (Gangloff *et al.*, 2000).

Analyses of the sequences of these nonhistone histone fold-containing proteins reveal that although the hydrophobic core needed for dimerization is conserved, the residues in the archaeal histones and the eukaryal core histones that specify the path of the DNA around the nucleosome are largely absent. Studies of the histone fold-containing TAFs suggested that a histone octamer-like complex may be present within transcription factor TFIID (Xie *et al.*, 1996), but subsequent structural studies have not documented such an arrangement. The $dTAF_{II}60-dTAF_{II}40$ heterodimer has been shown to form a tetramer that binds specifically to a conserved downstream promoter element (Burke and Kadonaga, 1997), but residues that contact the DNA have not been identified.

TAFs organized as specific heterodimers are found as components of several large, multisubunit transcription complexes, including the general transcription factor TFIID, and the SAGA and PCAF histone acetyltransferase complexes (reviewed by Green, 2000). The histone folds within these TAFs appear to aid in assembly of these large complexes rather than providing DNA binding domains.

TABLE II

Eukaryal Histone-Fold Proteins

Histone-fold pair[a]	Histone fold based on	Binds DNA[b]	Structure/function	Reference(s)
yCSE4·H4 hCENP-A·H4	Sequence homology	Yes	Substitutes for H3·H4 in centromere chromatin	Palmer et al. (1991) Stoler et al. (1995)
yH2A.Z·H2B	Sequence homology	ND	Substitutes for H2A·H2B in a subset of nucleosomes	Santisteban et al. (2000)
yHAP3·HAP5 hNF-YB·NF-YC hCBF-A·CBF-C	Sequence homology	No	Recognizes CCAAT box in promoters; binds to TBP	Mantovani (1999)
yTAF60·TAF17 hTAF80·TAF31 dTAF60·TAF40	Crystal structure	Yes	Components of TFIID and histone acetyltransferase	Xie et al. (1996) Ogryzko et al. (1998) Burke and Kadonaga (1997)
yTAF68·ND hTAF20·TAF135 dTAF30α·ND	Sequence homology	ND	Components of TFIID	Gangloff et al. (2000)
yTAF40·TAF19 hTAF28·TAF18 dTAF30β·ND	Crystal structure	ND	Components of TFIID	Birck et al. (1998)
yYDR1·BUR6 hDr1·DRAP hNC2α–NC2β dhNC2α–NC2β	Sequence homology	No	Represses and activates transcription	Mermelstein et al. (1996) Kim et al. (2000) Willy et al. (2000)
yTAF68·ADA1	Sequence homology	ND	Components of histone acetyltransferase	Gangloff et al. (2000)
ySPT3 (intramolecular)	Sequence homology	No	Component of histone acetyltransferase	Birck et al. (1998)
yDPB3·DPB4 hp12·p17	Sequence homology	ND	Components of DNA polymerase ε	Li et al. (2000c)
hp15·p17 dCHRAC14·CHRAC16	Sequence homology	Yes	Components of chromatin remodeling complex	Poot et al. (2000) Corona et al. (2000)

[a] The origin of the histone-fold pairs is indicated by y (yeast), h (human), d or (Drosophila).

[b] ND, not determined.

As has been observed with some archaeal histones, the SPT3 family of proteins (also components of SAGA) contain two histone-fold motifs in a single polypeptide. In these cases, however, the two histone folds are not directly adjacent but, rather, are separated by ~120 amino acid residues. In addition, in the second histone fold, $\alpha3$ is separated from $\alpha2$ by 36 residues, indicating that a histone fold can be formed from noncontiguous sequences (Birck *et al.*, 1998).

VII. Evolutionary Considerations

To date, the only prokaryotic histone folds known are archaeal histones. These are more closely related to the sequences of the four eukaryal nucleosome core histones than to the sequences of the other histone fold-containing proteins in Eukarya. It therefore seems reasonable to conclude that the histone fold originated in a prokaryote for the purpose that it now serves in contemporary Euryarchaea—to package and maintain the chromosome in a configuration that prevents DNA aggregation and preserves its flexibility. It is also intriguing to consider whether it may also have evolved to stabilize DNA in life evolving in a high-temperature environment. The DNA packaging problem must, however, have been encountered in all environments and solved early in cellular evolution. The diversity of extant prokaryotic architectural proteins argues that there were many successful solutions, but the dominance of histones in Eukarya suggests that the histone-based nucleosome solution was most readily adapted and modified to accommodate the increased size and complexity that occurred with eukaryotic evolution.

ACKNOWLEDGMENT

The authors thank Divya Soares for help in compiling the data in Table I.

REFERENCES

Agback, P., Baumann, H., Knapp, S., Ladenstein, R., and Härd, T. (1998). *Nature Struct. Biol.* **5**, 579–584.
Alilat, M., Sivolob, A., Revet, B., and Prunell, A. (1999). *J. Mol. Biol.* **291**, 815–841.
Arents, G., and Moudrianakis, E. N. (1995). *Proc. Natl. Acad. Sci. USA* **92**, 11170–11174.
Arents, G., Burlingame, R. W., Wang, B.-C., Love, W. E., and Moudrianakis, E. N. (1991). *Proc. Natl. Acad. Sci. USA* **88**, 10148–10152.
Aurora, R., Srinivasan, R., and Rose, G. D. (1994). *Science* **264**, 1126–1130.
Baer, B. W., and Rhodes, D. (1983). *Nature* **301**, 482–488.
Bailey, K. A., Chow, C. S., and Reeve, J. N. (1999). *Nucleic Acids Res.* **27**, 532–536.
Bailey, K. A., Pereira, S. L., Widom, J., and Reeve, J. N. (2000). *J. Mol. Biol.* **303**, 25–34.

Birck, C., Poch, O., Romier, C., Ruff, M., Mengus, G., Lavigne, A.-C., Davidson, I., and Moras, D. (1998). *Cell* **94**, 239–249.

Bloomfield, V. A. (1996). *Curr. Opin. Struct. Biol.* **6**, 334–341.

Bloomfield, V. A. (1997). *Biopolymers* **44**, 269–282.

Bult, C. J., White, O., Olsen, G. J., Zhou, L., Fleischmann, R. D., and Sutton, G. G., *et al.* (1996). *Science* **273**, 1058–1073.

Burke, T. W., and Kadonaga, J. T. (1997). *Genes Dev.* **11**, 3020–3031.

Chirinos, M., Hernandez, F., and Palacian, E. (1999). *Arch. Biochem. Biophys.* **370**, 222–230.

Corona, D. F. V., Budde, A., Deuring, R., Ferrari, S., Varga-Weisz, P. D., Wilm, M., Tamkun, J. W., and Becker, P. B. (2000). *EMBO J.* **19**, 3049–3059.

Darcy, T. J., Sandman, K., and Reeve, J. N. (1995). *J. Bacteriol.* **177**, 858–860.

Decanniere, K., Babu, A. M., Sandman, K., Reeve, J. N., and Heinemann, U. (2000). *J. Mol. Biol.* **303**, 35–47.

Dong, F., and van Holde, K. E. (1991). *Proc. Natl. Acad. Sci. USA* **88**, 10596–10600.

Gangloff, Y.-G., Werten, S., Romier, C., Carre, L., Poch, O., Moras, D., and Davidson, I. (2000). *Mol. Cell. Biol.* **20**, 340–351.

Grayling, R. A., Sandman, K., and Reeve, J. N. (1996). *Adv. Protein Chem.* **48**, 437–467.

Grayling, R. A., Bailey, K. A., and Reeve, J. N. (1997). *Extremophiles* **1**, 79–88.

Green, M. (2000). *Trends Biochem. Sci.* **25**, 59–63.

Hamiche, A., and Richard-Foy, H. (1998). *J. Biol. Chem.* **273**, 9261–9269.

Hamiche, A., Carot, V., Alilat, M., DeLucia, F., O'Donohue, M.-F., Révet, B., and Prunell, A. (1996). *Proc. Natl. Acad. Sci. USA* **93**, 7588–7593.

Hansen, J. C., and Wolffe, A. P. (1994). *Proc. Natl. Acad. Sci. USA* **91**, 2339–2343.

Hirschhorn, J. N., Brown, S. A., Clark, C. D., and Winston, F. (1992). *Genes Dev.* **6**, 2288–2298.

Holmes, V. F., and Cozzarelli, N. R. (2000). *Proc. Natl. Acad. Sci USA* **97**, 1322–1324.

Jackson, V. (1990). *Biochemistry* **29**, 719–731.

Kim, S., Cabane, K., Hampsey, M., and Reinberg, D. (2000). *Mol. Cell. Biol.* **20**, 2455–2465.

Li, J.-Y., Arnold-Schulz-Gahman, B., and Kellenberger, E. (1999). *Microbiology* **145**, 1–2.

Li, W.-T., Grayling, R. A., Sandman, K., Edmondson, S., Shriver, J. W., and Reeve, J. N. (1998). *Biochemistry* **37**, 10563–10572.

Li, W.-T., Sandman, K., Pereira, S. L., and Reeve, J. N. (2000a). *Extremophiles* **4**, 43–51.

Li, W.-T., Shriver, J. W., and Reeve, J. N. (2000b). *J. Bacteriol.* **182**, 812–817.

Li, Y., Pursell, Z. F., and Linn, S. (2000c). *J. Biol. Chem.* **275**, 23247–23252.

Liu, L. F., and Wang, J. C. (1987). *Proc. Natl. Acad. Sci. USA* **84**, 7024–7027.

Louters, L., and Chalkley, R. (1985). *Biochemistry* **24**, 3080–3085.

Lowary, P. T., and Widom, J. (1998). *J. Mol. Biol.* **276**, 19–42.

Luger, K., Mäder, A. W., Richmond, R. K., Sargent, D. F., and Richmond, T. J. (1997). *Nature* **389**, 251–260.

Mantovani, R. (1999). *Gene* **239**, 15–27.

Mermelstein, F., Yeung, K., Cao, J., Inostroza, J. A., Erdjument-Bromage, H., Eagelson, K., Landsman, D., Levitt, P., Tempst, P., and Reinberg, D. (1996). *Genes Dev.* **10**, 1033–1048.

Minsky, A., Ghirlando, R., and Reich, Z. (1997). *J. Theor. Biol.* **188**, 379–385.

Musgrave, D. R., Sandman, K. M., and Reeve, J. N. (1991). *Proc. Natl. Acad. Sci. USA* **88**, 10397–10401.

Ng, W. V., Kennedy, S. P., Mahairas, G. G., Berquist, B., and Pan, M., *et al.* (2000). *Proc. Natl. Acad. Sci. USA* **97**, 12176–12181.

Ogryzko, V. V., Kotani, T., Zhang, X., Schlitz, R. L., Howard, T., Yang, X.-J., Howard, B. H., Qin, J., and Nakatani, Y. (1998). *Cell* **94**, 35–44.

Orphanides, G., Wu, W.-H., Lane, W. S., Hampsey, M., and Reinberg, D. (1999). *Nature* **400**, 284–288.

Palmer, D. K., O'Day, K., Trong, H. L., Charbonneau, H., and Margolis, R. L. (1991). *Proc. Natl. Acad. Sci USA* **88**, 3734–3738.

Pereira, S. L., Grayling, R. A., Lurz, R., and Reeve, J. N. (1997). *Proc. Natl. Acad. Sci. USA* **94**, 12633–12637.

Pereira, S. L., and Reeve, J. N. (1998). *Extremophiles* **2**, 141–148.

Pereira, S. L., and Reeve, J. N. (1999). *J. Mol. Biol.* **289**, 675–681.

Poot, R. A., Dellaire, G., Hülsmann, B. B., Grimaldi, M. A., Corona, D. F. V., Becker, P. B., Bickmore, W. A., and Varga-Weisz, P. D. (2000). *EMBO J.* **19**, 3377–3387.

Protacio, R. U., Polach, K. J., and Widom, J. (1997). *J. Mol. Biol.* **274**, 708–721.

Prunell, A. (1998). *Biophys. J.* **74**, 2531–2544.

Rice, P. A., Yang, S., Mizuuchi, K., and Nash, H. A. (1996). *Cell* **87**, 1295–1306.

Robinson, H., Gao, Y.-G., McCrary, B. S., Edmondson, S. P., Shriver, J. W., and Wang, A.H.-J. (1998). *Nature* **392**, 202–205.

Ronimus, R., and Musgrave, D. R. (1996). *Biochim. Biophys. Acta* **1307**, 1–7.

Ruepp, A., Graml, W., Santos-Martinez, M.-L., Koretke, K. K., Volker, C., Mewes, H. W., Frishman, D., Stocker, S., Lupas, A. N., and Baumeister, W. (2000). *Nature* **407**, 508–513.

Sandman, K., and Reeve, J. N. (1999). *J. Bacteriol.* **181**, 1035–1038.

Sandman, K., and Reeve, J. N. (2000). *Arch. Microbiol.* **173**, 165–169.

Sandman, K., Krzycki, J. A., Dobrinski, B., Lurz, R., and Reeve, J. N. (1990). *Proc. Natl. Acad. Sci. USA* **87**, 5788–5791.

Sandman, K., Grayling, R. A., Dobrinski, B., Lurz, R., and Reeve, J. N. (1994a). *Proc. Natl. Acad. Sci. USA* **91**, 12624–12628.

Sandman, K., Perler, F. B., and Reeve, J. N. (1994b). *Gene* **150**, 207–208.

Sandman, K., Pereira, S. L., and Reeve, J. N. (1998a). *Cell. Mol. Life Sci.* **54**, 1350–1364.

Sandman, K., Zhu, W., Summers, M. F., and Reeve, J. N. (1998b). In "Thermophiles: The Key to Molecular Evolution and the Origin of Life?" (J. Weigel and M. W. W. Adams, eds.), pp. 243–253. Taylor & Francis, London.

Sandman, K., Bailey, K. A., Pereira, S. L., Soares, D., Li, W.-T., and Reeve, J. N. (2001). Archaeal histones and nucleosomes. *Methods Enzymol.* **334**, 116–129.

Santisteban, M. S., Kalashnikova, T., and Smith, M. M. (2000). *Cell* **103**, 411–422.

Shioda, M., Sugimori, K., Shiroya, T., and Takayanagi, S. (1989). *J. Bacteriol.* **171**, 4514–4517.

Slesarev, A. I., Belova, G. I., Kozyavkin, S. A., and Lake, J. A. (1998). *Nucleic Acids Res.* **26**, 427–430.

Soares, D. J., Sandman, K., and Reeve, J. N. (2000). *J. Mol. Biol.* **297**, 39–47.

Soppa, J. (1999). *Mol. Microbiol.* **31**, 1295–1305.

Spangenberg, C., Eisfeld, K., Stünkel, W., Luger, K., Flaus, A., Richmond, T. J., Truss, M., and Beato, M. (1998). *J. Mol. Biol.* **278**, 725–739.

Starich, M. R., Sandman, K., Reeve, J. N., and Summers, M. F. (1996). *J. Mol. Biol.* **255**, 187–203.

Stoler, S., Keith, K. C., Curnick, K. E., and Fitzgerald-Hayes, M. (1995). *Genes Dev.* **9**, 573–586.

Strahl, B. D., and Allis, C. D. (2000). *Nature* **403**, 41–45.

Struhl, K. (1999). *Cell* **98**, 1–4.

Takayanagi, S., Morimura, S., Kusaoke, H., Yokoyama, Y., Kano, K., and Shioda, M. (1992). *J. Bacteriol.* **174**, 7207–7216.

Thaström, A., Lowary, P. T., Widlund, H. R., Cao, H., Kubista, M., and Widom, J. (1999). *J. Mol. Biol.* **288**, 213–229.

Trun, N. J., and Marko, J. F. (1998). *ASM Newslett.* **64,** 276–283.

Tse, C., Fletcher, T. M., and Hansen, J. C. (1998). *Proc. Natl. Acad. Sci. USA* **95,** 12169–12173.

Tuerk, C., and Gold, L. (1990). *Science* **249,** 505–510.

van Holde, K., and Zlatanova, J. (1999). *BioEssays* **21,** 776–780.

Widlund, H. R., Cao, H., Simonsson, S., Magnusson, E., Simonsson, T., Nielsen, P. E., Kahn, J. D., Crothers, D. M., and Kubista, M. (1997). *J. Mol. Biol.* **267,** 807–817.

Willy, P. J., Kobayashi, R., and Kadonaga, J. T. (2000). *Science* **290,** 982–984.

Xie, X., Kokubo, T., Cohen, S. L., Mirza, U. A., Hoffmann, A., Chait, B. T., Roeder, R. G., Nakatani, Y., and Burley, S. K. (1996). *Nature* **380,** 316–322.

Zhu, W., Sandman, K., Lee, G. E., Reeve, J. N., and Summers, M. F. (1998). *Biochemistry* **37,** 10573–10580.

DNA Recombination and Repair in the Archaea

Erica M. Seitz, Cynthia A. Haseltine, and
Stephen C. Kowalczykowski

Sections of Microbiology and of Molecular and Cellular Biology
Center for Genetics and Development
University of California, Davis
Davis, California 95616-8665

I. Introduction

DNA is subjected daily to considerable environmental and endogenous damage, which challenges both the integrity of the essential information that it contains and its ability to be transferred to future generations. All cells, however, are prepared to handle damage to the genome through an extensive DNA repair system, thus underscoring the importance of this process in cell survival. The Archaea represent a rather diverse group of organisms, including many members who thrive under conditions that would be lethal for most bacteria and eukaryotes. These conditions, such as extreme temperatures, also present a new challenge to the Archaea and to their genomes, reinforcing the need to possess an efficient DNA repair system (DiRuggiero et al., 1999; Grogan, 2000). This, and the fact that the Archaea are a largely unexplored domain of life prompted

ADVANCES IN APPLIED MICROBIOLOGY, VOLUME 50

interest in the types of DNA repair mechanisms that operate within this domain.

Studies carried out in bacteria, especially in *Escherichia coli,* or in eukaryotes, particularly in the yeast *Saccharomyces cerevisiae,* revealed much of what is known about these processes. These studies showed that DNA repair occurs by several pathways (Lindahl and Wood, 1999); these include reversal of DNA damage, excision of damaged nucleotides (nucleotide excision repair or NER) or bases (base excision repair or BER), excision of misincorporated nucleotides (mismatch repair or MMR), and recombinational repair (Friedberg *et al.,* 1995). Although relatively little was known about DNA repair in Archaea, the recent sequencing of several archaeal genomes permitted the identification of structural homologues of many proteins involved in these different pathways. In this article, we review the most important features of DNA repair learned from studies of organisms such as *E. coli* and *S. cerevisiae.* In particular, we emphasize the elements which have been conserved throughout evolution, either at the level of global mechanisms or at the level of the protein effectors. We apply this knowledge to the third domain of life, the Archaea, and review what is known about DNA repair in this domain of life, with a specific emphasis on recombinational repair.

II. Recombinational Repair

One of the most serious types of damage that can be inflicted on the genome is a DNA break either in a single strand or in both strands of DNA [a double-stranded DNA (dsDNA) break; DSB]. DNA breaks of any type pose a particularly significant problem to the cell because they challenge the integrity of the DNA molecule and can lead, if not repaired, to loss of information, gross chromosomal rearrangements, and chromosome missegregation. Because of these potentially lethal consequences, both bacterial and eukaryal organisms have mechanisms for repairing this type of DNA lesion, although the manner by which each repairs the lesion differs. In the Bacteria, this type of damage is remedied primarily by the process of homologous DNA recombination (Kowalczykowski *et al.,* 1994; Kuzminov, 1999), whereas in the Eukarya, the DSB is repaired by either homologous recombination or nonhomologous end joining (NHEJ) (Pâques and Haber, 1999; Sung *et al.,* 2000). Recombination involves pairing of the damaged DNA with a homologous partner to copy any lost information from the homologue, thereby accurately repairing the DSB, whereas NHEJ involves ligation of the DSB without the need for significant homology, thus being inherently error-prone. Here we focus on DSB repair by homologous recombination, as NHEJ appears to be a uniquely eukaryal process.

A. An Overview of Homologous Recombination

Homologous DNA recombination is a primary means for the repair of DSBs. Although the general mechanism is similar in bacteria and eukaryotes, the proteins that are involved in this process differ (Fig. 1). Figure 1 depicts the DSB repair model (Resnick, 1976; Szostak et al., 1983) and the likely proteins that act at each step. After DSB formation, both ends of the break are resected to create single-stranded DNA (ssDNA), which then invades a homologous dsDNA molecule. After DNA strand invasion occurs, the 3′ ends of the invading strands serve as primers for the initiation of nascent DNA synthesis, which leads to the formation of two Holliday junctions that are cleaved in one of two orientations to generate two types of recombinant molecules (Fig. 1).

Biochemical studies have revealed the function of many enzymes that participate in the process of homologous recombination. In E. coli, it was determined that the process of homologous recombination involves the action of more than 25 different proteins (Kowalczykowski et al., 1994). Figure 1 shows some of the enzymes from E. coli and S. cerevisiae that act at each step in this process (Kowalczykowski et al., 1994; Pâques and Haber, 1999) and for which there are, or may be, either structural or functional homologues in the Archaea. The first step in the homologous DNA recombination pathway is an initiation or processing step, which involves processing of the broken DNA molecule so that a region with a partially ssDNA character is generated. This processing can be accomplished through the action of DNA helicases, nucleases, or both. The next step corresponds to the search for the homologous target DNA molecule, which is immediately followed by the exchange of their DNA strands. This step is accomplished by DNA strand exchange proteins, which bind to the ssDNA that was generated previously. The resultant nucleoprotein filament is the active form of these proteins, which acts both in the homology search process and in the invasion of the recipient DNA molecule. The consequence of this initial pairing event is a region of newly paired or heteroduplex DNA, which is also known as a joint molecule (Kowalczykowski and Eggleston, 1994). The third step involves the reciprocal exchange of the two DNA strands, creating a four-stranded structure known as a Holliday junction. The regions of heteroduplex DNA are extended by protein-promoted branch migration, which involves the action of either the DNA strand exchange protein or a specialized DNA helicase. The final step involves symmetric cleavage of the Holliday junction in one of two orientations by a Holliday junction-specific endonuclease to produce one of two alternative recombinant products (Kowalczykowski et al., 1994; West, 1994a,b; White et al., 1997; Lilley and White, 2000). Despite differences between the

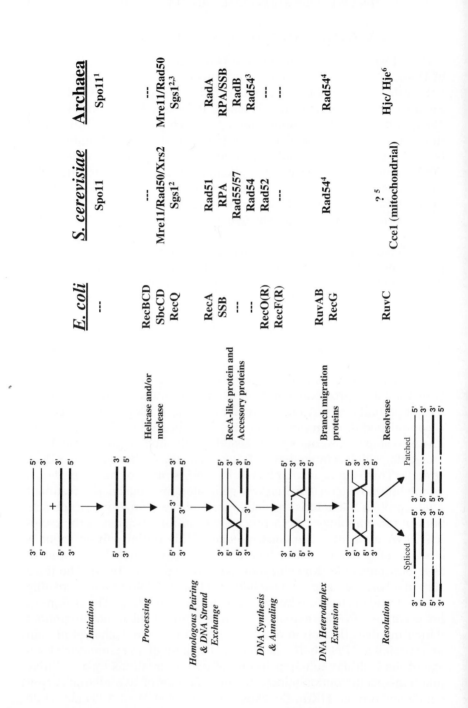

well-studied bacterial (namely, *E. coli*) and eukaryal systems (namely, *S. cerevisiae*), these basic steps remain mostly conserved.

1. Bacterial Homologous DNA Recombination

E. coli possesses two pathways for the repair of DNA strand breaks (Kowalczykowski *et al.*, 1994; Kuzminov, 1999): the RecBCD pathway, which repairs DSBs; and the RecF pathway, which repairs primarily single-strand gaps but can repair DSBs as well. Both of these pathways for recombinational repair depend on the action of the RecA protein. In the RecBCD pathway, the RecBCD helicase/nuclease both processes the DSB to create ssDNA and loads RecA protein onto this ssDNA in anticipation of DNA strand exchange. In the RecF pathway, RecQ helicase processes the broken DNA molecule to produce ssDNA, and the RecO and RecR proteins aid in loading RecA protein onto the ssDNA by mediating the removal of ssDNA binding (SSB) protein (Umezu *et al.*, 1993; Harmon and Kowalczykowski, 1998; Kuzminov, 1999).

2. Eukaryal Homologous DNA Recombination

Homologous DNA recombination is studied in the Eukarya most extensively with the yeast, *S. cerevisiae*, but recent studies in mammals demonstrate the commonality of this eukaryotic process (Pâques and Haber, 1999). As discussed later, some parallels can be drawn between the yeast and the bacterial systems, but for the most part, the system in yeast exists as a more complex process. The repair of DSBs by homologous recombination requires members of the yeast *RAD52* epistasis group, which consists of *RAD50, RAD51, RAD52, RAD54, RAD55, RAD57, RAD59, MRE11, XRS2,* and *RDH54/TID1* genes (Game, 1993; Pâques and Haber, 1999). The function of the proteins encoded by these genes has been studied both genetically and biochemically, but the precise function of some proteins is not yet fully understood (Fig. 1).

3. Archaeal Homologous DNA Recombination

The genome sequences of several archaeons has made it possible to identify structural homologues of many proteins involved in the process of homologous DNA recombination. In addition, some of these proteins

---

FIG. 1. Mechanism for double-stranded DNA break repair by homologous recombination, and the proteins involved. Shown are the proteins that are either known or proposed to act in each step of this process in *E. coli, S. cerevisiae,* and the Archaea. Notes: [1] The archaeal Spo11 protein is a subunit of TopoVI, and a direct role in DSB formation is not clearly defined. [2] A role for Sgs1 in initiation is unclear. [3] Assignment is based only on sequence homology. [4] The Rad54 protein is not a structural homologue of either the RuvAB or the RecG protein; however, it will promote DNA heteroduplex extension (J. Solinger *et al,* in press). [5] ? refers to the fact that an activity has been found in human cells but the responsible protein is unknown. [6] Hje refers to an activity only; the protein has not been identified.

have been studied biochemically, and there is some genetic evidence supporting the role of these genes in archaeal homologous DNA recombination. Evidence for stimulation of chromosomal marker exchange in the hyperthermophilic archaeon *Sulfolobus acidocaldarius* provides evidence for DNA repair, conjugation, and homologous recombination processes in these organisms (Schmidt *et al.,* 1999). Figure 1 and Table I present mainly the proteins involved in this process for which homologues have been found in the Eukarya and Archaea. For the most part, the proteins identified in the Archaea show greater structural and, in some cases, functional, similarity to eukaryal proteins than to their bacterial counterparts (Fig. 1).

B. Generation of DNA Breaks

DNA breaks can occur either in a single DNA strand, creating ssDNA gaps, or in both strands, DSBs. There are many routes for production of ssDNA gaps or DSBs, but DNA replication is a major mechanism for converting ssDNA lesions into larger gaps or DSBs (Kogoma, 1997; Kuzminov, 1999; Kowalczykowski, 2000; Michel, 2000). As illustrated in Fig. 2, ssDNA gaps can be created if a blocking lesion is not removed by repair processes prior to the arrival of the DNA replication machinery. If the lesion is on the lagging strand template, then Okazaki fragments cannot be joined; if the lesion is on the leading strand, then the replication fork halts and may initiate farther downstream. In either case, a region of single-stranded, unreplicated DNA is created. Lesions having the ability to halt the progression of replicative DNA polymerases are numerous and include the well-studied 6–4 thymine photoproducts and cyclobutane pyrimidine dimers caused by ultraviolet (UV) light (Edenberg, 1976).

DSBs can arise from several sources. Exogenously, DSBs are caused by ionizing radiation such as X-rays or γ rays or by various radiomimetic chemicals. Endogenously, DSBs can be created directly by reactive oxygen species and can also arise as a consequence of replicating a nicked DNA template (Fig. 2). Indeed, if a DNA replication fork encounters an interruption (nick or ssDNA gap) in one of the two DNA strands, this interruption will be converted to a DSB (Kuzminov, 1999; Pâques and Haber, 1999; Kowalczykowski, 2000). Nicks in DNA can result from numerous sources, including unsealed Okazaki fragments on the lagging strand and incision of a damaged DNA strand by another repair system, such as either nucleotide or base excision repair. DSBs can also be created as a consequence of the replication apparatus stalling or halting. Stalling can occur, for example, due to the presence of a chemical imperfection in the DNA or a protein complex tightly bound to DNA, either of which can block the progression of the fork. The stalled DNA replication

TABLE I

ARCHAEAL RECOMBINATION PROTEIN HOMOLOGUES[a]

	Mre11	Rad50	Xrs2	RPA/SSB	RadA paralogue							
					RadA	RadB	Other	Rad52	Rad54[b]	Rad59	Spo11	Hjc[c]
A. fulgidus	+	+	−	R	+	+	−	−	?	−	+	+
Halobacterium	+	+	−	R	+	+	−	−	?	−	+	?
M. jannaschii	+	+	−	R	+	+	?d	−	−	−	+	+
M. thermoautotrophicum	+	+	−	R	+	+	−	−	−	−	+	?
P. abyssii	+	+	−	R	+	+	−	−	?	−	+	+
P. furiosus	+	+	−	R	+	+	−	−	?	−	+	+
P. horikoshii	+	+	−	R	+	+	−	−	?	−	+	+
A. pernix	+	+	−	S	+	+	−	−	?	−	+	+
S. solfataricus	+	+	−	S	+	+	−	−	+e	−	+	+

[a] The potential recombination protein homologues from nine fully sequenced archaeal genomes are represented. A + indicates the presence of a single homologous protein sequence, while a − represents the failure to detect a homologue. A ? is shown for single protein sequences where a homologue may be present, but sufficiently high levels of homology to permit confident assignment are not apparent. An R represents the presence of an RPA-like structural protein homologue, while an S represents the presence of an SSB-like structural protein homologue.

[b] A single protein sequence with limited homology to Rad54 was identified in each of the organisms indicated by a ?.

[c] A Holliday-junction endonuclease activity distinct from Hjc was found S. shibatae and S. solfataricus and is called Hje.

[d] A single protein sequence was identified with homology to Rad55.

[e] This homologue is based on sequence similarity only.

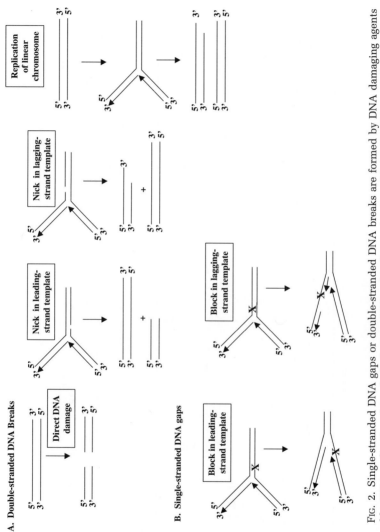

Fig. 2. Single-stranded DNA gaps or double-stranded DNA breaks are formed by DNA damaging agents and by DNA replication through the lesion. Depicted is the production of (A) a double-stranded DNA break, formed either by DNA damaging agents directly or by replication through a nicked template, and (B) a single-stranded DNA gap, formed by replication stopping at the lesion. Both ssDNA gaps and dsDNA breaks can be repaired by homologous recombination. [Adapted from Kowalczykowski (2000).]

forks must be restarted for the replication of the genome to be completed. This restart can be achieved through the introduction of a DSB at the regressed replication fork, followed by recombination-dependent replication (Kogoma, 1997; Michel *et al.,* 1997; Kuzminov, 1999; Pâques and Haber, 1999; Kowalczykowski, 2000; Marians, 2000; Michel, 2000).

In addition to these general mechanisms for DSB formation, DSBs in the Eukarya are also produced in a programmed and specific manner. For example, in meiotic cells, DSBs are enzymatically introduced during the initiation phase of meiosis, to ensure the crossing-over of homologues needed for their faithful segregation (Keeney *et al.,* 1997; Haber, 2000a,b).

1. DSBs in the Bacteria

In *E. coli,* DNA replication initiates at the chromosomal origin, OriC, and progresses bidirectionally along the two arms of the circular chromosome toward the replication terminus. The majority of these replication forks encounters an obstacle to their progression, leading to their stalling (Kogoma, 1997; Michel *et al.,* 1997; Kuzminov, 1999). These obstacles can be chemical lesions, DNA-bound protein complexes, or secondary DNA structure. Regardless of the obstacle, complete replication of the chromosome requires the origin-independent restart of the stalled replication fork. DNA recombination is responsible for this restart (Kogoma, 1996). Recent studies indicate that the first step in this process involves regression of the replication fork by reannealing of the two newly synthesized DNA strands after replication fork arrest. This creates an X-shaped Holliday junction that contains one accessible dsDNA end (Postow *et al.,* 2000; Flores *et al.,* 2001). The RecG protein, a DNA helicase involved in homologous recombination, can catalyze such Holliday junction formation by replication fork reversal (McGlynn and Lloyd, 2000). At this stage, this intermediate can be processed in either of two ways. The RecBCD enzyme, an enzyme involved in the initiation of DNA recombination in bacteria (see below), is a dsDNA nuclease that acts on the DSB created at the Holliday junction (which was formed by replication fork reversal) and starts degrading the DNA. This nucleolytic action effectively shortens the two newly synthesized strands and allows the replication fork to move back from the point where it initially stalled, giving it another opportunity to progress past the previous block after it reinitiates. Alternatively, the regressed replication fork/Holliday junction can be recognized and cleaved by the RuvABC complex to produce a DSB (Michel *et al.,* 1997; Seigneur *et al.,* 1998). The RuvAB complex is involved in the branch migration of Holliday junctions, and RuvC is an endonuclease that specifically cleaves these junctions, as discussed in more detail below. The DSB is then repaired by homologous recombination and is used to restart replication through the action of the PriA protein, which links recombination and replication restart (Kogoma, 1996, 1997; Kowalczykowski, 2000; Marians, 2000; Michel, 2000; Sandler and Marians, 2000).

2. DSBs in the Eukarya

The importance of the above findings is underlined by the fact that sites which are known to block DNA replication in mitotic eukaryal cells promote chromosomal instability due to an increased frequency of homologous recombination, suggesting that the relationship between replication blockage and recombination-dependent replication fork restart is universal (Rothstein et al., 2000). In yeast cells undergoing meiosis, DSBs have long been observed to coincide with known meiotic recombination hot spots (Nicolas et al., 1989; Sun et al., 1989; Debrauwere et al., 1999). These meiotic DSBs were mapped at nucleotide resolution along the entire length of chromosome III and were found to cluster in intergenic promoter-containing intervals, but their occurrence did not require transcription (Baudat and Nicolas, 1997; Borde et al., 1999). Because some breaks were found to have the Spo11 protein covalently linked to the 5' ends of the break sites (Liu et al., 1995; Keeney et al., 1997), it was hypothesized that this protein is the endonuclease responsible for the formation of the meiotic DSB. Mutation of a conserved tyrosine residue in this protein (the residue that attacks the phosphodiester bond and results in a transient covalent DNA–protein complex) eliminated the DSBs and meiotic recombination (Bergerat et al., 1997). Following this discovery, Spo11 homologues were discovered in Schizosaccharomyces pombe, Drosophila melanogaster, Caenorhabditis elegans, and Mus musculus and were found to be essential for meiotic recombination (Dernburg et al., 1998; McKim and Hayashi-Hagihara, 1998; Celerin et al., 2000; Cervantes et al., 2000). In mice, knockouts of the Spo11 gene result in drastic gonadal abnormalities due to defective meiosis, and this gene is additionally required for meiotic synapsis (Baudat et al., 2000; Romanienko and Camerini-Otero, 2000). Overall, these studies demonstrate that homologous DNA recombination during meiosis is initiated by the formation of specific DSBs. Recent results demonstrate that the formation of these breaks in yeast is carefully controlled by the cell and is coupled to the last round of meiotic DNA replication (Borde et al., 2000).

3. Spo11 in the Archaea

An archaeal type II topoisomerase from the hyperthermophile Sulfolobus shibatae that showed homology to the S. cerevisiae Spo11 protein was discovered and is referred to as topoisomerase VI (TopoVI) (Bergerat et al., 1994, 1997). TopoVI is a type II topoisomerase, and these enzymes help regulate DNA topology during transcription, replication, and recombination by catalyzing DNA strand transfer through transient DSBs. This particular topoisomerase is composed of two subunits, A and B, and defines a new family of topoisomerases. The A subunit showed significant homology to the Spo11 protein in *S. cerevisiae* and to the Spo11 homologue in *S. pombe,* the Rec12 protein. Upon inspection of the nine fully sequenced archaeal genomes, we identified several additional homologues, and Fig. 3 shows an alignment of these proteins from eight archaeal organisms. A Spo11 protein homologue was not found in *P. furiosus.* Overall, these proteins share 28–35% similarity to the *S. cerevisiae* Spo11 protein, and each has five conserved DNA gyrase motifs, labeled I–V (Figs. 3 and 4). The *S. shibatae* TopoVI can relax both positive and negative supercoils and has a strong decatenase activity, implying a function in the maintenance of chromosome topology (Bergerat *et al.,* 1997).

C. Initiation of Homologous DNA Recombination: DSB End Processing

After the formation of a DSB, processing of the DNA ends must occur to create a suitable substrate for the next step in homologous recombination, which is catalyzed by a DNA strand exchange protein (Fig. 1). In *E. coli,* the RecBCD enzyme is responsible for this end-processing event (for review, see Kowalczykowski *et al.,* 1994; Kuzminov, 1999; Arnold and Kowalczykowski, 1999), but in the Eukarya and Archaea the mechanism by which this initial processing event occurs is largely unknown. There are, however, enzymes involved in some aspect of DNA end processing that are homologous between the Eukarya and the Archaea; these are the Rad50 and Mre11 proteins (Pâques and Haber, 1999; Sung *et al.,* 2000), which, interestingly, also share homology with a DNA nuclease in *E. coli,* comprised of the SbcC and SbcD proteins (Sharples and Leach, 1995).

The RecBCD enzyme is not the only protein capable of initiating recombination in *E. coli.* In a *recBC⁻ sbcBC⁻* background, recombination proceeds by an alternate pathway known as the RecF pathway. In the absence of the RecBCD enzyme, another helicase, RecQ, processes the DSB (Clark and Sandler, 1994; Mendonca *et al.,* 1995). Interestingly, the Eukarya also have structural homologues of the RecQ helicase; in *S. cerevisiae* it is the Sgs1 protein, and it also affects recombination, but its precise function is unclear (Gangloff *et al.,* 1994; Watt *et al.,* 1995). In humans, there are five proteins that in their conserved helicase domains show significant amino acid similarity to the *E. coli* RecQ helicase: Blm, Wrn, RecQL, RecQ4, and RecQ5 (Puranam and Blackshear, 1994; Seki *et al.,* 1994; Ellis *et al.,* 1995; Yu *et al.,* 1996; Kitao *et al.,* 1998; Shen and Loeb, 2000). Mutations at the *BLM, WRN,* and *RECQ4* loci lead to Bloom's, Werner's, or Rothmund–Thomson

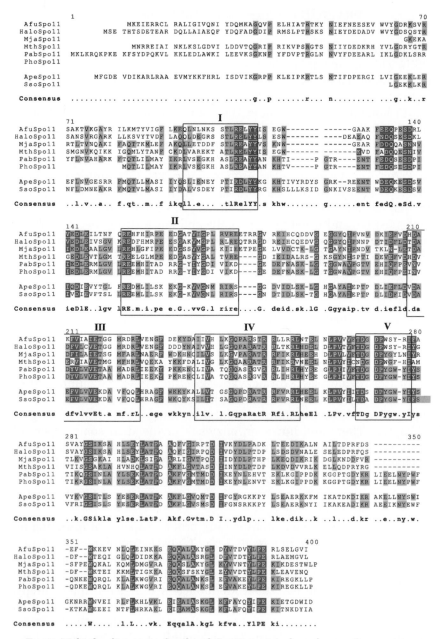

Fig. 3. Multiple alignment of archaeal Spo11 protein homologues. Sequences were as follows: *A. fulgidus* (Afu), gi2649657; *Halobacterium* sp. NRC-1 (Halo), gi10580448; *M. jannaschii* (Mja), mj0369; *M. thermoautotrophicum* (Mth), gi2622109; *P. abyssii* (Pab), gi5458027; *P. horikoshii* (Pho), ph1563; *A. pernix* (Ape), gi5104364; and *S. solfataricus* (Sso), bac04_042. The sequences were aligned using MULTALIN at http://www.toulouse.inra.fr/multalin.html. Highly conserved residues are shaded in dark gray, while moderately conserved residues are shaded in light gray. DNA gyrase motifs I–V are indicated.

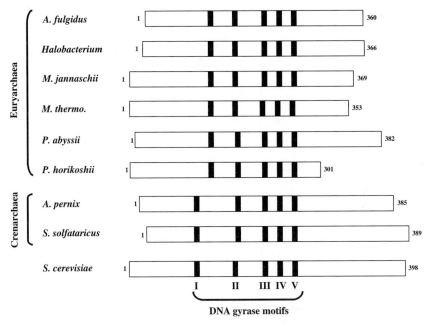

FIG. 4. Schematic representation of archaeal Spo11 protein homologues. Also shown, for comparison, is the *S. cerevisiae* Spo11 protein. DNA gyrase motifs I–V are indicated.

syndromes, respectively, which are rare, inherited diseases that result in DNA replication abnormalities and genomic instability (Kitao *et al.,* 1999a,b; Chakraverty and Hickson, 1999; Shen and Loeb, 2000). Interestingly, a member of the RecQ helicase family was identified in the crenarchaeote *A. pernix* (Kawarabayasi *et al.,* 1999).

1. Bacterial RecBCD-like Enzymes

DNA processing in wild-type *E. coli* is carried out by the RecBCD enzyme, a heterotrimeric protein complex that possesses DNA helicase activity, as well as dsDNA and ssDNA exonuclease activities (Kowalczykowski *et al.,* 1994; Arnold and Kowalczykowski, 1999; Kuzminov, 1999). The exonuclease activity of the RecBCD enzyme initially degrades DNA in a preferential 3′-to-5′ direction (Fig. 5). This destructive activity is regulated by the interaction of the RecBCD enzyme with an eight-nucleotide DNA hot-spot sequence called χ (Lam *et al.,* 1974; Smith *et al.,* 1980; Dixon and Kowalczykowski, 1993; Anderson and Kowalczykowski, 1997a; Bianco and Kowalczykowski, 1997). When the RecBCD enzyme encounters a properly oriented χ site, the 3′-to-5′ exonuclease activity is attenuated, while a weaker 5′-to-3′ exonuclease is activated (Fig. 5). Since the helicase activity

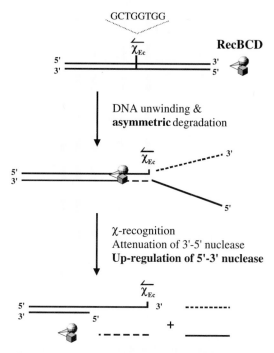

FIG. 5. RecBCD helicase/exonuclease activity is regulated by the recombination hot spot, χ. The RecBCD enzyme enters the DSB and both unwinds and degrades the DNA (the 3'-strand is degraded more extensively than the 5'-strand). Recognition of χ (5'-GCTGGTGG-3') is followed by both attenuation of the 3'–5' nuclease activity and a switch in the polarity of nuclease degradation (to 5'–3'), resulting in degradation of the opposite DNA strand. Also (not shown), the RecA protein is loaded by the RecBCD enzyme onto the χ-containing strand. [Adapted from Anderson and Kowalczykowski (1997a).]

is unaffected, these changes result in a switch in polarity of DNA strand degradation: before χ, the RecBCD enzyme preferentially degrades the 3'-ending strand, whereas after encountering a χ site, the RecBCD enzyme degrades the 5'-ending strand (Fig. 5) (Dixon and Kowalczykowski, 1993; Anderson and Kowalczykowski, 1997a). This processing results in a DNA molecule containing a 3'-ssDNA overhang, onto which the RecBCD enzyme also facilitates the loading of the RecA protein. The RecA nucleoprotein filament then promotes homologous pairing and DNA strand exchange (Anderson and Kowalczykowski, 1997b). Indeed, this facilitated loading of the RecA protein by the RecBCD enzyme is essential to the RecBCD-mediated recombination pathway (Arnold and Kowalczykowski, 2000). Functional homologues of the RecBCD enzyme exist in other bacteria, and although their mechanism of action differs somewhat, the net effect is to process DSBs into 3'-tailed ssDNA (Chédin et al., 2000).

There is no known homologue of the RecBCD enzyme in either the Eukarya or the Archaea at this time, but a structural homologue of the SbcCD enzyme of *E. coli* exists in both of these phylogenetic domains (Fig. 1) (Connelly *et al.*, 1999). The SbcC and SbcD proteins form a complex that possesses ATP-independent ssDNA endonuclease and ATP-dependent dsDNA exonuclease activities (Connelly and Leach, 1996; Connelly *et al.*, 1997). The SbcC protein contains an ATP-binding motif, and the SbcD protein contains a nuclease domain. This complex can also recognize and cleave DNA hairpins (Connelly *et al.*, 1998, 1999; Cromie *et al.*, 2000).

2. E. coli RecQ Helicase

The RecQ helicase is responsible for processing DSBs in the absence of a functional RecBCD enzyme, and it functions in the RecF pathway of recombination. Null mutations in *recQ*, in combination with other mutations, result in a 100-fold reduction in homologous recombination proficiency and cause an increase in sensitivity to UV irradiation (Nakayama *et al.*, 1984, 1985). RecQ is a 3′-to-5′ DNA helicase that can initiate homologous recombination either at a DSB or at ssDNA regions (Lanzov *et al.*, 1991; Lloyd and Buckman, 1995) and can unwind a variety of DNA substrates, including intermediates formed by homologous pairing events (Harmon and Kowalczykowski, 1998). RecQ helicase, in the presence of RecA and SSB proteins, can also initiate homologous recombination *in vitro* (Harmon and Kowalczykowski, 1998). Another function for RecQ helicase comes from evidence that it acts together with topoisomerase III to control recombination (Harmon *et al.*, 1999).

3. Eukaryal Sgs1 Helicase

The *S. cerevisiae* Sgs1 helicase is a member of the RecQ helicase family that is involved in the segregation of chromosomes, control of aging, and regulation of recombination. Mutation of *SGS1* results in premature aging in yeast cells and the accumulation of extrachromosomal rDNA circles (Gangloff *et al.*, 1994; Watt *et al.*, 1995, 1996; Sinclair and Guarente, 1997; Saffi *et al.*, 2000). The Sgs1 protein is also a 3′-to-5′ helicase (Bennett *et al.*, 1998). Additionally, like the *E. coli* system, the Sgs1 protein interacts with *S. cerevisiae* TopoIII to control recombination events (Gangloff *et al.*, 1994; Bennett *et al.*, 2000; Duno *et al.*, 2000; Fricke *et al.*, 2000).

Five additional members of the RecQ helicase family exist in humans, and three are responsible for causing diseases, known as Werner's, Bloom's, and Rothmund–Thomson syndromes (Ellis *et al.*, 1995; Yu *et al.*, 1996; Kitao *et al.*, 1998,1999a,b). These diseases are characterized by the premature onset of aging and an increased incidence of chromosomal abnormalities (Epstein and Motulsky, 1996; Lindor *et al.*, 2000).

4. Archaeal Sgs1 Helicase

A putative Sgs1 protein homologue exists in the crenarchaeote
A. pernix (gi5105033) (Kawarabayasi *et al.*, 1999). Searching the rest of
the fully sequenced archaeal genomes has not yet resulted in convincing Sgs1 protein homologues. The *A. pernix* Sgs1 protein homologue is
similar in size to the S. cerevisiae Sgs1 protein and shows 42% similarity to the S. cerevisiae Sgs1 protein and 47% similarity to the *E. coli*
RecQ protein in the region containing the helicase domains.

5. Eukaryal MRE11/RAD50/ XRS2 (NBS1) Proteins

The genes involved in DNA end processing in *S. cerevisiae* are called
RAD50, MRE11, and *XRS2,* and their gene products form a complex.
This complex is involved in many DNA repair processes, which include
homologous recombination, nonhomologous end joining, telomere
maintenance, and the generation of DSBs in meiosis (Pâques and Haber,
1999; Sung *et al.*, 2000). The Rad50 protein shows homology to the
E. coli SbcC protein, while the Mre11 protein shows homology to the
E. coli SbcD protein (Sharples and Leach, 1995). The Rad50 protein
is a member of a family of proteins called the *s*tructural *m*aintenance
of *c*hromosomes (SMC) family (Hirano, 1999). This protein has
ATP-dependent DNA binding and partial DNA unwinding activities
(Raymond and Kleckner, 1993). Several mutations near the nucleotide
binding site additionally cause defects in meiotic but not in mitotic
DSB repair (Alani *et al.*, 1990). The Mre11 protein is homologous to
a family of phosphodiesterases (Ogawa *et al.*, 1995). In accordance
with this fact, both the *S. cerevisiae* and the human Mre11 proteins
have ssDNA endonuclease activity and a 3'-to-5' exonuclease activity
(Furuse *et al.*, 1998; Paull and Gellert, 1998; Usui *et al.*, 1998). The
Mre11 and Rad50 proteins from humans and yeast form a complex,
which results in enhanced exonuclease activity. These proteins, like the
bacterial SbcD protein, specifically require manganese for activation
of nuclease activity (Furuse *et al.*, 1998). Processing of DSBs during
meiotic recombination is dependent on the nuclease activity of Mre11,
which is proposed to remove the DSB-promoting protein, Spo11, from
the 5' terminus of the DSB to which it is covalently attached (Sung
et al., 2000). The Rad50/Mre11 complex interacts with a third protein
called Xrs2. This interaction takes place via the Mre11 subunit (Johzuka
and Ogawa, 1995), although the role of Xrs2 in changing the function
of the Mre11/Rad50 complex remains undefined.

In humans, the Rad50/Mre11 complex interacts with a third protein,
called p95 or NBS1 (named due to its involvement in Nijemegen breakage syndrome) (Dolganov *et al.*, 1996). Although this third subunit
appears to be analogous to the yeast Xrs2 protein, there is essentially

no sequence homology between these two proteins (Petrini, 1999). This third protein confers upon the complex the ability to open DNA hairpins efficiently, as well as an ATP-dependent endonuclease activity that acts on 3'-ssDNA tails adjacent to a duplex region (Paull and Gellert, 1999). This complex can also unwind duplex DNA to a limited extent, causing strand separation that is stimulated by ATP (Paull and Gellert, 1999).

6. Archaeal RAD50/MRE11 Proteins

Rad50 and Mre11 protein homologues exist in at least nine archaeons to date (Table I and Figs. 6–8). The archaeal Rad50 proteins share 30–38% similarity with the *S. cerevisiae* Rad50 protein and 5–13% similarity with *E. coli* SbcC protein and have conserved Walker-A and -B domains (Fig. 6). We also identified archaeal Mre11 protein homologues in each of the fully sequenced genomes available; these share 20–25% similarity with the *S. cerevisiae* Mre11 protein and 8–20% similarity with *E. coli* SbcD protein. The archaeal Mre11 proteins all contain the four domains that were proposed to be essential for nuclease activity (I–IV in Figs. 7 and 8). A homologue of either the Xrs2 or the NBS1 subunit has not yet been detected, raising the possibility that the Archaea lack this third subunit.

Mre11 (*pfMre11*) and *Rad50* (*pfRad50*) from the euryarchaeote *Pyrococcus furiosus* were recently cloned, and their gene products purified (Hopfner *et al.*, 2000a). This Mre11 homologue, the pfMre11 protein, showed sequence similarity with other members of the Mre11 protein family and had 29% identity and 42% similarity with the human Mre11 protein in the conserved N-terminal domains of the two proteins. The pfMre11 protein, alone, digests ssDNA in a Mn^{2+}-dependent manner. The *pfRad50* gene is located next to the *pfMre11* gene in the *P. furiosus* genome, which is similar to the genetic organization of the *E. coli sbcC* and *sbcD* genes. The pfRad50 protein displays only 19% homology to the human Rad50 protein, although the key residues of the Walker-A and -B ATP binding motifs are conserved between the pfRad50 protein and other members of this protein family (Hopfner *et al.*, 2000a).

The pfMre11 and pfRad50 proteins form a stable complex (pfMRE11/ Rad50), which can digest linear plasmid DNA in an ATP-dependent manner. pfMRE11/Rad50 shows 3'-to-5' ssDNA exonuclease activity, and this activity is ATP dependent, like the bacterial SbcCD complex and the eukaryal Mre11/Rad50 complex. These activities were observed at elevated temperatures of 50°C (Hopfner *et al.*, 2000a). The high-resolution X-ray crystal structures of the ATP-bound and ATP-free Rad50 catalytic domains were determined for *pfRad50*. The two Rad50 catalytic domains associate in an ATP-dependent manner and form a putative DNA binding groove at the interface of this interaction

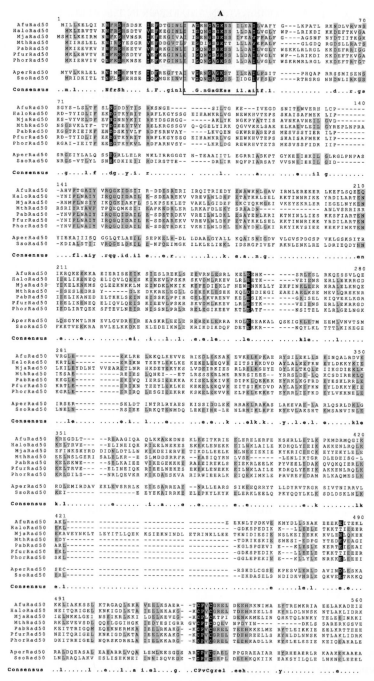

FIG. 6. Multiple alignment of archaeal Rad50 protein homologues. Sequences were as follows: *A. fulgidus* (Afu), gi2649562; *Halobacterium* sp. NRC-1 (Halo), gi10580117; *M. jannaschii* (Mja), mj1322; *M. thermoautotrophicum* (Mth), gi2621615; *P. abyssii* (Pab),

118

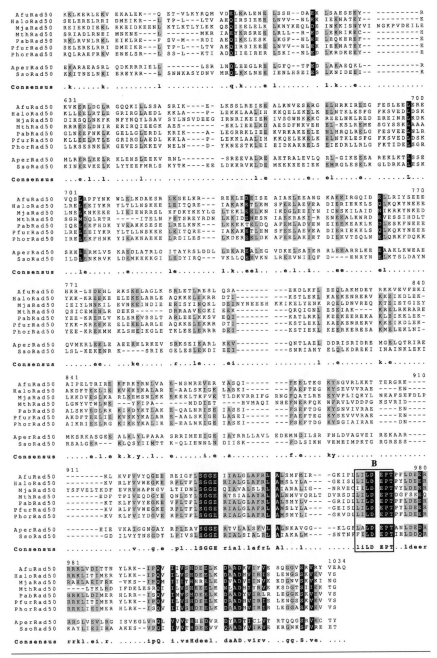

gi5458643; *P. furiosus* (Pfu), orf 1474; *P. horikoshii* (Pho), gi3257342; *A. pernix* (Ape), gi5103499; and *S. solfataricus* (Sso), bac26_052. The sequences were aligned using MULTALIN at http://www.toulouse.inra.fr/multalin.html. Highly conserved residues are shaded in dark gray, while moderately conserved residues are shaded in light gray. The two conserved Walker-A and -B ATP binding domains are indicated as A and B.

119

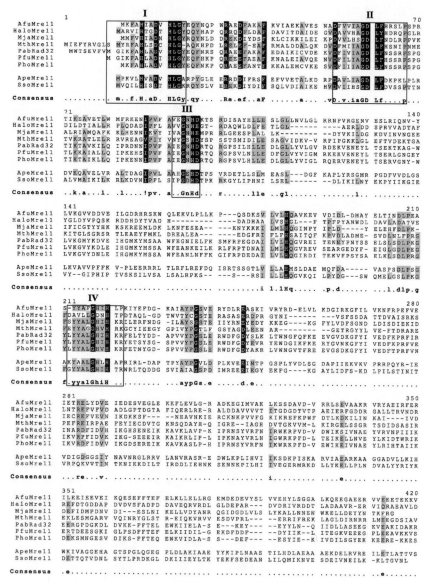

FIG. 7. Multiple alignment of archaeal Mre11 protein homologues. Sequences were as follows: *A. fulgidus* (Afu), G69378; *Halobacterium* sp. NRC-1 (Halo), gi10580116; *M. jannaschii* (Mja), B64465; *M. thermoautotrophicum* (Mth), E69171; *P. abyssii* (Pab), E75103; *P. furiosus* (Pfu), orf1475; *P. horikoshii* (Pho), D71083; *A. pernix* (Ape), E72765; and *S. solfataricus* (Sso), bac26_053. The sequences were aligned using MULTALIN at http://www.toulouse.inra.fr/multalin.html. Highly conserved residues are shaded in dark gray, while moderately conserved residues are shaded in light gray. Conserved nuclease domains I–IV as described for the Mre11 family are indicated.

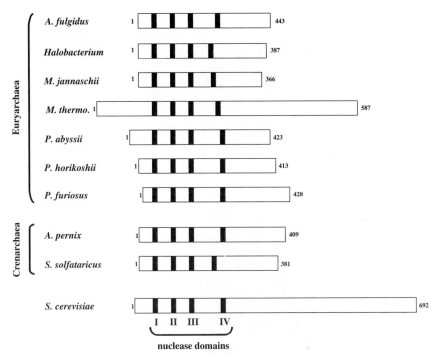

FIG. 8. Schematic representation of archaeal Mre11 protein homologues. Also, shown, for comparison, is the *S. cerevisiae* Mre11 protein. Conserved nuclease domains I–IV are indicated.

(Fig. 9). This suggests that the Rad50 protein may regulate DNA binding and release after DNA end processing through its association with the Mre11 protein (Hopfner *et al.,* 2000b). The fact that the Archaea possess both a Mre11/Rad50 protein homologue and a Spo11 protein homologue suggests that this group of organisms may both form and process DSBs more similarly to the Eukarya than to the Bacteria.

D. DNA PAIRING AND STRAND EXCHANGE

Perhaps the most crucial step in homologous recombination is that of homologous pairing and DNA strand exchange (Fig. 10) (Kowalczykowski and Eggleston, 1994; Bianco *et al.,* 1998; Kuzminov, 1999). The first archaeal recombination protein identified was a DNA strand exchange protein. This protein was discovered based upon its homology to both the bacterial and the eukaryal DNA strand exchange proteins, although it displayed more homology to the eukaryal DNA strand exchange protein (Sandler *et al.,* 1996). In the Bacteria, the role of

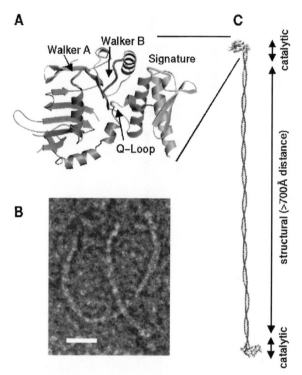

FIG. 9. Structure of the catalytic domain of the *P. furiosus* Rad50 protein. (A) The bilobal
ABC type ATPase fold of the Rad50 protein catalytic domain, which is created by association of the N-terminal and C-terminal ATPase segments of Rad50 protein. The Walker-A
and -B motifs, as well as other important catalytic domains, are indicated. (B) Electron
micrograph of the elongated rods of the 600-residue coiled-coil domain of the Rad50 protein homodimer. The scale bar is 10 nm. (C) Proposed structure of a Rad50 homodimer.
(Courtesy of J. A. Tainer, Scripps Research Institute.)

homologous pairing and DNA strand exchange is fulfilled by the RecA
protein (Bianco and Kowalczykowski, 1999). In the Eukarya, the Rad51
protein, which is homologous to the RecA protein, assumes this role
(Ogawa *et al.,* 1993), and in the Archaea, this DNA strand exchange
step is mediated by the RadA protein (Seitz *et al.,* 1998).

1. Bacterial DNA Strand Exchange: The RecA Protein

Pioneering work on the *E. coli* RecA protein helped to define its
role as the prototypical DNA strand exchange protein. The *recA* gene
was originally isolated in *E. coli* over 30 years ago as a mutation responsible for a dramatic reduction in recombination levels, and its involvement was eventually established for almost all pathways of bacterial recombination (Clark and Margulies, 1965). Subsequently, the

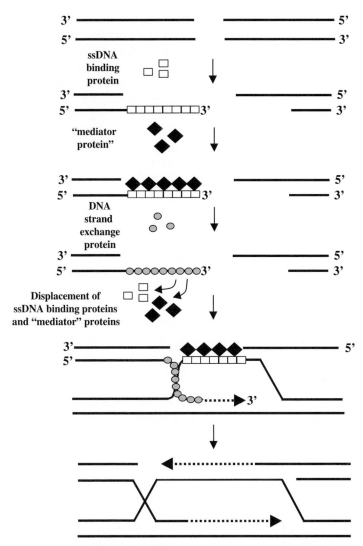

FIG. 10. Biochemical mechanism for the homologous pairing and DNA strand exchange step of homologous recombination. Shown is the DNA strand exchange protein-mediated homologous pairing event between a dsDNA molecule with a DSB and an intact target DNA molecule. After processing of the DSB, ssDNA tails are created, to which a ssDNA binding protein binds. To bind the ssDNA, the DNA strand exchange protein must then displace the ssDNA binding protein; this replacement is aided by mediator or exchange proteins. Next the DNA strand exchange protein catalyzes a homology search and pairs the two DNA molecules. The opposite end of the DSB, after processing, pairs either by the same process or by annealing of the displaced ssDNA in the joint molecule with the repair of ssDNA in the DSB. After DNA strand invasion, the 3′ end serves as a primer for DNA replication (dashed line).

RecA protein was found to possess many biochemical activities: ss- and dsDNA-dependent ATPase, DNA- and ATP-dependent coprotease, ATP-stimulated DNA annealing and ATP-dependent DNA strand exchange activities (Radding, 1989; Cox, 1999; Bianco and Kowalczykowski, 1999). After initial processing of the DSB ends by the RecBCD or RecQ enzymes (Anderson and Kowalczykowski, 1997b; Harmon and Kowalczykowski, 1998), the RecA protein begins a search for homology and catalyzes the pairing and exchange of a DNA strand between each of the two DNA molecules (Fig. 10). RecA protein-mediated homologous pairing and DNA strand exchange occur through a series of distinct steps: presynapsis, synapsis, and DNA heteroduplex extension. During presynapsis, the RecA protein binds to ssDNA in a stoichiometric fashion, with one RecA monomer bound per three nucleotides of ssDNA. The RecA protein interacts with ssDNA in a nonspecific, cooperative manner but does display a preference for binding and pairing DNA sequences rich in G and T residues (Tracy and Kowalczykowski, 1996). RecA protein assembly on ssDNA is polar and occurs in a 5'-to-3' direction to yield a continuous right-handed helical nucleoprotein filament of RecA protein termed the "presynaptic complex" (Stasiak *et al.*, 1984; Egelman and Stasiak, 1986; Stasiak and Egelman, 1986, 1994). Formation of this presynaptic complex occurs much more readily in the presence of a single-stranded DNA binding protein, the SSB protein. Because the RecA protein binds poorly to dsDNA, the presence of secondary structure in ssDNA impedes the formation of a contiguous RecA protein filament. The SSB protein removes this block by disrupting the secondary structure and is subsequently displaced by the RecA protein. Removal of this ssDNA secondary structure permits contiguous filament formation by the RecA protein (Kowalczykowski and Krupp, 1987). The formation of the active RecA nucleoprotein filament typically depends on the presence of a cofactor such as ATP or dATP, and in this ATP-bound form, the RecA protein is in a state that has a high affinity for binding to DNA. The RecA protein hydrolyzes ATP at a rate (k_{cat}) of 25–30 min^{-1}. Although this ATP hydrolysis is not required for the homologous pairing and DNA strand exchange step, it is important in converting the RecA protein from a high-affinity ATP-bound form to an ADP-bound form that has a low affinity for DNA (Kowalczykowski, 1991). This allows the RecA protein both to bind tightly to DNA and to dissociate readily from DNA. Within the filament lies the ssDNA molecule, which has been extended by binding of the RecA protein to 1.5 times the axial spacing of regular B-form DNA (Stasiak *et al.*, 1981; Egelman and Stasiak, 1986, 1998; Stasiak and Egelman, 1986, 1984; Egelman and Yu, 1989).

   During the synaptic step of this process, the RecA nucleoprotein filament catalyzes the search for homology within another dsDNA molecule and exchanges DNA strands between the two molecules. First, the RecA filament makes a series of random, nonhomologous contacts with the target duplex DNA molecule before finding the homologous sequence. Next, the RecA protein catalyzes the exchange of DNA strands, producing a joint molecule. Subsequent to the formation of this joint molecule, the heteroduplex DNA can be extended by the RecA protein through a branch migration step that occurs in only one direction (5' to 3' relative to the displaced ssDNA) (Cox and Lehman, 1981); however, *in vivo,* the RuvAB proteins likely assume this function (West, 1997). The SSB protein also plays a second function in DNA strand exchange at this postsynaptic step, by binding to the displaced ssDNA strand and preventing RecA protein-dependent reinvasion of the duplex DNA molecule by the displaced strand (Kowalczykowski *et al.,* 1994).

2. Eukaryal DNA Strand Exchange: The Rad51 Protein

The existence of a RecA protein homologue in the Eukarya was discovered almost 10 years ago (Shinohara *et al.,* 1992). Mutants of *S. cerevisiae* were isolated on the basis of their sensitivity to ionizing radiation and their inability to undergo meiosis. Of the corresponding genes, studies showed that a *rad51* null mutant is defective in both mitotic and meiotic recombination and is impaired in DSB repair (Game, 1993). Additionally, it was found that the Rad51 protein showed a strong amino acid similarity to the RecA protein (Shinohara *et al.,* 1992). The Rad51 protein possesses many of the same biochemical activities as the RecA protein: stoichiometric binding to DNA (one Rad51 protein monomer per three nucleotides of DNA), ssDNA-dependent ATPase activity, and catalysis of DNA strand exchange (Sung, 1994). The Rad51 protein also forms a right-handed helical nucleoprotein filament on DNA, similar to that of the RecA protein (Ogawa *et al.,* 1993). Interesting differences do exist between these two homologues, however: the Rad51 protein hydrolyzes ATP at a much slower rate (0.7 min^{-1}), has a greater affinity for dsDNA binding, and catalyzes DNA strand exchange much less efficiently, even in the presence of the eukaryotic SSB protein, replication protein-A (RPA), than the RecA protein. Rad51 protein-promoted DNA strand exchange is almost entirely dependent on the presence of a ssDNA binding protein, in contrast to the RecA protein-promoted reaction (Sung and Robberson, 1995; Sugiyama *et al.,* 1997). The ready binding of Rad51 protein to dsDNA poses a unique problem, in that it blocks DNA strand exchange *in vitro* (Sung and Robberson, 1995). Interestingly, the Rad51 protein also shows a pairing bias that is opposite to that of the RecA protein (Mazin *et al.,* 2000b), suggesting that the biochemical properties of the two nucleoprotein filaments may be different. Additionally, the Rad51 protein interacts with other members of the *RAD52* epistasis group, some of which stimulate activities of the Rad51 protein (Sung *et al.*, 2000) (Fig. 10; see below).

3. Archaeal DNA Strand Exchange: The RadA Protein

A role for the RadA protein (Sandler *et al.*, 1996) in DNA repair via homologous recombination came from genetic analysis showing that deletion of the *radA* gene in *Haloferax volcanii* (Woods and Dyall-Smith, 1997) resulted in an archaeon that exhibited a decreased growth rate and an increased sensitivity to DNA damaging agents such as UV irradiation and ethylmethane sulfonate (EMS). The RecA protein homologue from the hyperthermophilic crenarchaeote *Sulfolobus solfataricus* was the first to be purified and studied biochemically (Seitz *et al.*, 1998). It shares many of the same biochemical characteristics of the RecA and Rad51 proteins: the RadA protein is a DNA-dependent ATPase, forms a helical nucleoprotein filament on DNA (Fig. 11), and catalyzes DNA strand exchange. The RadA protein also binds ssDNA with the same stoichiometry as do the RecA and Rad51 proteins, one RadA monomer per three nucleotides of DNA, and it shows a preference for binding to and pairing DNA sequences that are rich in G and T residues (Seitz and Kowalczykowski, 2000). These biochemical

RadA protein   RecA protein

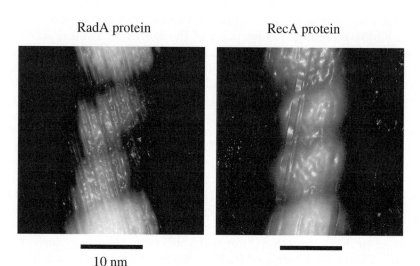

10 nm

FIG. 11. Nucleoprotein filaments of RecA and RadA proteins imaged by atomic force microscopy. Shown are complexes of the RadA and RecA proteins assembled on pBR322 dsDNA in the presence of the ATP analogue, ADP · Al · F₄. As shown, the RadA protein forms a right-handed helical structure that is similar to the structure formed by the RecA protein. [Adapted from Seitz *et al.* (1998).]

activities were seen only at elevated temperatures, close to those at which *S. solfataricus* thrives. The nucleoprotein filament formed by the archaeal RadA protein is the same right-handed helical structure formed by the *E. coli* RecA and the *S. cerevisiae* Rad51 proteins (Egelman and Stasiak, 1986; Ogawa *et al.*, 1993; Seitz *et al.*, 1998) (Fig. 11). The RadA protein's biochemical activities seem more akin to those of the Rad51 protein, however, in that the rate of ATPase activity is rather low (k_{cat} = 0.2 min^{-1}), and the efficiency of DNA strand exchange is also rather poor (Seitz *et al.*, 1998).

The RadA proteins from other hyperthermophilic archaeons, *Desulfurococcus amylolyticus, Pyrobaculum islandicum,* and *P. furiosus,* possess similar biochemical activities, also at elevated temperatures (Kil *et al.*, 2000; Komori *et al.*, 2000b; Spies *et al.*, 2000). Figure 12 shows an alignment of nine archaeal RadA protein sequences, demonstrating the extensive sequence conservation; the well-conserved Walker-A and -B nucleoside triphosphate binding motifs are indicated. In accord with its biochemical similarity to the eukaryal Rad51 protein, the amino acid sequences show that the archaeal RadA proteins are structurally more closely related to the eukaryal Rad51 protein (34–42% identical and 53–63% similar) than to their bacterial counterpart (14–17% identical and 25–31% similar). Domain analysis of the RadA protein from *P. furiosus* demonstrates that the C-terminal portion of the protein, which contains the central core domain (Domain II), possesses DNA-dependent ATPase activity and DNA strand exchange activity, although much reduced in comparison to those of the native RadA protein. Addition of the missing N-terminal peptide to the C-terminal portion restored RadA protein activity to 60% of the wild-type level as measured by ATPase and DNA strand exchange activities, which suggests that the N terminus is needed for the protein to achieve the proper structure for optimal activity (Komori *et al.*, 2000a).

E. Single-Stranded DNA Binding Proteins

As stated previously, DNA strand exchange takes place in essentially three stages. During the steps of presynapsis and postsynapsis, ssDNA binding proteins help to alleviate ssDNA secondary structure and to prevent reinvasion of the displaced single strand of DNA after synapsis, respectively (Kowalczykowski *et al.*, 1994). These functions are fulfilled in bacteria by the ssDNA binding (SSB) protein and in eukaryotes by replication protein-A (RPA) (Fig. 13). Several ssDNA binding proteins have also been identified in the Archaea. Although single-stranded DNA binding proteins are conserved throughout the Archaea, Bacteria, and Eukarya, their protein architectures are quite different.

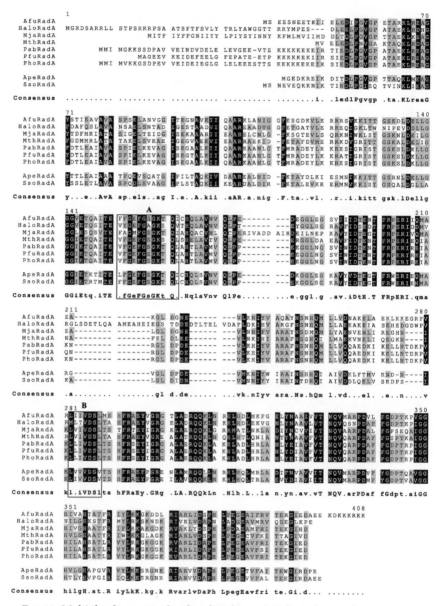

FIG. 12. Multiple alignment of archaeal RadA protein homologues. Sequences were as follows: *A. fulgidus* (Afu), gi2649602; *Halobacterium* sp. NRC-1 (Halo), gi10581871; *M. jannaschii* (Mja), gi2146708; *M. thermoautotrophicum* (Mth), gi2622493; *P. abyssi* (Pab), gi7448305; *P. furiosus* (Pfu), gi3560537; *P. horikoshii* (Pho), gi3256652; *A. pernix* (Ape), gi5103509; and *S. solfataricus* (Sso), gi2129447. The sequences were aligned using MULTALIN at http://www.toulouse.inra.fr/multalin.html. Highly conserved residues are shaded in dark gray, while moderately conserved residues are shaded in light gray. The two conserved Walker-A and -B domains are indicated as A and B.

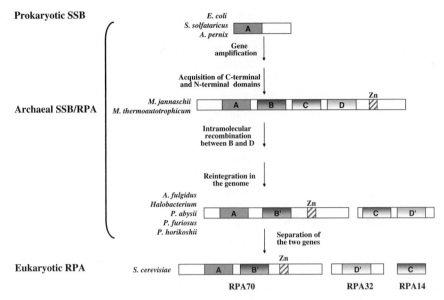

FIG. 13. A model for the evolutionary relationship between the single-stranded DNA binding proteins. Shown is a possible scheme for the evolution of the heterotrimeric eukaryal RPA protein from the single subunit of the bacterial and archaeal SSB proteins. The path illustrated is the simplest and does not necessarily imply the actual evolutionary mechanism. [Adapted from Chédin *et al.* (1998a).]

1. The Bacterial SSB Protein

The *E. coli* SSB protein is important in the processes of replication, recombination, mutagenesis, transposition, repair, and response to DNA damage (Meyer and Laine, 1990). This protein binds preferentially and cooperatively to ssDNA (Lohman and Ferrari, 1994). The *E. coli* SSB protein is encoded by a single gene, while the active form of the protein is a homotetramer in which each monomer contains one ssDNA binding domain (Lohman and Ferrari, 1994). During the process of homologous recombination, the SSB protein is involved in stimulation of RecA protein-mediated DNA strand exchange and in protecting ssDNA from nucleolytic degradation (Kowalczykowski *et al.*, 1994; Anderson and Kowalczykowski, 1998).

2. The Eukaryal RPA

The eukaryal RPA complex is composed of three distinct subunits (Gomes and Wold, 1995, 1996; Wold, 1997). The large subunit of this protein, RPA70, has several domains. The N terminus mediates interactions between RPA and many cellular proteins, while the middle region contains two functional and homologous ssDNA binding sites. The C terminus is involved in interactions with the other subunits of this heterotrimeric complex and also contains a zinc-finger domain, which is important for RPA function (Wold, 1997). RPA32 carries a third functional ssDNA binding site and is phosphorylated in a cell cycle-dependent manner (Bochkareva et al., 1998). Finally, the smallest subunit, RPA14, has an additional ssDNA binding domain. Although the bacterial and eukaryal proteins have completely different protein architectures and share little homology overall, a significant amount of homology is found between their ssDNA binding domain motifs. For example, the ssDNA binding domain A of the RPA70 subunit shows similarity to the E. coli SSB protein. This homology also extends to phage-encoded SSB's and, now, to the archaeal ssDNA binding proteins (Philipova et al., 1996; Chédin et al., 1998b; Kelly et al., 1998; Haseltine, 2001; Wadsworth and White, 2001).

3. Archaeal ssDNA Binding Proteins

A ssDNA binding protein was initially found by sequence analysis in each of three archaeons: *Methanococcus jannaschii, Methanobacterium thermoautotrophicum,* and *Archaeoglobus fulgidus* (Fig. 13 and Table I) (Chédin et al., 1998b). These proteins are homologous to the eukaryal *RFA1* gene, which corresponds to the RPA70 subunit, the largest subunit of the RPA heterotrimeric complex. Interestingly, the ssDNA binding proteins discovered in these three archaeons possessed completely different architectures from either the SSB protein or RPA (Chédin et al., 1998b; Kelly et al., 1998). The euryarchaeal *M. jannaschii* and *M. thermoautotrophicum* proteins exist as a single polypeptide chain and encompass four ssDNA binding domains in tandem, all of which show homology to each other (Fig. 13). Additionally, these ssDNA binding domains contain amino acids that are conserved in the eukaryal RPA70 subunit and are known to make contacts with DNA. Furthermore, a strongly conserved zinc-finger domain was also found within these proteins. This finding implies that these proteins function as a single subunit that does not require multimerization, as in the case of the SSB protein, or association with other subunits, as in the case of the eukaryal RPA.

Investigation into other members of the Archaea, however, revealed ssDNA binding proteins with varied architectures (Chédin et al., 1998b). For example, in *A. fulgidus,* a protein containing two subunits with two DNA binding domains in each was discovered. The second subunit also contained a putative zinc-finger motif. This organization proved to be true for *Pyrococcus abysii, Pyrococcus horikoshii, P. furiosus,* and *Halobacterium* sp. NRC-1 as well (Fig. 13). Finally, the genomes of *Aeropyrum pernix* and *S. solfataricus,* two members of the Crenarchaeota, possess proteins with a completely different architecture (Haseltine, 2001; Wadsworth and White, 2001). These proteins contain a single subunit with a single ssDNA binding domain and an acidic C terminus, which are hallmarks of an *E. coli* SSB protein-like structure. This suggests that the ssDNA binding proteins from members of the Crenarchaeota and Euryarchaeota must have diverged early in evolution and that representatives of each type of ssDNA binding protein still exist in members of the Archaea.

The ssDNA binding proteins from *M. jannaschii* (Kelly *et al.,* 1998; E. M. Seitz and S. C. Kowalczykowski, unpublished observation) and, most recently, *S. solfataricus* (Haseltine, 2001; Wadsworth and White, 2001) were purified. Both proteins show ssDNA binding activity at elevated temperatures, but neither stimulate the ATPase activity or DNA strand exchange activities of the RadA protein. Since secondary structure is not stable in ssDNA at elevated (75–80°C) temperatures, there may be little need for an SSB protein in the presynaptic step of archaeal recombination. Consequently, these ssDNA binding proteins might be needed only for postsynaptic steps.

F. Additional Proteins Involved in DNA Strand Exchange

During the process of DNA strand exchange, the RecA, Rad51, and RadA proteins may encounter obstacles that prevent them from binding to ssDNA or from efficiently completing the DNA strand exchange or DNA heteroduplex extension step. In some instances, ssDNA binding proteins can actually serve as competitors to binding of the DNA strand exchange proteins to ssDNA. This competition is overcome by "mediator" proteins that can facilitate the binding of the DNA strand exchange protein to ssDNA (Fig. 10). In *E. coli,* the RecF, RecO, and RecR proteins serve this function by facilitating binding of the RecA protein to a SSB protein-coated ssDNA gap (Umezu *et al.,* 1993; Webb *et al.,* 1997; Kuzminov, 1999). While there is no structural homologue of either the RecF, the RecO, or the RecR protein in the Eukarya, two factors, the Rad52 protein and Rad55/57 proteins, help the Rad51 protein to overcome the competition imposed by the binding of RPA to ssDNA (Pâques and Haber, 1999; Sung *et al.,* 2000). The Rad55/57 proteins share homology to the Rad51 protein and are, therefore, referred to as Rad51 protein paralogues. Homologues of the RecF, RecO, RecR or Rad52 protein have not been identified in the Archaea. However, there exists a RadA protein paralogue, the RadB protein (Komori *et al.,* 2000b), whose function is unclear, but it may also serve a "mediator" role during DNA strand exchange.

1. Recombination Mediator/DNA Annealing Proteins

a. Bacterial RecFOR Proteins. In both the Bacteria and the Eukarya, there exist proteins that aid the DNA strand exchange protein. In wild-type *E. coli,* the need for these "accessory" proteins is revealed when the DNA lesion is a daughter strand gap, whose repair occurs via the RecF pathway of recombinational repair (Horii and Clark, 1973; Kuzminov, 1999). In this pathway, three proteins facilitate aspects of RecA nucleoprotein filament formation: RecF, RecO, and RecR (Fig. 1) (Kolodner *et al.,* 1985). In the course of daughter strand gap repair, SSB protein is the first protein to bind to the ssDNA within the gap. To facilitate the exchange of RecA protein for SSB protein, the RecOR protein complex binds to the SSB protein–ssDNA complex and facilitates the polymerization of the RecA protein filament at the expense of the SSB protein-coated ssDNA. The RecA protein can now pair the ssDNA gap with a homologous sequence to permit repair of the ssDNA gap. In this capacity, the RecO and -R proteins help both to direct the RecA protein to the gap and to displace the SSB protein that is coating the ssDNA. The RecF protein forms a complex with the RecR protein, and this complex binds randomly to dsDNA to stop RecA nucleoprotein filament extension (Webb *et al.,* 1997). The RecO protein can also anneal complementary ssDNA (Luisi-DeLuca and Kolodner, 1994) and, in fact, can anneal ssDNA that is complexed with the SSB protein (N. Kantake, M. V. V. M. Madiraju, T. Sugiyama, and S. Kowalczykowski, in preparation). To date, no structural homologues of RecF, RecO, or RecR have been uncovered in eukaryal or archaeal organisms, although these proteins are conserved throughout the Bacteria; however, functional homologues exist.

b. The Eukaryal Rad52 Protein. The importance of *S. cerevisiae* *RAD52* in recombination is underscored by the fact that null mutations in *RAD52* eliminate the cell's ability to carry out all homologous recombination events (Game, 1993; Rattray and Symington, 1994). *RAD52* has therefore been implicated in multiple recombination pathways: homologous recombination, ssDNA annealing (SSA), and break-induced replication (BIR) (Pâques and Haber, 1999; Sung *et al.,* 2000). The Rad52 protein bears no structural homology to any known recombination factors in the Bacteria; however, it appears to be a functional homologue of the RecO(R) protein. Additionally, no Rad52 protein homologues have been identified in the Archaea.

The Rad52 protein binds ssDNA and mediates DNA strand annealing between two homologous DNA molecules; this activity is stimulated by the presence of RPA bound to the DNA (Mortensen *et al.,* 1996; Shinohara *et al.,* 1998; Sugiyama *et al.,* 1998). The Rad52 protein binds to DNA by forming ring-shaped multimers (Shinohara *et al.,* 1998; Van

Dyck *et al.*, 1999) and binds to ssDNA with a higher affinity than to ds-DNA (Mortensen *et al.*, 1996; Van Dyck *et al.*, 1999). The Rad52 protein forms a complex with the Rad51 protein, as shown by immunoprecipitation (Sung, 1997b). The Rad52 protein is also able to form a complex with RPA or with RPA–ssDNA complexes (Shinohara *et al.*, 1998; Sugiyama *et al.*, 1998). During DNA strand exchange, the Rad52 protein is able to overcome the inhibition to Rad51 protein posed by the binding of RPA to ssDNA (New *et al.*, 1998; Shinohara *et al.*, 1998). While the Rad52 protein can bind ssDNA, it does not displace RPA from ssDNA; rather it mediates an efficient exchange between the Rad51 protein and RPA (Sung, 1997b; New *et al.*, 1998; Shinohara and Ogawa, 1998). The mechanism by which the Rad52 protein carries out this role as "mediator" may be through its ability to target the Rad51 protein to ssDNA, although presently the exact mechanism is not entirely clear.

2. Rad51 and RadA Protein Paralogues

a. Eukaryal Rad55/57 Proteins (Rad51 Protein Paralogues). Additional members of the yeast *RAD52* epistasis group function in conjunction with the Rad51 protein, and some of these members exist in archaeal genomes. Two proteins in *S. cerevisiae* show limited homology to both the RecA and the Rad51 proteins and are called the Rad55 and Rad57 proteins (Sung *et al.*, 2000). The homology between these proteins and either the RecA or the Rad51 protein resides mainly in the sequence motifs that are involved in nucleoside triphosphate binding. In yeast, mutations in these genes result in cells that are cold-sensitive for both recombination and sensitivity to ionizing radiation. The recombination defect of a *rad55 rad57* double mutant is no greater than that of either single mutation alone, which suggests an epistatic relationship between the two genes (Lovett and Mortimer, 1987). The two proteins interact with one another, as evidenced by yeast two-hybrid experiments and coimmunoprecipitation (Johnson and Symington, 1995). The Rad55/57 complex aids the Rad51 protein in forming a more continuous filament on ssDNA that is complexed with RPA during the presynaptic step of DNA strand exchange (Sung, 1997a).

Human cells contain five Rad51 paralogues of unknown function, known as XRCC2, XRCC3, Rad51B, Rad51C, and Rad51D. These human Rad51 paralogues are all mitotically expressed (Albala *et al.*, 1997; Rice *et al.*, 1997; Cartwright *et al.*, 1998a,b; Dosanjh *et al.*, 1998; Liu *et al.*, 1998), and share 20–30% amino acid homology with the human Rad51 protein and with each other. The *XRCC2* and *XRCC3* genes are important for chromosome stability in mammalian cells (Fuller and Painter, 1988; Tucker *et al.*, 1991; Cui *et al.*, 1999), and *XRCC2* and *XRCC3* are important for efficient repair of DSBs by homologous recombination (Cui *et al.,* 1999; Pierce *et al.,* 1999). Additionally, these five human Rad51 paralogues interact with one other (Schild *et al.,* 2000).

b. Archaeal RadA Protein Paralogues. The Archaea possess proteins homologous to RadA protein as well, and they may serve the same sort of presynaptic role in homologous recombination as demonstrated for Rad55/57 (Fig. 1). The RadA protein paralogue in the Archaea is referred to as RadB. Figure 14 shows an alignment of nine RadB proteins and the conserved Walker-A and -B motifs. These RadB proteins differ from RadA in two ways: first, the RadB proteins are smaller than the RadA protein, lacking both an N- and a C-terminal extension (Fig. 15). Second, while the sequences are homologous, they share only about 30–40% similarity with the RadA protein. In addition, there is a difference between the euryarchaeal and the crenarchaeal RadB protein sequences. The crenarchaeal RadB proteins show more sequence similarity to the *E. coli* RecA protein and, in fact, cannot be identified through a Blast search with the *S. cerevisiae* Rad51 protein sequence. Figure 16a shows an alignment between the *E. coli* RecA protein and the RadB proteins from the crenarchaeotes *S. solfataricus* and *A. pernix.* The crenarchaeal RadB protein is truncated on both the N and the C termini in comparison to the RecA protein but shows 25–27% amino acid similarity over the entire protein. Conversely, RadB proteins from euryarchaeotes show more sequence similarity to the *S. cerevisiae* Rad51 protein (Bult *et al.,* 1996; Klenk *et al.,* 1997; Smith *et al.,* 1997; Kawarabayasi *et al.,* 1998; Komori *et al.,* 2000b). Figure 16b shows an alignment of euryarchaeal RadB proteins with the *S. cerevisiae* Rad51 protein. These euryarchaeal RadB proteins share 38–54% amino acid similarity, across the entire protein, to the Rad51 protein.

The *radB* gene from *P. furiosus* was cloned and its gene product purified (Komori *et al.,* 2000b). This protein possesses a weak DNA-independent ATPase activity, and, interestingly, a higher affinity for binding to ssDNA than does the RadA protein. The RadB protein inhibits RadA protein-promoted D-loop formation under all conditions examined. Inhibition is also seen in RadA protein-promoted DNA strand exchange unless the RadB protein is added after the RadA protein is allowed to bind the ssDNA. Electron microscopy reveals that the RadB protein forms a filamentous structure on ssDNA. The RadB protein did not show any interaction with the RadA protein, which differs from the situation with Rad51 and Rad55/57. Interestingly, this protein coimmunoprecipitates with the Hjc enzyme from *P. furiosus,* a Holliday junction-resolving enzyme (see below), and the RadB protein inhibited Holliday junction cleavage by the Hjc protein. The fact that the RadB

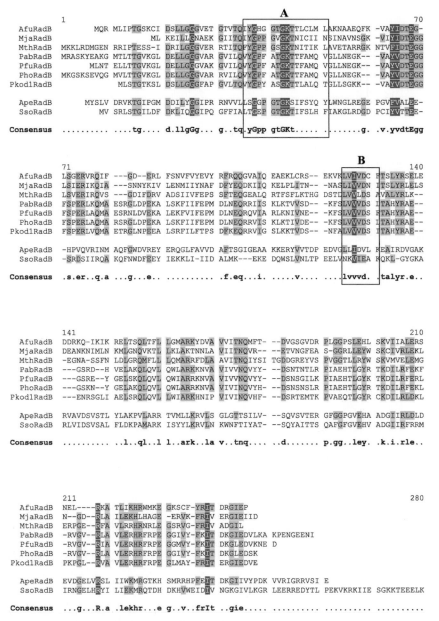

FIG. 14. Multiple alignment of archaeal RadB protein homologues. Sequences were as follows: *A. fulgidus* (Afu), gi_2648436; *M. jannaschii* (Mja), mj0254; *M. thermoautotrophicum* (Mth), gi_2622824; *P. abyssii* (Pab), gi5457551; *P. furiosus* (Pfu), orf527; *P. horikoshii* (Pho), gi3256505, *P.* KOD1 (Pkod), gi6009935; *A. pernix* (Ape), gi5105190; and *S. solfataricus* (Sso), c62_008. The sequences were aligned using MULTALIN at http://www.toulouse.inra.fr/multalin.html. Highly conserved residues are shaded in dark gray, while moderately conserved residues are shaded in light gray. The two conserved Walker-A and -B domains are indicated as A and B.

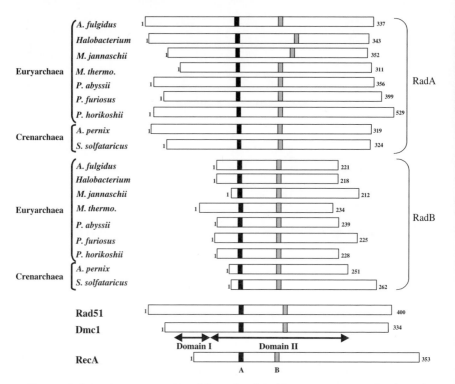

FIG. 15. Schematic representation of the archaeal RadB proteins compared to the RadA proteins and to the RecA/Rad51 proteins. Shown for comparison are the *S. cerevisiae* proteins, Rad51 and Dmc1, and the *E. coli* protein, RecA. RadA proteins are approximately 100 amino acids longer than RadB proteins at the N terminus (Domain I). RadB proteins consist primarily of a central core domain (Domain II). The two conserved Walker-A and -B domains are indicated as A and B.

protein did not stimulate any RadA protein activity could be due to the fact that, to function properly, it must form a heterodimer with another unknown protein, like the *S. cerevisiae* Rad55/57 protein complex (Komori *et al.*, 2000b).

3. Rad54 Proteins

a. The Yeast Rad54 Protein. Another member of the *RAD52* epistasis group, the Rad54 protein, was shown in *S. cerevisiae* to enhance Rad51 protein function during the synaptic phase of DNA strand exchange (Petukhova *et al.*, 1999; Mazin *et al.*, 2000a; Van Komen *et al.*, 2000). This protein belongs to a group of proteins known as the Swi2/Snf2 family, which are involved in a variety of chromosomal processes (Eisen *et al.*, 1995). The Rad54 protein has dsDNA-dependent ATPase activity, and it can induce a conformational change in dsDNA, which is manifest as a change in the linking number of covalently closed dsDNA (Petukhova *et al.*, 1999; Tan *et al.*, 1999). The Rad54 protein interacts with the Rad51 protein in both yeast two-hybrid and *in vitro* analyses (Petukhova *et al.*, 1998), and the Rad54 protein stimulates, by more than 10-fold, Rad51 protein-dependent homologous DNA pairing (Petukhova *et al.*, 1999; Mazin *et al.*, 2000a; Van Komen *et al.*, 2000).

b. Archaeal Rad54 Protein Homologues. A putative Rad54 protein homologue exists in the crenarchaeote *S. solfataricus* (Table I and Figs. 16 and 17). The *S. solfataricus* Rad54 homologue shows conservation of the seven helicase motifs that are found in the yeast Rad54 protein, and it is about 30 amino acids longer than the yeast protein. Figure 17 shows an alignment of the *S. solfataricus* Rad54 protein with the *S. cerevisiae* Rad54 protein, and the conserved helicase motifs are labeled. Also indicated are conserved leucine residues that may constitue a leucine zipper motif. Figure 18 is a schematic comparison of these two proteins. The *S. solfataricus* Rad54 protein lacks the nuclear localization signal (NLS) of the *S. cerevisiae* Rad54 protein but has 47 and 25% amino acid similarity and identity, respectively, to the first 200 amino acids immediately following the yeast Rad54 NLS. This 200-amino acid region makes the Rad54 protein family distinct from other Swi2/ Snf2 DNA-dependent ATPases (Kanaar *et al.*, 1996). Additionally, the *S. solfataricus* Rad54 protein has a conserved leucine zipper motif that is found in the *S. cerevisiae* Rad54 protein. Homologues of the Rad54 protein cannot be identified unequivocally in other archaeons due to weak sequence conservation, and currently there is no biochemistry available for any putative archaeal Rad54 protein.

G. Holliday-Junction Cleaving Enzymes

When first proposed, the Holliday model for recombination envisioned that exchange of both single strands of dsDNA with a homologous duplex DNA would produce a four-way junction, termed the Holliday (1964) junction. This four-way Holliday-junction is central to many models of homologous recombination, and physical evidence for this junction in meiotic recombination was demonstrated (Schwacha and Kleckner, 1995). The formation of this four-way junction is followed by branch migration, which includes the progressive exchange of base-pairing between the homologous duplex DNA molecules (West, 1992; White *et al.*, 1997). Cleavage of this junction by the introduction of two symmetric phosphodiester cleavages (Fig. 19) in one of two possible orientations results in two possible recombinant

(a)

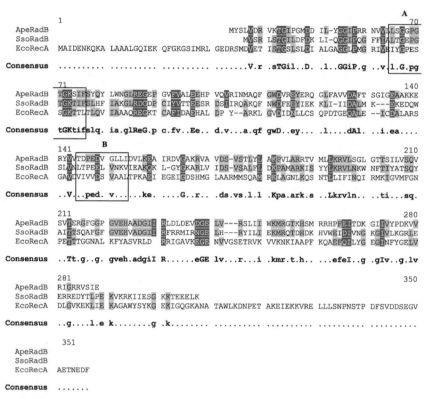

FIG. 16. Multiple alignment of RadB protein homologues. (a) Alignment of crenarchaeal RadB proteins with *E. coli* RecA protein. Sequences were as follows: *A. pernix* (Ape), gi5105190; *S. solfataricus* (Sso), c62_008; and *E. coli* (Eco), gi1789051. (b) Alignment of euryarchaeal RadB proteins with *S. cerevisiae* Rad51 protein. Sequences were as follows: *A. fulgidus* (Afu), gi_2648436; *M. jannaschii* (Mja), mj0254; *M. thermoautotrophicum*, gi_2622824; *P. abyssii* (Pab), gi5457551; *P. furiosus* (Pfu), orf527; *P. horikoshii* (Pho), gi3256505, P. KOD1 (Pkod), gi6009935; and *S. cerevisiae* (Sce), gi603333. The sequences were aligned using MULTALIN at http://www.toulouse.inra.fr/multalin.html. Highly conserved residues are shaded in dark gray, while moderately conserved residues are shaded in light gray. The two conserved Walker-A and -B domains are indicated as A and B.

DNA products: spliced, which results in exchange of genetic markers; and patched, which results in heteroduplex DNA but no exchange of the flanking genetic markers.

The branch migration step (Fig. 1) can be catalyzed by a DNA strand exchange protein; however, in *E. coli* two proteins, RuvA and RuvB, which form the heterodimer called RuvAB, promote particularly

(b)

```
                                                                      A
          1                                                                       70
AfuRadB                MQRMLI PTGSKCIDSL LGCGVETGTV TQIYCHGGTC KTTLCLMLAK NA------AE
MjaRadB                       MLKEI ILCNAEKGII TQIYCPPGVC KTNICIINSI NA------VN
MthRadB   MKK LRDMGENRRI PTESS-IDRI LGCGVERRTI TQIYCPPGSC KTNITIKLAV ET------AR
PabRadB   MRA SKYEAKGMTL TTGVKGLDEL LGCGVARGVI LQIYCPFATC KTTFAMQVGL L---------
PfuRadB             MLNTELL TTGVKGLDEL LGCGVAKGVI LQIYCPFATC KTTFAMQVGL L---------
PhoRadB   MKG SKSEVQGMVL TTGVKGLDEL LGCGVARRVI LQIYCPFATC KTTFAMQVGL L---------
Pkod1RadB                ML STGTKSLDSL LGCGFAPGVL TQIYCPYASC KTTLALQTGL L---------
SceRad51  PMGFVTAADF HMRRSELICL TTGSKNLDTL LGCGVETGST TELFCEFRTC KSQLCHTLAV TCQIPLDIGG

Consensus ..........  .........1 .tg.k.1d.1 LgGgve.g.i tq.yGp..tG Ktt...... . ..........
```

```
                                                                              B
          71                                                                    140
AfuRadB   QFK-VAYIDT EC-LSGERVR QIF---GD-- ERLFSNVFVY EVYRFRQCGV AIQEAEKLCR S--EKVKLVI
MjaRadB   SGK-VIYIDT ECGLSIERIK QIA---SNNY KIVLENMIIY NAFDFYECDK IIQKELPLIT N---NASLIV
MthRadB   RGKNTVPIDT ECGLSVERIR QVS---GDIF DRVADSIIVF EPSSFTECGE ALQRTFSFLK THGDSTDLVV
PabRadB   NEGKVAYVDT ECGFSPERLA QMAESRGLDP EKALSKFIIF EPMDLNECRR IISKLKTVVS D---KFSLVV
PfuRadB   NEGKVAYVDT ECGFSPERLA QMAESRNLDV EKALEKFVIF EPMDLNECRQ VIARLKNIVN E---KFSLVV
PhoRadB   NEGKVAYVDT ECGFSPERLA QMARSRGLDP EKALSKFIIF EPMDLNECRR VISKLKTIVD E---KFSLVV
Pkod1RadB SGKKVAYVDT ECGFSPERLV QMAETRGLNP EEALSRFILF TPSDFKECRR VIGSLKKTVD S---NFAYVV
SceRad51  GEGKCLYIDT ECTFRPVRLV SIAQRFGLDP DDALNNVAYA RAYNADHQLR LLDAAAQMMS E--SRFSLIV

Consensus ..kkvayiDT EGgfspeRl. q.a...gl.. e.al...i.f ep.df.eQ.. .i........ . ...fsLvv
```

```
          141                                                                   210
AfuRadB   VDCFTSLYRS ELEDDRKQ-I KIKREITSQL TFLLGMARKY DVAVVITNQM FT---DVGS- --GVDRPLGG
MjaRadB   VDNITSIYRL ELSDEANKNI MLNKMIGNQV KTILKLAKTN NLAVIITNCV R----ETVN- --GFEAS-GG
MthRadB   LDSAVALYRL K---EGNA-S SFNLDIGRQM FLLLQMARRF DLAAVITNCI YSITGDDGR- --EYVSPVGG
PabRadB   VDSLTAHYRA E----GSR-D --HVELAKQL QVLQWLARRK NVAVLVVNCV YYDSNTNTL- -----RPIAE
PfuRadB   VDSFTAHYRA E----GSR-E --YGELSKQL QVLQWIARRK NVAVLVVNCV YYDSNSGIL- -----KPIAE
PhoRadB   VDSLTAHYRA E----GSK-N --YGELAKQL QVLQWLARRK NVAVLVVNCV YHDSNSNSL- -----RPIAE
Pkod1RadB VDSITAHYRA E----ENR-S GLIAELSRQL QVLLWIARKH NIPVLVINCV HFDSRTEMT- -----KPVAE
SceRad51  VDSVMALYRT DFSGRGEL-S ARQMHLAKFM RALQRLADQF GVAVVVTNCV VAQVDGGMAF NPDPKKPIGG

Consensus vDs.talYR. e....g.... ....eL..ql ..Ll..Ar.. nvavivtNQv .......... .P.gg
```

```
          211                                                                   272
AfuRadB   PSLEHLSKVI IALERSNEL- -RKATLIKHR WMKEGKSCFY RITDRCIEP
MjaRadB   RLLEYWSKCI VRLEKLN--G DRLAILEKHL HAGE-ERVKE RIVERCIEII D
MthRadB   TLLRYWSKVM VELEMGERPG ERFAVLRRHH NRLEGSRVGE RIVADCIL
PabRadB   HTLGYRTKDI LRFEKF-RVG VRLAVLERHR FRPEGGIVYE KITDKCIEDV LKAKPENGEE NI
PfuRadB   HTLGYKTKDI LRFERL-RVG VRIAVLERHR FRPEGGMVYE KITDKCIEDV KNED
PhoRadB   HTLGYRTKDI LRFEKL-RVG VRLAVLERHR FRPEGGIVYE KITDKCIEDS K
Pkod1RadB QTLGYRCKDI LRLDKLPKPG LRVAVLERHR FRPEGLMAYE RITERCIEDV E
SceRad51  NIVAHSSTTR LGFKKG-KGC QRLCKVVDSP CLPEAECV-E AIYEDCVGDP REEDE

Consensus ..l.y.sk.i lrlek....g .R.avlerhr .rpEg..v.f rIt..Gied. ....................
```

FIG. 16. (*Contd.*)

efficient branch migration (Iwasaki *et al.*, 1992; West, 1997). In addition, the RecG protein has DNA unwinding activity that can promote branch migration (Lloyd and Sharples, 1993; Whitby and Lloyd, 1998).

Holliday-junction cleaving or resolving enzymes are found throughout all three domains of life (Aravind *et al.*, 2000) and are also present in bacteriophages (White *et al.*, 1997). These nucleases are specific for DNA molecules that contain branch points and, in particular, four-way junctions. Holliday-junction resolving enzymes can be divided into three types. Type 1 enzymes cleave Holliday junctions at specific dinucleotide sequences, and members include *E. coli* RuvC, yeast mitochondrial Cce1, *E. coli* RusA (White *et al.*, 1997), and perhaps archaeal

```
SceRad54:   189 DNKEEESKKMIKSTQEKDNINKEKNSQEERPTQRIGRHPALMTNGVRNKP--LRE-LL 244
                D  EEE K++  +     +        +E+  Q+I       +  +   +K   LRE LL
SsoRad54:   356 DISEEEFMKLVSENRTIVELGGNLVEIDEKSLQKIKDLLYKIKSKKIDKIDILRESLL 414

                                                                          I
SceRad54:   245 GDSENSAE--NKKKFASVPVVIDPKLAKI-LRPHQVEG---VRFLNRGAYGCIMADEM 297
                GD E + E  ++ +      +++P   K  LRP+Q++G    +RF+N+  G  +AD+M
SsoRad54:   415 GDIEINDELLDRLRGNKSFQLLEPYNIKANLRPYQIKGFSWMRFMNKLGFGICLADDM 473

                                              IA
SceRad54:   298 GLGKTLQCIALMWTLLRQGPQGKRLIDKCIIVCPSSLVNNWANELIKWLGPNTLTPLA 356
                GLGKTLQ IA+     ++         +   +++CP S++ NW  EL K+
SsoRad54:   474 GLGKTLQTIAVFSDAKKENE-----LTPSLVICPLSVLKNWEEELSKFAPHLRFAVFH 527

                                                                          II
SceRad54:   357 VDGKKSSMGGGNTTVSQAIHAWAQAQGRNIVKPVQIISYETQRRNVDQLKNCNVGQML 415
                 D   K   +       +             +++ +Y  L  R+   +LK     ++
SsoRad54:   528 EDRSKIKLEDYD-------------------IIQTTYAVLQRDT-RLKLVEWKYIV 565

                                              III
SceRad54:   416 ADEGHRLKNGDSLTFTALDSISCHRRVILSGTPIQNDLSEYFALLSFSNPGLLGSRAE 474
                   DE   +KN  +  F A+  +     R+ L+GTPI+N + + ++++++F NPGLLGS +E
SsoRad54:   566 IDEAQNIKNPQTKIFKAVKELKSKYRIALTGTPIENKVDDLWSIMTFLNPGLLGSYSE 624

SceRad54:   475 FRKNFENPILRGRDADATDKEITKGEAQLQKLSTIVSKFIIRRT--NDILAKYLPCKY 531
                F++ F  PI +G   D    KE        +L  I+S FI+RRT    +  +LP K
SsoRad54:   625 FKSKFATPIKKG---DNMAKE---------ELKAIISPFILRRTKYDKAIINDLPDKI 671

SceRad54:   532 EHVIFVNLKPLQNELYNKLIKSREVKKVVKGVGGSQPLRAIGI-------LKKLCNHP 583
                E ++    NL P Q  +Y      EV+ +  +      ++  G+        LK++ +HP
SsoRad54:   672 ETNVYCNLTPEQAAMYKA-----EVENLFNNIDSVTGIKRKGMILSTLLKLKQIVDHP 725

SceRad54:   584 NLLNFEDEFDDEDD-----LELPDDYNMPGSKARDVQTKYSAKFSILERFLHKIKTES 637
                LL    ++        +E+ ++   G K   + T++     I+    + K   E
SsoRad54:   726 ALLKGGEQSVRRSGKMIRTMEIIEEALDEGDKIA-IFTQFVDMGKIIRNIIEK---EL 780

                                              IV
SceRad54:   638 DDKIVLISNYTQTLDLIEKMCRYKHYSAVRLDGTMSINKRQKLVDRFNDPEGQEFIFL 696
                +  +  ++       +    ++ C +     AV L   +       ++ +F +   +FI +
SsoRad54:   781 NTEVPFLYGELSKKERDDRECSH----AVILFDIIMRTLPDDIISKFQNNPSVKFI-V 834

                       V                      VI
SceRad54:   697 LSSKAGGCGINLIGANRLILMDPDWNPAADQQALARVWRDGQKKDCFIYRFISTGTIE 755
                LS  KAGG GINL  ANR+I  D   WNPA + QA   RV+R GQ ++  +++ IS GT+E
SsoRad54:   835 LSVKAGGFGINLTSANRVIHFDRWWNPAVEDQATDRVYRIGQTRNVIVHKLISVGTLE 893

SceRad54:   756 EKIFQRQSMKMSLSSCVVDAKEDVERLFSSDNLRQLFQ 794
                EKI Q  + K SL   ++ + +       S++ LR++ +
SsoRad54:   894 EKIDQLLAFKRSLFKDIISSGDSWITELSTEELRKVIE 932
```

FIG. 17. Comparison of the *S. solfataricus* Rad54 protein homologue with *S. cerevisiae* Rad54. Protein sequences (Sso Rad54 homologue sh13a0224_002&004 and Sce Rad54 protein gi6321275) were aligned using BLAST at http://www.ncbi.nlm.nih.gov/BLAST/ . The seven helicase domains characteristic of Swi2/Snf2 DNA-dependent ATPases are indicated, although the homology in motif IV is weak. Identical residues are represented by the single-letter amino acid code, while highly conserved residues are indicated by a +. Residues that may constitute a leucine zipper motif are circled.

Hjc (Kvaratskhelia and White, 2000a). This sequence requirement is probably important to limit cleavage to the Holliday junction. Type 2 enzymes, on the other hand, which include the bacteriophage enzymes T4 endo VII and T7 endo I, have little or no substrate specificity. These endonucleases can cleave a wide variety of other DNA structures, such

S. cerevisiae

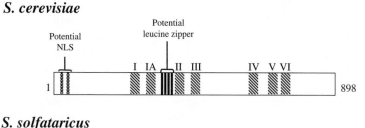

S. solfataricus

FIG. 18. Schematic representation of the *S. solfataricus* Rad54 protein homologue. Potential nuclear localization signal (NLS) and potential leucine zipper regions are indicated. The seven helicase domains characteristic of Swi2/Snf2 DNA-dependent ATPases are represented by cross-hatched boxes, although the homology in motif IV is weak.

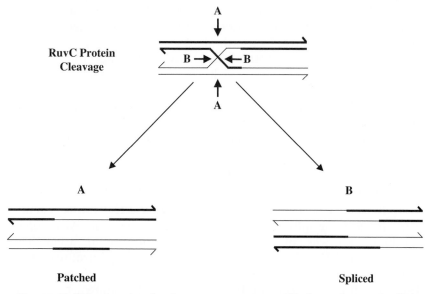

FIG. 19. Holliday-junction cleaving enzymes are responsible for resolution of Holliday junctions in one of two possible orientations. Shown are the products of the endonucleolytic cleavage by the RuvC protein of a Holliday junction in either of two possible orientations, A or B. Cleavage in the A orientation results in a patched recombinant product, while cleavage in the B orientation results in a spliced recombinant product.

as three-way junctions, bulged duplexes, mismatches, and cisplatin adducts (White *et al.,* 1997). The third type of Holliday-junction resolvases is defined by a newly discovered archaeal Hje enzyme. Like type 1, this enzyme shows substrate specificity, but like type 2, it does not exhibit sequence specificity for cleavage (Kvaratskhelia and White, 2000b). Although these Holliday-junction resolving enzymes show the same type of specificity for binding to and cleaving four-way junctions, at the amino acid level these proteins show little or no conservation. Indeed, while a Holliday-junction cleaving activity is detected in yeast nuclei and mammalian extracts, no proteins have been assigned to this activity as of yet (Constantinou *et al.,* 2001).

1. The Bacterial RuvC Protein

The *E. coli* RuvC protein is the prototypic Holliday-junction cleaving enzyme (Bennett and West, 1996; Shah *et al.,* 1997, West, 1997; Eggleston and West, 2000). The crystal structure of RuvC was determined at atomic resolution and demonstrates that the catalytic center, comprising four acidic residues, lies at the bottom of a cleft that fits a DNA duplex (Ariyoshi *et al.,* 1994a,b). The RuvC protein specifically binds four-way Holliday junctions as a dimer and cleaves the strands in a magnesium- and homology-dependent manner. The ssDNA nicks made by RuvC are symmetric; they are found in strands of similar polarity, exclusively on the 3′ site of thymine residues. Strand cleavage by the RuvC dimer occurs in a sequence-specific manner, and the optimal sequence for cleavage is (A~T)TT↓(C >G~A) (Fogg *et al.,* 1999).

2. Archaeal Holliday-Junction Cleaving Enzymes

The first archaeal Holliday-junction cleaving activity was detected in the hyperthermophilic archaeon, *P. furiosus;* the gene was cloned, and the protein was subsequently purified (Komori *et al.,* 1999). This protein, named Hjc (for Holliday-junction cleavage), introduces symmetrically related nicks into two DNA strands of similar polarity, as observed with the *E. coli* RuvC enzyme and other known resolvases. This *P. furiosus* Hjc enzyme resolves Holliday junctions by introducing paired cuts, 3′ to the point of strand exchange, without discernible sequence specificity. The *P. furiosus* Hjc protein does not share any sequence similarity with any of the other known resolvases, although this sequence is highly conserved in the genomes of other archaeons (Table I and Fig. 20). *P. furiosus* Hjc protein cleaves the recombination intermediates that are formed by the *E. coli* RecA protein as efficiently as does the *E. coli* RuvC enzyme (Komori *et al.,* 1999).

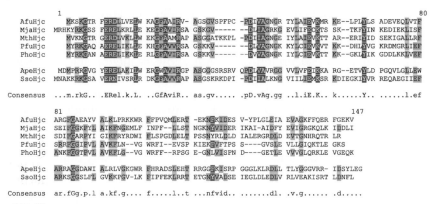

FIG. 20. Multiple alignment of archaeal Holliday-junction cleavage protein homologues. Sequences were as follows: A. fulgidus (Afu), gi2648580; M. jannaschii (Mja),
gi2496010; M. thermoautotrophicum (Mth), gi2622382; P. furiosus (Pfu), gi5689160;
P. horikoshii (Pho), gi5689160; A. pernix (Ape), gi5104108; and S. solfataricus (Sso),
gi6015898. The sequences were aligned using MULTALIN at http://www.toulouse.
inra.fr/multalin.html. Highly conserved residues are shaded in dark gray, while moderately conserved residues are shaded in light gray. The P. aerophilum homologue is not
shown because the genome sequence has not been publicly released.

The S. solfataricus Hjc protein was identified based on homology to
the P. furiosus Hjc protein and showed 34% amino acid sequence identity to this protein. Additional homologues of the Hjc enzyme were identified in the archaea shown in Fig. 20, plus Pyrobaculum aerophilum.
These proteins show 35% amino acid identity between them, including
13 totally conserved residues that may function in binding the catalytic
metal ions (Fig. 20). This conserved catalytic metal ion binding domain
was identified previously in several restriction enzymes and is part of
the active site of the type II restriction enzyme EcoRV (Kvaratskhelia
et al., 2000). Domain analysis of the P. furiosus Hjc enzyme also revealed the importance of several residues that confer enzymatic activity
to this protein, three of which were found to be conserved in the motif
found in type II restriction endonuclease family proteins (Komori et al.,
2000a). The S. solfataricus Hjc enzyme binds specifically to four-way
DNA junctions in a Mg^{2+}-dependent manner, cleaves the junction 3' to
the center of the junction, and may show some sequence specificity for
cleavage (Kvaratskhelia and White, 2000a).

Another archaeal Holliday-junction resolving enzyme, Hje (for
Holliday-junction endonuclease), was found in two members of the
crenarchaeota, S. solfataricus and S. shibatae (Table I) (Kvaratskhelia
and White, 2000b). The partial purification of these enzymes showed

that these endonucleases resolve Holliday junctions in a Mg^{2+}-dependent manner by introducing paired nicks in opposing strands, thereby releasing nicked duplex DNA products. Further experiments showed that the Hje protein does not show sequence specificity for junction cleavage, suggesting that Hje does not belong to the type 1 class of sequence-specific junction resolving enzymes, such as the E. coli RuvC and yeast mitochondrial Cce1 proteins. The Hje proteins do not cleave three-way junctions as does the T4 endonucleaseVII enzyme but do discriminate between the continuous and the exchanging strands of the four-way DNA junction to a greater extent than any other known Holliday-junction cleavage enzyme (Kvaratskhelia and White, 2000b). The archaeal Hje enzyme may therefore use this type of discrimination for recognition and resolution of Holliday junctions to achieve specificity without having to rely on the local nucleotide sequence, like the RuvC enzyme. The Hje enzyme introduces a new class of Holliday-junction resolving enzymes that is unlike any of the previously studied enzymes (Kvaratskhelia and White, 2000a). The S. solfataricus Hje enzyme produces a cleavage pattern completely different from that of the Hjc enzyme, which suggests that there are two Holliday-junction resolving enzymes in this archaeon (Kvaratskhelia and White, 2000a).

H. Summary: Archaeal Recombinational Repair

The process of homologous DNA recombination in the Archaea has only just begun to be explored. This nascent analysis has been greatly facilitated by the relatively recent sequencing of several different archaeal genomes, since the ability to perform genetic screens in these organisms is still rather difficult due to unusual growth requirements, as well as the inability to transform genetically many members of this group.

The picture emerging for this process in the Archaea is one that shows much more similarity to the pathway of eukaryal homologous DNA recombination than to that of bacterial recombination. Homologues of the eukaryal Spo11 protein, which is involved in the creation of DSBs in meiosis, exist in nearly all members of the Archaea, although it is unclear at this point whether this protein plays a direct role in the initiation of homologous recombination in the Archaea, since it is a subunit of topoisomerase VI. The lack of a bacterial RecBCD enzyme homologue to process the DSB suggests that there is a different initiation or DNA end processing mechanism in the Archaea. Homologues of another eukaryal/bacterial nuclease complex that can process DNA ends are, however, found in the Archaea: the Rad50 and Mre11 proteins (Fig. 1 and Table I). Although their precise role in recombination is uknown, perhaps in conjunction with a DNA helicase, appropriate DSB processing can be effected. Interestingly, there also exists at least one example of an archaeal homologue of the RecQ/Sgs1 helicase family. Therefore, related mechanisms of DSB processing are likely for the Archaea and Eukarya.

The archaeal homologous DNA strand exchange protein, RadA, clearly shows more homology to the eukaryal Rad51 protein rather than to the bacterial RecA protein, both structurally and functionally. The fact that RadA protein homologues exist in more than 14 archaeons illustrates the importance of this protein in archaeal cellular function, and given the ubiquity of the Rad51 and RecA proteins, all Archaea are expected to have a RadA homologue.

The Archaea also possess an interesting family of ssDNA binding proteins, which likely serves an important function in the processes of DNA replication, recombination, and repair. These proteins are also more similar at the sequence level to the eukaryal RPA, but they display very diverse structural forms. The euryarchaeal proteins closely resemble RPA in that they also incorporate a zinc binding domain within the protein; however, these proteins exist in one- or two-subunit structural variants, rather than the three-subunit quaternary structure of RPA (Fig. 13). In contrast, however, the crenarchaeal protein resembles the bacterial SSB protein in structural form (a single ssDNA binding domain with an acidic tail, which assembles into a tetramer) while retaining sequence similarity to the binding domains of eukaryal RPA.

The existence of *RAD52* epistasis group homologues in the Archaea also substantiates this similarity to the eukaryal process. These homologues include members, known as RadB proteins, that bear similarity to RecA or Rad51 proteins but are distinct from RadA proteins. RadB proteins, which are RadA protein paralogues, may be homologues of Rad55 or Rad57 proteins. A putative Rad54 protein homologue is also present.

Finally, Holliday-junction resolvases exist in the Archaea. While these enzymes do not show homology to any known resolvases, they are able to bind to four-way Holliday junctions and promote their cleavage in a Mg^{2+}-dependent manner, as shown for all other Holliday-junction cleaving enzymes. The Hjc enzyme, present in most archaeons, is a Holliday-junction resolving enzyme, which may show some sequence specificity for cleavage. The Hje enzymes seem to define their own class of Holliday-junction resolvases, in that they do not display any sequence specificity for cleavage of the Holliday junction but do discriminate between stacked four-way junctions that contain continuous and those that contain exchanging strands, which is different from any Holliday-junction resolving enzyme known to date. Until the identification of the eukaryal Holliday-junction resolvases responsible for this step in homologous recombination, it is impossible to say whether the archaeal resolvases resemble eukaryal resolvases.

Thus, the archaeal system does seem to represent a "simpler" version of the complex eukaryal process, but with unique features, and with some features that bear resemblance to those of Bacteria.

III. DNA Repair Pathways

All living cells have many mechanisms for repairing the various types of DNA damage encountered (Lindahl and Wood, 1999). The multiple pathways employed can be divided into several distinct groups: direct reversal of DNA damage, which chemically reverses DNA damage; base excision repair (BER), which removes the damaged base; nucleotide excision repair (NER), which removes lesions in oligonucleotide form; mismatch repair (MMR), which corrects mispaired bases in DNA; and bypass pathways, which involve specialized DNA polymerases that can insert residues opposite damaged sites so that DNA replication can continue. In this article, we focus mainly on the pathways where homologues have been identified or studied in the Archaea. These processes include direct reversal of DNA damage, NER, and BER (Fig. 21). Toward the end of the article we discuss what is known in the other

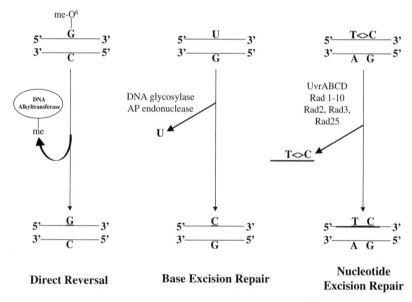

FIG. 21. Three DNA repair pathways common to all phylogenetic domains. Direct reversal chemically reverses the modification and includes the removal of a methyl group from O^6-methylguanine. Base excision repair corrects modifications, such as the incorporation of a uracil residue, by removing a single base. Nucleotide excision repair involves the removal of intact nucleotides, such as a T–C pyrimidine dimer; the lesion is excised as an oligonucleotide, whose length differs for bacterial and eukaryal NER systems.

pathways of MMR and error-prone DNA repair in this phylogenetic domain.

A. Direct DNA Damage Reversal

The first DNA repair mode to be discovered was photoreactivation of DNA (Friedberg *et al.,* 1995). Photoproducts in DNA are created by exposure to UV radiation at wavelengths near the absorption maximum of DNA. To repair the major photoproduct formed, a pyrimidine dimer, organisms have a photoreactivation system to reverse the base damage directly. Photoreactivation is a light-dependent process involving the enzyme-catalyzed monomerization of *cis-syn*-cyclobutyl pyrimidine dimers (Fig. 22), and the enzymes that catalyze the photoreactivation of pyrimidine dimers in DNA are referred to as DNA photolyases or photoreactivating enzymes (Friedberg *et al.,* 1995). This activity is widely distributed in nature and exists in the Bacteria, Eukarya, and Archaea (Friedberg *et al.,* 1995; DiRuggiero *et al.,* 1999; Grogan, 2000).

1. Photolyase

Photolyase is able to split dimers using visible light as the source of energy. This enzyme is able to absorb visible or near-UV light because it contains a photochemically active chromophore (reduced FAD) as well as another chromophore which transduces the absorbed energy to the FAD cofactor. In bacteria, such as *E. coli,* the *phrB* gene encodes the DNA photolyase; in lower eukaryotes, such as *S. cerevisiae,* this gene is referred to as *PHR1.* The *E. coli* and *S. cerevisiae* photolyases contain 5,10-methenyltetrahydrofolate (MTHF) as the second chromophore and have an absorption maximum at 380 nm (Sancar *et al.,* 1987; Johnson *et al.,* 1988). However, the Gram-positive bacterium *Streptomyces griseus* and the cyanobacterium *Anacystis nidulans* contain 8-hydroxy-5-deazaflavin as a second chromophore, which has an absorption maximum at 440 nm (Eker *et al.,* 1981, 1990; Yasui *et al.,* 1988; Sack *et al.,* 1998). Photoreactivation activity has been detected in four archaeons *in vivo: H. halobium, M. thermoautotrophicum, S. solfataricus,* and *S. acidocaldarius* (Fig. 22) (Grogan, 2000). The DNA photolyase from *M. thermoautotrophicum* was purified and characterized and was found to have an absorption maximum at 440 nm (Kiener *et al.,* 1989).

2. DNA Alkyltransferases

Another mechanism of DNA damage repair occurs in response to certain mutagenic alkylating agents, which react with DNA to produce both O-alkylated and N-alkylated products. O^6-alkylguanine and O^4-alkylthymine are potentially mutagenic lesions because they can

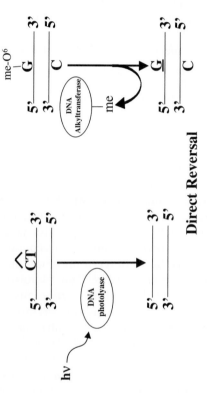

Direct Reversal

Protein Function	E. coli	S. cerevisiae	Euryarchaeota	Crenarchaeota
DNA Photolyase	PhrB	Phr1	**Activity:** *H. halobium, M. thermoautotrophicum* **Homologue:** *M. thermoautotrophicum* (gi2507184)	**Activity:** *S. acidocaldarius, S. solfataricus*
DNA alkyl-transferase	Ogt	Mgt1	**Activity:** *T. litoralis, P. furiosus* **Ogt Homologues:** *A. aeolicus* (gi2983880), *A. fulgidus* (gi2648205), *M. thermoautotrophicum* (gi2621699), *P. abyssii* (gi5457822), *P. horikoshii* (gi3258272) **Mgt1 homologue:** *Pyrococcus* KOD1 (gbD86335), *M. jannaschii* (mj1529)	**Activity:** *P. islandicum, S. acidocaldarius* **Ogt Homologues:** *A. pernix* (gi5104628), *S. solfataricus* (bac03_008)

FIG. 22. Proteins involved in direct reversal DNA repair that are common to all phylogenetic domains. The table compares proteins involved in the photoreactivation and DNA alkyl transfer processes for the Bacteria, Eukarya, and Archaea.

mispair during semiconservative DNA synthesis. The DNA repair protein, O^6-alkylguanine DNA alkyltransferase (ATase), functions by transferring the problematic alkyl groups from the O^6 position of guanine and the O^4 position of thymine to a cysteine residue at the active site of the protein (Foote *et al.*, 1980; Olsson and Lindahl, 1980). This irreversible process results in the stoichiometric inactivation of the protein.

The *E. coli* enzyme that is responsible for transferring methyl groups from the O^6 position of O^6-methylguanine was originally called O^6-methylguanine DNA methyltransferase, but it is also known as Ada due to its importance in the adaptive response to alkylation damage (Friedberg *et al.*, 1995). This protein is able to recognize methyl groups and larger alkyl groups as substrates. *E. coli* possesses an additional protein, however, called Ogt (a DNA alkyltransferase encoded by the *ogt* gene), which transfers the alkyl groups from O^4-methylthymine and O^6-methylguanine to a cysteine residue in the ATase (Goodtzova *et al.*, 1997). The protein responsible for O^6-alkylguanine DNA alkyltransferase activity in *S. cerevisiae* is the product of the *MGT1* gene and is known as Mgt1 protein. This protein shows conservation with the *E. coli* Ada and Ogt proteins and with the human and mammalian Mgt1 proteins as well (Xiao and Samson, 1992).

In the Archaea, DNA alkyltransferases and DNA methyltransferases were found in several members. The protein MGMT (for O^6- methylguanine DNA methyltransferase) was isolated from the hyperthermophilic archaeon *Pyrococcus* sp. KOD1 and possesses methyltransferase activity at temperatures as high as 90°C (Leclere *et al.*, 1998). Additionally, alkyltransferase activity was detected in cell extracts from two euryarchaeotes, *Thermococcus litoralis* and *P. furiosus*, and two crenarchaeotes, *S. acidocaldarius* and *P. islandicum*. The principal activity of these extracts resembled that of the *E. coli* Ogt protein (Skorvaga *et al.*, 1998). Subsequent analysis of sequenced archaeal genomes revealed Ogt homologues also in *A. aeolicus, A. fulgidus, A. pernix, M. thermoautotrophicum, M. jannaschii, P. abysii, P. horikoshii,* and *S. solfataricus* (Fig. 23) (Grogan, 2000). Figure 23 shows an alignment of eight archaeal Ogt protein homologues with the bacterial Ogt protein from *T. maritima*. These proteins all have a conserved methyl-acceptor cysteine residue. The conservation of these alkyltransferases throughout evolution suggests a strong need for this function, which is most likely due to the toxic and mutagenic consequences of this type of DNA damage.

B. BASE EXCISION REPAIR

Base excision repair (BER) involves the removal of nonbulky DNA lesions such as uracil, thymine glycols and hydrates, and 8-oxo-guanine in essentially two steps (Fig. 24). First, a DNA glycosylase releases the

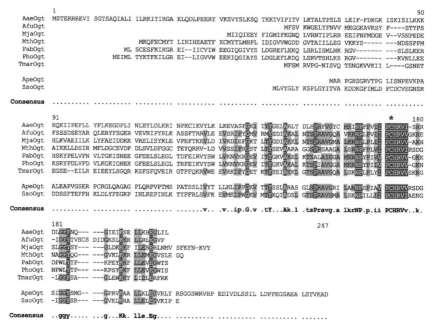

FIG. 23. Multiple alignment of Ogt protein homologues. Sequences were as follows: *Aquifex aeolicus* (Aae), gi2983880; *A. fulgidus* (Afu), gi2648205; *M. jannaschii* (Mja), mj1529; *M. thermoautotrophicum* (Mth), gi2621699; *P. abyssii* (Pab), gi5457822; *P. horikoshii* (Pho), gi3258272; *A. pernix* (Ape), gi5104628; *S. solfataricus* (Sso), bac03–008; and *Thermotoga maritima* (Tmar), gi4981422. *T. maritima* is a member of the Bacteria. The sequences were aligned using MULTALIN at http://www.toulouse.inra.fr/multalin.html. Highly conserved residues are shaded in dark gray, while moderately conserved residues are shaded in light gray. The methyl acceptor cysteine is marked by an asterisk.

base by cleaving the glycosidic bond that connects the base to the deoxyribose. Next, the abasic sugar (apurinic/apyrimidinic (AP) site) is released by the combined actions of AP lyase and AP endonucleases (Friedberg *et al.*, 1995; Sancar, 1996; Wood, 1996).

1. DNA Glycosylases

DNA glycosylases recognize only a certain form of base damage, such as a specific inappropriate base (*e.g.*, uracil) or a specific base mispairing.

FIG. 24. Proteins involved in base excision repair (BER) that are common to all phylogenetic domains. The table compares proteins involved in BER for the Bacteria, Eukarya, and Archaea, showing conserved homologues of a uracil DNA glycosylase, a mismatch glycosylase, an 8-oxoguanine DNA glycosylase, and an apurinic nuclease. Notes: [1]Although reported as UDG homologues (Sandigursky and Franklin, 2000), these sequences are annotated in their respective genomes as DNA polymerase homologues. [2]This protein has also been suggested to be a novel mismatch glycosylase (Horst and Fritz, 1996; Begley *et al.*, 1999) and has been categorized here as a MutY homologue for simplicity.

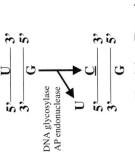

5'——U——3'
3'——G——5'

DNA glycosylase
AP endonuclease

5'——C——3'
3'——G——5'   U

Base Excision Repair

Protein Function	Bacteria	Eucarya	Euryarchaeota	Crenarchaeota
Uracil DNA Glycosylase	UDG	UDG	Activity: A. fulgidus, P. islandicum, P. furiosus, T. litoralis Homologues: A. fulgidus (gi2648243)[1], P. abyssii (gi3257896)[1], P. horikoshii (gi5458117)[1]	Activity: S. shibatae, S. solfataricus Homologue: A. pernix (gi5104069)[1]
Mismatch Glycosylase	Nth	--	Homologues: A. fulgidus (gi2648861), Halobacterium (gi10580185), M. jannaschii (mj1434), M. thermoformicicum (gi232205), P. aerophilum, P. abyssii (gi5458097), P. furiosus (orf1411), P. horikoshii (gi3257923)	Homologue: S. solfataricus (c04_006)
	MutY	--	Homologue: Halobacterium (gi10581009), M. thermoautotrophicum MIG (gi2621835)[2]	Homologue: A. pernix (gi5104542)
8-oxoguanine DNA Glycosylase	oxoG	Ogg1	Activity: M. jannaschii (mjOgg) Homologue: M. jannaschii (gi2833558)	--
Apurinic Endonuclease	Endo IV	Apn	Endo IV Homologues: M. jannaschii (mj1614), M. thermoautotrophicum (gi8928109)	--
	Endo V	--	Homologue: M. thermoautotrophicum (gi2622612)	

DNA glycosylases were first identified in *E. coli* but are ubiquitous in nature. Generally speaking, DNA glycosylases are small, single-subunit proteins that have no cofactor requirement. These enzymes recognize the presence of damaged or mismatched bases and catalyze the breakage of the glycosyl bond between the base and the DNA sugar–phosphate backbone. Some of these enzymes have an associated AP lyase activity that produces $3'$-α,β-unsaturated aldehyde and $5'$-phosphate products (McCullough *et al.*, 1999). Glycosylase action, or the loss of purines or pyrimidines, results in the production of a common intermediate, the AP site. These sites are further processed by the AP endonucleases or AP lyases that cleave the phosphodiester bond either $5'$ or $3'$ to the AP site, respectively. This site is then processed further to yield a $3'$-OH suitable for polymerization and ligation (Sancar, 1996).

a. Uracil DNA Glycosylases. Deamination of cytosine results in the formation of a uracil base. Since uracil will base-pair with adenine, cytosine deamination results in a transition mutation from G–C to A–T, if the uracil-containing strand is used as a replication template (Friedberg *et al.*, 1995). DNA glycosylases that excise uracil or thymine at the *N*-glycosidic bond can be classified into two major types according to amino acid sequence and function. The first type is uracil DNA glycosylase (UDG), which excises uracil from both ss- and dsDNA (U/G and U/A mispairs). This type of enzyme does not, however, excise thymine from T/G mismatches. UDG is found in all organisms, and there is 56% amino acid sequence identity between *E. coli* UDG and human UDG (Olsen *et al.*, 1989; Krokan *et al.*, 1997). The second type of DNA glycosylase includes a mismatch-specific uracil DNA glycosylase (MUG), found in *E. coli* and *Serratia marcescens*, and thymine DNA glycosylase (TDG) from humans (Neddermann *et al.*, 1996). MUG and TDG recognize the mismatched basepairs in dsDNA and remove both mismatched uracil and thymine. TDG recognizes and repairs U/G and T/G mispairs equally, while MUG is mostly U/G mispair specific. MUG has 32% amino acid identity with the central part of human TDG.

A uracil DNA glycosylase (UDG) was first described based on protein activity in the archaea *S. shibatae, S. solfataricus, P. islandicum, P. furiosus,* and *T. litoralis* (Fig. 24) (Koulis *et al.*, 1996). Subsequent to this discovery, a UDG from the archaeon *A. fulgidus* was isolated (Sandigursky and Franklin, 2000). These enzymes showed biochemical characteristics similar to those of the *E. coli* enzyme, as well as the same enzyme from the thermophilic bacterium *T. maritima* (Sandigursky and Franklin, 1999). This archaeal UDG enzyme can remove uracil opposite guanine, as would occur in DNA after cytosine deamination. However, this glycosylase was not able to remove thymine from a similar substrate containing a T–G base pair, which is similar to the activity of the *T. maritima* UDG (Sandigursky and Franklin, 1999). Additional homologues of this protein exist in *P. horikoshii, P. abysii,* and *A. pernix* and were identified based on amino acid sequence homology (Fig. 24) (Sandigursky and Franklin, 2000).

b. Mismatch Glycosylases. A mismatch glycosylase (Mth-MIG) that shows functional similarity to MUG/TDG glycosylases was discovered encoded on the cryptic plasmid pV1 of *M. thermoautotrophicum* (Fig. 24). Mth-MIG processes U/G and T/G but not U on a single strand of DNA (Horst and Fritz, 1996; Begley *et al.,* 1999). Mth-MIG shows little amino acid similarity to MUG/TDG and UDG but shows significant sequence similarity to the [4Fe–4S]-containing Nth/MutY DNA glycosylase family, which catalyzes *N*-glycosylic reactions on DNA substrates other than U/G and T/G mispairs and which are conserved in both the Bacteria and the Eukarya. These types of DNA glycosylases include DNA endonuclease III (Nth; thymine glycol DNA glycosylase), MutY DNA glycosylase (A/G-specific adenine glycosylase), UV endonuclease (UV endo), and methylpurine DNA glycosylase II (MpgII). The unique structural and functional characteristics of Mth-MIG suggest that it is a new type of U/G and T/G mismatch-specific glycosylase. Another putative homologue of this protein was identified in the archaeon *M. jannaschii* based upon sequence homology to endonuclease III (Fig. 24) (Begley *et al.,* 1999).

An additional DNA glycosylase with significant sequence homology to [4Fe–4S]-containing Nth/MutY DNA glycosylases was discovered in the hyperthermophilic archaeon *P. aerophilum* (Fig. 24) (Yang *et al.,* 2000). This protein, Pa-MIG, shows 34% amino acid identity to the *M. thermoformicicum* Mth-MIG protein and 30% amino acid identity to the *E. coli* MutY protein. This protein also has amino acid residues that are generally conserved in the [4Fe–4S]-containing Nth/MutY DNA glycosylase family (Lu and Fawcett, 1998; Yang *et al.,* 2000). The Pa-MIG protein also has a conserved tyrosine residue that is conserved among all Nth proteins and is critical for associated AP lyase activity. Biochemically, the Pa-MIG protein processes both U/G and T/G mismatches and may have a weak AP lyase activity associated with the enzyme, as does the *E. coli* MutY enzyme. This protein could also process T/7,8-dihydro-8-oxoguanine (GO) and U/GO substrates but could not process A/G and A/GO mispairs, which are substrates for the MutY protein, or G/G and G/GO mispairs. Members of this Nth/MutY/MIG/MpgII/UV endoglycosylase superfamily can also be found in *A. pernix, A. fulgidus, M. jannaschii,* and *P. horikoshii* (Yang *et al.,* 2000). Figure 25 shows an alignment of nine archaeal members of this DNA glycosylase family. The conserved lysine residue within the Nth protein family is indicated, and the cysteine residues involved in the [4Fe–4S] binding cluster are also indicated. *M. thermoformicicum* Mth-MIG is not indicated due

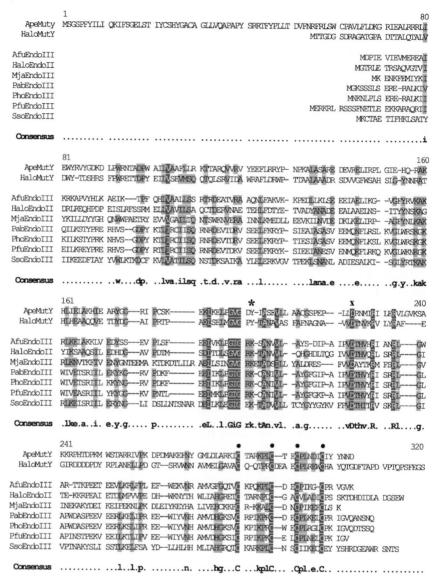

FIG. 25. Multiple alignment of archaeal MutY and endonuclease III protein homologues. Sequences were as follows: *A. pernix* (Ape), gi5104542; *Halobacterium* (HaloMutY), gi10581009; *A. fulgidus* (Afu), gi2648861; *Halobacterium* (HaloEndoIII), gi10580185; *M. jannaschii* (Mja), mj1434; *P. abyssii* (Pab), gi5458097; *P. furiosus* (Pfu), orf1411; *P. horikoshii* (Pho); gi3257923; and *S. solfataricus* (Sso), gi3257923. The sequences were aligned using MULTALIN at http://www.toulouse.inra.fr/multalin.html. Highly conserved residues are shaded in dark gray, while moderately conserved are residues are shaded in light gray. The conserved lysine residue within the Nth family is marked with an asterisk. The strictly conserved aspartic acid residue is indicated by an "x." The cysteine residues involved in binding the [4Fe–4S] cluster are marked with dots.

to the incompletion of this genome sequencing project at this date, and *P. aerophilum* is not indicated due to restrictions on obtaining the sequences. The archaeal MIG family is remotely related to the human MBD4 thymine glycosylase (Pa-MIG shows 21% amino acid identity in the glycosylase domain to human MBD4 protein), which also repairs T/G and U/G mismatches in dsDNA. The C-terminal catalytic domain of the human MBD4 protein shows homology to *E. coli* endonuclease III and MutY proteins (Petronzelli *et al.*, 2000).

2. 8-Oxoguanine DNA Glycosylases

Another member of the DNA glycosylase family that has a homologue in the Archaea is 8-oxoguanine DNA glycosylase (Gogos and Clarke, 1999). 8-Oxoguanine (oxoG) is caused by oxidizing agents or ionizing radiation and can be highly mutagenic if not repaired properly. DNA glycosylases that are specific for this oxoG type of lesion were discovered throughout the Bacteria and Eukarya, although they do not appear to belong to the same family. The eukaryal oxoG DNA glycosylases of yeast and mammals (Ogg 1 protein in *S. cerevisiae* and humans) belong to a protein sequence-related family of DNA glycosylases whose members have a wide range of specificities. The bacterial enzymes, however, such as the *E. coli* MutM enzyme (or Fpg), make up their own distinct family that share sequence conservation, require zinc for activity, and have a strong δ-elimination activity (Girard *et al.*, 1997). An oxoG DNA glycosylase was identified, based on sequence homology to the DNA glycosylase superfamily, in the euryarchaeote *M. jannaschii,* and its gene product purified (Fig. 24). This protein, called mjOgg, is distantly related to other known oxoG-specific enzymes belonging to the same glycosylase superfamily and shows no greater sequence homology with the eukaryal Ogg1 protein than other members. mjOgg shows DNA glycosylase activity and a specificity for oxoG. This enzyme also has an associated DNA lyase activity (Gogos and Clarke, 1999).

3. AP Endonucleases

The AP endo/endonuclease IV family is another class of enzymes involved in BER that have putative representatives in the Archaea, based on sequence analysis. Homologues have been found in *M. jannaschii* and *M. thermoautotrophicum* (Fig. 24). Following the release of free, damaged, or inappropriate bases by DNA glycosylases, AP sites are produced. The repair of these lesions is initiated by AP endonucleases, which catalyze the incision of DNA exclusively at AP sites, and this prepares the DNA for subsequent excision, repair synthesis, and DNA ligation. Endonuclease IV, encoded by the *nfo* gene in *E. coli,* catalyzes the formation of ssDNA breaks at sites of base loss in duplex DNA.

Endo IV attacks phosphodiester bonds 5' to the sites of base loss in DNA, leaving 3'-OH groups. The bacterial endo IV protein is a homologue of eukaryotic apurinic endonucleases (Aravind *et al.*, 1999). Additionally, a homologue of *E. coli* Nfi, or endonuclease V, was tentatively identified, based on sequence homology, in *M. thermoautotrophicum* (Fig. 24) (Aravind *et al.*, 1999). These putative protein homologues have yet to be studied biochemically.

C. Nucleotide Excision Repair

Another ubiquitous repair pathway is the nucleotide excision repair (NER) pathway (Friedberg *et al.*, 1995; Sancar, 1996). During NER, damaged bases such as pyrimidine dimers and (6–4) photoproducts are enzymatically excised from DNA as intact nucleotides that are a part of an olignonucleotide fragment (Fig. 26). There are two excision mechanisms. One is via an endonuclease–exonuclease mechanism, where an endonuclease makes an incision at a phosphodiester bond either 5' or 3' to the lesion, and then an exonuclease digests the damaged strand past the lesion. The second mechanism involves the action of an excision nuclease (excinuclease), which incises the phosphodiester bonds on either side of the lesion and at some distance away from the lesion, to excise the lesion in a nucleotide fragment of a unique length. The fragment and UvrC protein are then released by the action of a DNA helicase (UvrD protein, or helicase II, in *E. coli*) (Fig. 26) (Friedberg *et al.*, 1995; Sancar, 1996).

NER has been characterized in detail in both the Bacteria and the Eukarya, where the damage to the DNA is excised by the combined actions of several proteins in an ATP-dependent manner. The multisubunit complex that comprises the excinuclease in *E. coli* is made up of the UvrA, UvrB, and UvrC proteins (Sancar, 1996). The UvrA protein functions in recognizing the site of DNA damage, while the UvrB and UvrC proteins catalyze the excision reaction, hydrolyzing the eighth phosphodiester bond on the 5' side of the damaged base or bases and the fourth or fifth phosphodiester bond on the 3' side of the damaged base or bases. This leads to the excision of the lesion in the form of a 12- to 13-nucleotide fragment (Sancar, 1996). The UvrD protein (helicase II) then releases the oligonucleotide fragment as well as the DNA-bound UvrC protein. The eukaryal excinuclease incises the 20th–25th phosphodiester bond 5' and the 3rd–8th phosphodiester bond 3' to the lesion to generate 24- to 32-nucleotide fragments (Fig. 26). This NER system, however, involves the action of many more proteins than the bacterial process and is, thus, much more complex. None of the protein subunits that make up the eukaryal excinuclease show any significant homology

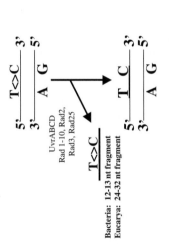

5' ——T<>C—— 3'
3' ——A   G—— 5'

UvrABCD
Rad 1–10, Rad2,
Rad3, Rad25

——T<>C——

5' ——T   C—— 3'
3' ——A   G—— 5'

Bacteria: 12–13 nt fragment
Eucarya: 24–32 nt fragment

Nucleotide Excision Repair

Protein Function	Bacteria	Eucarya	Euryarchaeota	Crenarchaeota
	UvrABCD	...	Activity: *M. thermoautotrophicum*	
	UvrA	...	Homologue: *M. thermoautotrophicum* (MT443)	
	UvrB	...	Homologue: *M. thermoautotrophicum* (MT442)	
Excinuclease	...	Rad1	Homologues: *A. fulgidus* (AF0264), *M. jannaschii* (MJ1505), *M. thermoautotrophicum* (MT1415)	---
	...	Rad2	Homologues: *A. fulgidus* (AF0264), *M. thermoautotrophicum* (MT1633), *P. abyssii* (PAB1877)	
	...	Rad3	Homologue: *P. abyssii* (PAB2385)	

FIG. 26. Proteins involved in nucleotide excision repair (NER) that are common to all phylogenetic domains. The table compares proteins involved in NER for the Bacteria, Eukarya, and Archaea, showing the conserved excinucleases involved in this process.

to the bacterial enzyme. The eukaryal system, however, is conserved throughout the Eukarya (Wood, 1996).

When the Archaea were explored for the presence of NER activity, the activity was found to be more similar to that of the bacterial system. The first experiments, using a cell extract from *M. thermoautotroph-icum*, demonstrated the release of an oligomer containing the lesion that was 10–11 nucleotides in length (Ogrunc *et al.*, 1998). This finding paralleled the results with the purified *E. coli* excinuclease, which released a 12-mer fragment, whereas the mammalian excinuclease released a 27-mer fragment. The archaeal reaction was ATP dependent, in accordance with the behavior of both the bacterial and the eukaryal excinucleases. This archaeon also has UvrA and UvrB homologues, based on sequence homology (Fig. 26) (Grogan, 2000).

The mechanism of NER seems to differ, however, for other members of the Archaea, and homologues of the eucaryal NER system were detected. These include homologues of Rad1, Rad2, Rad3, Rad25, and Rad27, as well as mouse ERCC1 and human XP-F proteins (Fig. 26) (Aravind *et al.*, 1999; Grogan, 2000). In the Eukarya, two nucleases are used to create the dual incisions during NER. In *S. cerevisiae*, the nucleases are the Rad2 protein and the Rad1–10 protein complex (Game, 1993, 2000; Prakash and Prakash, 2000). The Rad1 and Rad10 proteins form a complex that has a ssDNA endonuclease activity which cleaves 3′-ended ssDNA at the junction with duplex DNA (Rad1–10). The Rad2 protein also has ssDNA endonuclease activity. Homologues of the yeast Rad1 protein were uncovered in the archaea *M. jannaschii*, *A. fulgidus*, and *M. thermoautotrophicum*, although none was found in bacteria (Aravind *et al.*, 1999). All of the nucleases from this Rad1 family of proteins contain a conserved ERKX$_2$SD motif and a conserved aspartate residue. The archaeal homologues predict, interestingly, an N-terminal helicase domain that is normally inactive in eukaryotes (Aravind *et al.*, 1999). Putative homologues of Rad2 were identified in *P. abysii*, *A. fulgidus*, and *M. thermoautotrophicum* (Fig. 26) (Aravind *et al.*, 1999; Grogan, 2000). Two helicases in *S. cerevisiae*, Rad3 and Rad25, are also involved in NER. These helicases are responsible for creating a bubble structure during NER (Prakash and Prakash, 2000), and a homologue of the Rad3 helicase was identified in *P. abysii* (Fig. 25) (Grogan, 2000).

D. MISMATCH REPAIR

Both bacterial and eukaryal organisms can repair mismatched DNA base pairs. Mismatches arise by several mechanisms, including errors generated during the process of DNA replication, the formation of heteroduplex DNA as part of the recombination process, and through the deamination of 5-methylcytosine. This type of modified base can be found in the DNA of many organisms, from bacteria to eukaryotes. Deamination causes the conversion of a G–5-mC base pair to a G–T base pair (Friedberg *et al.*, 1995; Yang, 2000).

The basic enzymology of the major MMR processes is very similar in bacteria and eukaryotes. MMR in *E. coli* has been studied extensively and occurs via a methyl-directed MutHLS system. MutS protein initiates this process by binding, as a homodimer, to base–base mismatches and loop insertions–deletions that may have arisen due to polymerase misincorporation and slippage errors, respectively. This MutS repair complex then recruits a MutL protein homodimer, which activates the endonuclease activity of MutH. The ATP binding and hydrolysis activities of MutS and MutL proteins may cause conformational changes to regulate binding to mismatches and subsequent interactions with other factors such as MutH. Once MutH is activated, its endonuclease activity is directed to incise the newly replicated DNA strand at hemimethylated sites formed after the passage of the replication fork. The nicked strand is then unwound by the activity of helicase II and degraded back past the mismatch, either by 5′-to-3′ or by 3′-to-5′ exonucleases, and repair synthesis fills in the resulting gap (Modrich, 1991; Yang, 2000).

Unlike the system in *E. coli*, *S. cerevisiae* has six MutS protein homologues, which are referred to as MutS homologue (MSH) proteins (Kolodner and Marsischky, 1999). In yeast, MMR begins with a MSH2 protein recognizing the mismatch and forming a heterodimer with either a MSH3 or a MSH6 protein to bind the mismatches; each of the latter provides specificity for the type of error that is recognized (Eisen, 1998; Kolodner and Marsischky, 1999). The roles of the other MutS homologues in yeast are not as well understood. MSH1 protein is involved in MMR in mitochondrial DNA, although the function of this protein has not yet been completely characterized (Chi and Kolodner, 1994). The MSH4 and MSH5 proteins are not involved in MMR but, instead, function during meiotic crossing-over and chromosome segregation (Pochart *et al.*, 1997). Mismatch recognition and repair mechanisms in humans and other higher eukaryotes show similarity to those that exist in yeast (Fishel and Wilson, 1997; Kolodner and Marsischky, 1999).

The Archaea, so far, have been shown to possess only a single MutS protein homologue (Eisen, 1998; Aravind *et al.*, 1999). The putative MutS protein homologue was detected in only one member of the Archaea, *M. thermoautotrophicum* (Eisen, 1998), based on sequence homology to the *E. coli* MutS protein; however, this MutS protein homologue was shown to group closer to a subgroup of MutS protein homologues that includes MSH4 and MSH5, which are chromosome crossover and segregation proteins (Eisen, 1998). There is no biochemical characterization of this protein as of yet.

E. Flap Endonuclease Protein Homologues

DNA structures containing single-stranded branches or "flaps" are found as intermediates of DNA replication, recombination, or repair (DeMott et al., 1996; Bambara et al., 1997). Degradation of these flap structures during these processes is carried out by a protein known as FEN-1 (flap endonuclease-1). This protein possesses 5'-to-3' exonuclease activity and can act as an endonuclease for 5' ssDNA flaps. FEN-1 protein homologues were discovered in several members of the Archaea: A. fulgidus, P. furiosus, M. jannaschii, and P. horikoshii (Hosfield et al., 1998a; Rao et al., 1998; Matsui et al., 1999). These proteins show a high level of sequence homology with the human FEN-1 protein; the M. jannaschii FEN-1 homologue shows 76% amino sequence similarity, and the homologues from A. fulgidus and P. furiosus show 72 and 74% amino sequence similarity, respectively. The A. fulgidus, P. furiosus, M. jannaschii, and P. horikoshii FEN-1 protein homologues were purified, and they show specificity for flap DNA structures (Hosfield et al., 1998a; Rao et al., 1998; Matsui et al., 1999). The FEN-1 protein from P. furiosus was crystallized, and the structure was determined (Hosfield et al., 1998b).

F. Translesion DNA Synthesis and Mutagenesis

In the bacterium E. coli, mutagenesis that occurs after exposure to DNA-damaging agents requires a distinct system (the SOS-induced mutagenesis system), which processes DNA damage in an error-prone manner. Several genes in E. coli are regulated by the SOS system, and two of these are error-prone DNA polymerases: UmuD$'_2$C, which is also referred to as PolV (Tang et al., 1999; Goodman, 2000); and DinB, which is referred to as PolIV (Wagner et al., 1999). Homologues of the E. coli DinB protein were discovered in S. cerevisiae, C. elegans, M. musculus, and H. sapiens (Gerlach et al., 1999; Woodgate, 1999). In yeast, the Rad30 protein is homologous to both the UmuC and the DinB proteins and is a DNA polymerase (DNA pol η) that can replicate thymine dimers in template DNA (Johnson et al., 1999). Additionally, a human homologue of yeast Rad30 (Xeroderma pigmentosum variant; XPV) shows activities similar to those of the yeast pol η (Masutani et al., 1999a; Masutani et al., 1999b).

A DinB/UmuC protein homologue was identified by sequence analysis in the archaeon, S. solfataricus (Kulaeva et al., 1996). This protein homologue shows 32% sequence similarity to the DinB protein and 22% sequence similarity to the UmuC protein. Additionally, DNA mutagenesis induced by exposure to UV radiation was detected in the *Pyrococcus* species of archaea (Watrin and Prieur, 1996). Biochemical characterization of this archaeal protein homologue is yet to be reported.

G. SUMMARY: DNA REPAIR MECHANISMS IN THE ARCHAEA

As discussed above, recombinational repair in the Archaea shares more orthologous protein components with the eukaryal system than with the bacterial system, based on the similarities with many components of the yeast *RAD52* epistasis group.

However, the comparison of other DNA repair pathways has not produced a simple conclusion. Proteins involved in the direct reversal of DNA damage are similar in both bacteria and eukaryotes, and the archaeal protein homologues show similarities to both as well. The archaeal DNA alkyltransferases, however, show homology to the bacterial Ogt protein.

The archaeal DNA glycosylases involved in BER show homology to both bacterial and eukaryal enzymes, a consequence of the fact that many bacterial DNA glycosylases are also conserved in the Eukarya. The archaeal UDG protein displays both biochemical and sequence similarity to bacterial UDG proteins. The Archaea have a mismatch glycosylase with homology to the Nth/MutY/MIG/MpgII/UV endoglycosylase superfamily, which is also conserved in both the Bacteria and the Eukarya. An archaeal 8-oxoguanine DNA glycosylase exists in *M. jannaschii,* but the sequence of this enzyme differs greatly from those of both its eukaryal and its bacterial counterparts. Finally, the members of the AP endo/endonuclease IV family in the Archaea are similar in sequence to the bacterial proteins.

In the case of NER, the archaeal proteins show similarities in some cases to the bacterial proteins and in other species to the eukaryal proteins. An activity was identified in *M. thermoautotrophicum* that mimics the action of the UvrABCD proteins, and UvrA and B protein homologues exist, based on sequence similarity, in this archaeon. However, in other archaea, protein homologues of the eukaryal NER machinery were detected.

Less information is available on the processes of mismatch repair and error-prone DNA repair in this third domain of life. So far, only one MutS homologue was found; although this homologue was discovered based on sequence homology to the *E. coli* MutS protein, it groups closer to a subgroup that includes eukaryal MutS protein homologues. Another protein involved in DNA replication, recombination, and repair, FEN-1 protein, has homologues in several archaea, and these show a high degree of sequence homology with the human FEN-1 protein. Finally, a homologue of a bacterial protein involved in error-prone DNA replication, DinB/UmuC, was found in just one member of the Archaea.

In conclusion, it appears that the Archaea possess proteins involved in DNA repair that are similar to both bacterial and eukaryal components and some proteins that are only distantly related to either. For this reason, it is difficult to classify the entire archaeal domain as being "more" bacterial or eukaryal in its means for repairing damage to its DNA. Further investigation into the processes by which the Archaea are able to repair DNA damage will reveal mechanisms by which this unique domain of life deals with the classic problem of DNA damage and should lend insight into the evolution of DNA repair processes.

ACKNOWLEDGMENTS

We thank the following members of the Kowalczykowski lab for providing comments on the manuscript: Piero Bianco, Carole Bornarth, Joel Brockman, Frédéric Chédin, Mark Dillingham, Naofumi Handa, Alex Mazin, Jim New, and Yun Wu. We also thank Dr. John A. Tainer for providing the Rad50 protein structural figure. This work was supported by NIH Training Grant GM07377 to E.M.S., NSF Postdoctoral Fellowship in Microbial Biology 0074380 to C.A.H., and NIH Grants GM62653 and GM 41347 and Human Frontiers Science Program Grant HFSP-RG63 to S.C.K.

REFERENCES

Alani, E., Padmore, R., and Kleckner, N. (1990). *Cell* **61**, 419–436.
Albala, J. S., Thelen, M. P., Prange, C., Fan, W., Christensen, M., Thompson, L. H., and Lennon, G. G. (1997). *Genomics* **46**, 476–479.
Anderson, D. G., and Kowalczykowski, S. C. (1997a). *Genes Dev.* **11**, 571–581.
Anderson, D. G., and Kowalczykowski, S. C. (1997b). *Cell* **90**, 77–86.
Anderson, D. G., and Kowalczykowski, S. C. (1998). *J. Mol. Biol.* **282**, 275–285.
Aravind, L., Walker, D. R., and Koonin, E. V. (1999). *Nucleic Acids Res.* **27**, 1223–1242.
Aravind, L., Makarova, K. S., and Koonin, E. V. (2000). *Nucleic Acids Res.* **28**, 3417–3432.
Ariyoshi, M., Vassylyev, D. G., Iwasaki, H., Fujishima, A., Shinagawa, H., and Morikawa, K. (1994a). *J. Mol. Biol.* **241**, 281–282.
Ariyoshi, M., Vassylyev, D. G., Iwasaki, H., Nakamura, H., Shinagawa, H., and Morikawa, K. (1994b). *Cell* **78**, 1063–1072.
Arnold, D. A., and Kowalczykowski, S. C. (1999). *In* "Encyclopedia of Life Sciences." Nature Publishing Group, London, www.els.net.
Arnold, D. A., and Kowalczykowski, S. C. (2000). *J. Biol. Chem.* **275**, 12261–12265.
Bambara, R. A., Murante, R. S., and Henricksen, L. A. (1997). *J. Biol. Chem.* **272**, 4647–4650.
Baudat, F., and Nicolas, A. (1997). *Proc. Natl. Acad. Sci. USA* **94**, 5213–5218.
Baudat, F., Manova, K., Yuen, J. P., Jasin, M., and Keeney, S. (2000). *Mol. Cell* **6**, 989–998.
Begley, T. J., Haas, B. J., Noel, J., Shekhtman, A., Williams, W. A., and Cunningham, R. P. (1999). *Curr. Biol.* **9**, 653–656.

Bennett, R. J., and West, S. C. (1996). *Proc. Natl. Acad. Sci. USA* **93**, 12217–12222.

Bennett, R. J., Sharp, J. A., and Wang, J. C. (1998). *J. Biol. Chem.* **273**, 9644–9650.

Bennett, R. J., Noirot-Gros, M. F., and Wang, J. C. (2000). *J. Biol. Chem.* **275**, 26898–26905.

Bergerat, A., Gadelle, D., and Forterre, P. (1994). *J. Biol. Chem.* **269**, 27663–27669.

Bergerat, A., de Massy, B., Gadelle, D., Varoutas, P. C., Nicolas, A., and Forterre, P. (1997). *Nature* **386**, 414–417.

Bianco, P. R., and Kowalczykowski, S. C. (1997). *Proc. Natl. Acad. Sci. USA* **94**, 6706–6711.

Bianco, P. R., and Kowalczykowski, S. C. (1999). *In* "Encyclopedia of Life Sciences." Nature Publishing Group, London, www.els.net.

Bianco, P. R., Tracy, R. B., and Kowalczykowski, S. C. (1998). *Front. Biosci.* **3**, D570–D603.

Bochkareva, E., Frappier, L., Edwards, A. M., and Bochkarev, A. (1998). *J. Biol. Chem.* **273**, 3932–3936.

Borde, V., Wu, T. C., and Lichten, M. (1999). *Mol. Cell. Biol.* **19**, 4832–4842.

Borde, V., Goldman, A. S., and Lichten, M. (2000). *Science* **290**, 806–809.

Bult, C. J., White, O., Olsen, G. J., Zhou, L., Fleischmann, R. D., Sutton, G. G., Blake, J. A., Fitzgerald, L. M., Clayton, R. A., and Gocayne, J. D. (1996). *Science* **273**, 1058–1073.

Cartwright, R., Dunn, A. M., Simpson, P. J., Tambini, C. E., and Thacker, J. (1998a). *Nucleic Acids Res.* **26**, 1653–1659.

Cartwright, R., Tambini, C. E., Simpson, P. J., and Thacker, J. (1998b). *Nucleic Acids Res.* **26**, 3084–3089.

Celerin, M., Merino, S. T., Stone, J. E., Menzie, A. M., and Zolan, M. E. (2000). *EMBO J.* **19**, 2739–2750.

Cervantes, M. D., Farah, J. A., and Smith, G. R. (2000). *Mol. Cell* **5**, 883–888.

Chakraverty, R. K., and Hickson, I. D. (1999). *Bioessays* **21**, 286–294.

Chédin, F., Noirot, P., Biaudet, V., and Ehrlich, S. D. (1998a). *Mol. Microbiol.* **29**, 1369–1377.

Chédin, F., Seitz, E. M., and Kowalczykowski, S. C. (1998b). *Trends Biochem. Sci.* **23**, 273–277.

Chédin, F., Ehrlich, S. D., and Kowalczykowski, S. C. (2000). *J. Mol. Biol.* **298**, 7–20.

Chi, N. W., and Kolodner, R. D. (1994). *J. Biol. Chem.* **269**, 29984–29992.

Clark, A. J., and Margulies, A. D. (1965). *Proc. Natl. Acad. Sci. USA* **53**, 451–459.

Clark, A. J., and Sandler, S. J. (1994). *Crit. Rev. Microbiol.* **20**, 125–142.

Connelly, J. C., and Leach, D. R. (1996). *Genes Cells* **1**, 285–291.

Connelly, J. C., de Leau, E. S., Okely, E. A., and Leach, D. R. (1997). *J. Biol. Chem.* **272**, 19819–19826.

Connelly, J. C., Kirkham, L. A., and Leach, D. R. (1998). *Proc. Natl. Acad. Sci. USA* **95**, 7969–7974.

Connelly, J. C., de Leau, E. S., and Leach, D. R. (1999). *Nucleic Acids Res.* **27**, 1039–1046.

Constantinou, A., Davies, A. A., and West, S. C. (2001). *Cell* **104**, 259–268.

Cox, M. M. (1999). *Prog. Nucleic Acid Res. Mol. Biol.* **63**, 311–366.

Cox, M. M., and Lehman, I. R. (1981). *Proc. Natl. Acad. Sci. USA* **78**, 6018–6022.

Cromie, G. A., Millar, C. B., Schmidt, K. H., and Leach, D. R. (2000). *Genetics* **154**, 513–522.

Cui, X., Brenneman, M., Meyne, J., Oshimura, M., Goodwin, E. H., and Chen, D. J. (1999). *Mutat. Res.* **434**, 75–88.

Debrauwere, H., Buard, J., Tessier, J., Aubert, D., Vergnaud, G., and Nicolas, A. (1999). *Nature Genet.* **23**, 367–371.

DeMott, M. S., Shen, B., Park, M. S., Bambara, R. A., and Zigman, S. (1996). *J. Biol. Chem.* **271**, 30068–30076.

Dernburg, A. F., McDonald, K., Moulder, G., Barstead, R., Dresser, M., and Villeneuve, A. M. (1998). *Cell* **94**, 387–398.

DiRuggiero, J., Brown, J. R., Bogert, A. P., and Robb, F. T. (1999). *J. Mol. Evol.* **49**, 474–484.

Dixon, D. A., and Kowalczykowski, S. C. (1993). *Cell* **73**, 87–96.

Dolganov, G. M., Maser, R. S., Novikov, A., Tosto, L., Chong, S., Bressan, D. A., and Petrini, J. H. (1996). *Mol. Cell. Biol.* **16**, 4832–4841.

Dosanjh, M. K., Collins, D. W., Fan, W., Lennon, G. G., Albala, J. S., Shen, Z., and Schild, D. (1998). *Nucleic Acids Res.* **26**, 1179–1184.

Duno, M., Thomsen, B., Westergaard, O., Krejci, L., and Bendixen, C. (2000). *Mol. Gen. Genet.* **264**, 89–97.

Edenberg, H. J. (1976). *Biophys. J.* **16**, 849–860.

Egelman, E. H., and Stasiak, A. (1986). *J. Mol. Biol.* **191**, 677–697.

Egelman, E. H., and Stasiak, A. (1988). *J. Mol. Biol.* **200**, 329–349.

Egelman, E. H., and Yu, X. (1989). *Science* **245**, 404–407.

Eggleston, A. K., and West, S. C. (2000). *J. Biol. Chem.* **275**, 26467–26476.

Eisen, J. A. (1998). *Nucleic Acids Res.* **26**, 4291–4300.

Eisen, J. A., Sweder, K. S., and Hanawalt, P. C. (1995). *Nucleic Acids Res.* **23**, 2715–2723.

Eker, A. P., Dekker, R. H., and Berends, W. (1981). *Photochem. Photobiol.* **33**, 65–72.

Eker, A. P., Kooiman, P., Hessels, J. K., and Yasui, A. (1990). *J. Biol. Chem.* **265**, 8009–8015.

Ellis, N. A., Groden, J., Ye, T. Z., Straughen, J., Lennon, D. J., Ciocci, S., Proytcheva, M., and German, J. (1995). *Cell* **83**, 655–666.

Epstein, C. J., and Motulsky, A. G. (1996). *Bioessays* **18**, 1025–1027.

Fishel, R., and Wilson, T. (1997). *Curr. Opin. Genet. Dev.* **7**, 105–113.

Flores, M. J., Bierne, H., Ehrlich, S. D., and Michel, B. (2001). *EMBO J.* **20**, 619–629.

Fogg, J. M., Schofield, M. J., White, M. F., and Lilley, D. M. (1999). *Biochemistry* **38**, 11349–11358.

Foote, R. S., Mitra, S., and Pal, B. C. (1980). *Biochem. Biophys. Res. Commun.* **97**, 654–659.

Fricke, W. M., Kaliraman, V., and Brill, S. J. (2000). *J. Biol. Chem.* (in press).

Friedberg, E. C., Walker, G. C., and Siede, W. (1995). "DNA Repair and Mutagenesis." ASM Press, Washington DC.

Fuller, L. F., and Painter, R. B. (1988). *Mutat. Res.* **193**, 109–121.

Furuse, M., Nagase, Y., Tsubouchi, H., Murakami-Murofushi, K., Shibata, T., and Ohta, K. (1998). *EMBO J.* **17**, 6412–6425.

Game, J. C. (1993). *Semin. Cancer Biol.* **4**, 73–83.

Game, J. C. (2000). *Mutat. Res.* **451**, 277–293.

Gangloff, S., McDonald, J. P., Bendixen, C., Arthur, L., and Rothstein, R. (1994). *Mol. Cell. Biol.* **14**, 8391–8398.

Gerlach, V. L., Aravind, L., Gotway, G., Schultz, R. A., Koonin, E. V., and Friedberg, E. C. (1999). *Proc. Natl. Acad. Sci. USA* **96**, 11922–11927.

Girard, P. M., Guibourt, N., and Boiteux, S. (1997). *Nucleic Acids Res.* **25**, 3204–3211.

Gogos, A., and Clarke, N. D. (1999). *J. Biol. Chem.* **274**, 30447–30450.

Gomes, X. V., and Wold, M. S. (1995). *J. Biol. Chem.* **270**, 4534–4543.

Gomes, X. V., and Wold, M. S. (1996). *Biochemistry* **35**, 10558–10568.

Goodman, M. F. (2000). *Trends Biochem. Sci.* **25**, 189–195.

Goodtzova, K., Kanugula, S., Edara, S., Pauly, G. T., Moschel, R. C., and Pegg, A. E. (1997). *J. Biol. Chem.* **272**, 8332–8339.

Grogan, D. W. (2000). *Trends Microbiol.* **8**, 180–185.

Haber, J. E. (2000a). *Mutat. Res.* **451**, 53–69.

Haber, J. E. (2000b). *Curr. Opin. Cell. Biol.* **12**, 286–292.

Harmon, F. G., and Kowalczykowski, S. C. (1998). *Genes Dev.* **12**, 1134–1144.

Harmon, F. G., DiGate, R. J., and Kowalczykowski, S. C. (1999). *Mol. Cell* **3**, 611–620.

Haseltine, C. A., and Kowalczykowski, S. C. (2001). Submitted for publication.

Hirano, T. (1999). *Genes Dev.* **13**, 11–19.

Holliday, R. (1964). *Genet. Res.* **5**, 282–304.

Hopfner, K. P., Karcher, A., Shin, D., Fairley, C., Tainer, J. A., and Carney, J. P. (2000a). *J. Bacteriol.* **182**, 6036–6041.

Hopfner, K. P., Karcher, A., Shin, D. S., Craig, L., Arthur, L. M., Carney, J. P., and Tainer, J. A. (2000b). *Cell* **101**, 789–800.

Horii, Z.-I., and Clark, A. J. (1973). *J. Mol. Biol.* **80**, 327–344.

Horst, J. P., and Fritz, H. J. (1996). *EMBO J.* **15**, 5459–5469.

Hosfield, D. J., Frank, G., Weng, Y., Tainer, J. A., and Shen, B. (1998a). *J. Biol. Chem.* **273**, 27154–27161.

Hosfield, D. J., Mol, C. D., Shen, B., and Tainer, J. A. (1998b). *Cell* **95**, 135–146.

Iwasaki, H., Takahagi, M., Nakata, A., and Shinagawa, H. (1992). *Genes Dev.* **6**, 2214–2220.

Johnson, J. L., Hamm-Alvarez, S., Payne, G., Sancar, G. B., Rajagopalan, K. V., and Sancar, A. (1988). *Proc. Natl. Acad. Sci. USA* **85**, 2046–2050.

Johnson, R. D., and Symington, L. S. (1995). *Mol. Cell. Biol.* **15**, 4843–4850.

Johnson, R. E., Prakash, S., and Prakash, L. (1999). *Science* **283**, 1001–1004.

Johzuka, K., and Ogawa, H. (1995). *Genetics* **139**, 1521–1532.

Kanaar, R., Troelstra, C., Swagemakers, S. M., Essers, J., Smit, B., Franssen, J. H., Pastink, A., Bezzubova, O. Y., Buerstedde, J. M., Clever, B., Heyer, W. D., and Hoeijmakers, J. H. (1996). *Curr. Biol.* **6**, 828–838.

Kawarabayasi, Y., Sawada, M., and Horikawa, H. *et al.* (1998). *DNA Res.* **5**, 55–76.

Kawarabayasi, Y., Hino, Y., and Horikawa, H., *et al.* (1999). *DNA Res.* **6**, 83–101, 145–152.

Keeney, S., Giroux, C. N., and Kleckner, N. (1997). *Cell* **88**, 375–384.

Kelly, T. J., Simancek, P., and Brush, G. S. (1998). *Proc. Natl. Acad. Sci. USA* **95**, 14634–14639.

Kiener, A., Husain, I., Sancar, A., and Walsh, C. (1989). *J. Biol. Chem.* **264**, 13880–13887.

Kil, Y. V., Baitin, D. M., Masui, R., Bonch-Osmolovskaya, E. A., Kuramitsu, S., and Lanzov, V. A. (2000). *J. Bacteriol.* **182**, 130–134.

Kitao, S., Ohsugi, I., Ichikawa, K., Goto, M., Furuichi, Y., and Shimamoto, A. (1998). *Genomics* **54**, 443–452.

Kitao, S., Lindor, N. M., Shiratori, M., Furuichi, Y., and Shimamoto, A. (1999a). *Genomics* **61**, 268–276.

Kitao, S., Shimamoto, A., Goto, M., Miller, R. W., Smithson, W. A., Lindor, N. M., and Furuichi, Y. (1999b). *Nature Genet.* **22**, 82–84.

Klenk, H. P., Clayton, R. A., and Tomb, J. F., *et al.* (1997). *Nature* **390**, 364–370.

Kogoma, T. (1996). *Cell* **85**, 625–627.

Kogoma, T. (1997). *Microbiol. Mol. Biol. Rev.* **61**, 212–238.

Kolodner, R. D., and Marsischky, G. T. (1999). *Curr. Opin. Genet. Dev.* **9**, 89–96.

Kolodner, R., Fishel, R. A., and Howard, M. (1985). *J. Bacteriol.* **163**, 1060–1066.

Komori, K., Sakae, S., Shinagawa, H., Morikawa, K., and Ishino, Y. (1999). *Proc. Natl. Acad. Sci. USA* **96**, 8873–8878.

Komori, K., Miyata, T., Daiyasu, H., Toh, H., Shinagawa, H., and Ishino, Y. (2000a). *J. Biol. Chem.* **275**, 33791–33797.

Komori, K., Miyata, T., DiRuggiero, J., Holley-Shanks, R., Hayashi, I., Cann, I. K., Mayanagi, K., Shinagawa, H., and Ishino, Y. (2000b). *J. Biol. Chem.* **275**, 33782–33790.

Koulis, A., Cowan, D. A., Pearl, L. H., and Savva, R. (1996). *FEMS Microbiol. Lett.* **143**, 267–271.

Kowalczykowski, S. C. (1991). *Annu. Rev. Biophys. Biophys. Chem.* **20**, 539–575.

Kowalczykowski, S. C. (2000). *Trends Biochem. Sci.* **25**, 156–165.

Kowalczykowski, S. C., and Eggleston, A. K. (1994). *Annu. Rev. Biochem.* **63**, 991–1043.

Kowalczykowski, S. C., and Krupp, R. A. (1987). *J. Mol. Biol.* **193**, 97–113.

Kowalczykowski, S. C., Dixon, D. A., Eggleston, A.K., Lauder, S.D., and Rehrauer, W. M. (1994). *Microbiol. Rev.* **58**, 401–465.

Krokan, H. E., Standal, R., and Slupphaug, G. (1997). *Biochem. J.* **325**, 1–16.

Kulaeva, O. I., Koonin, E. V., McDonald, J. P., Randall, S. K., Rabinovich, N., Connaughton, J. F., Levine, A. S., and Woodgate, R. (1996). *Mutat. Res.* **357**, 245–253.

Kuzminov, A. (1999). *Microbiol. Mol. Biol. Rev.* **63**, 751–813.

Kvaratskhelia, M., and White, M. F. (2000a). *J. Mol. Biol.* **297**, 923–932.

Kvaratskhelia, M., and White, M. F. (2000b). *J. Mol. Biol.* **295**, 193–202.

Kvaratskhelia, M., Wardleworth, B. N., Norman, D. G., and White, M. F. (2000). *J. Biol. Chem.* **275**, 25540–25546.

Lam, S. T., Stahl, M. M., McMilin, K. D., and Stahl, F. W. (1974). *Genetics* **77**, 425–433.

Lanzov, V., Stepanova, I., and Vinogradskaja, G. (1991). *Biochimie* **73**, 305–312.

Leclere, M. M., Nishioka, M., Yuasa, T., Fujiwara, S., Takagi, M., and Imanaka, T. (1998). *Mol. Gen. Genet.* **258**, 69–77.

Lilley, D. M., and White, M. F. (2000). *Proc. Natl. Acad. Sci. USA* **97**, 9351–9353.

Lindahl, T., and Wood, R. D. (1999). *Science* **286**, 1897–1905.

Lindor, N. M., Furuichi, Y., Kitao, S., Shimamoto, A., Arndt, C., and Jalal, S. (2000). *Am. J. Med. Genet.* **90**, 223–228.

Liu, J., Wu, T. C., and Lichten, M. (1995). *EMBO J.* **14**, 4599–4608.

Liu, N., Lamerdin, J. E., Tebbs, R. S., Schild, D., Tucker, J. D., Shen, M. R., Brookman, K. W., Siciliano, M. J., Walter, C. A., Fan, W., Narayana, L. S., Zhou, Z. Q., Adamson, A. W., Sorensen, K. J., Chen, D. J., Jones, N. J., and Thompson, L. H. (1998). *Mol. Cell* **1**, 783–793.

Lloyd, R. G., and Buckman, C. (1995). *Genetics* **139**, 1123–1148.

Lloyd, R. G., and Sharples, G. J. (1993). *EMBO J.* **12**, 17–22.

Lohman, T. M., and Ferrari, M. E. (1994). *Annu. Rev. Biochem.* **63**, 527–570.

Lovett, S. T., and Mortimer, R. K. (1987). *Genetics* **116**, 547–553.

Lu, A. L., and Fawcett, W. P. (1998). *J. Biol. Chem.* **273**, 25098–25105.

Luisi-DeLuca, C., and Kolodner, R. D. (1994). *J. Mol. Biol.* **236**, 124–138.

Marians, K. J. (2000). *Trends Biochem. Sci.* **25**, 185–189.

Masutani, C., Araki, M., Yamada, A., Kusumoto, R., Nogimori, T., Maekawa, T., Iwai, S., and Hanaoka, F. (1999a). *EMBO J.* **18**, 3491–3501.

Masutani, C., Kusumoto, R., Yamada, A., Dohmae, N., Yokoi, M., Yuasa, M., Araki, M., Iwai, S., Takio, K., and Hanaoka, F. (1999b). *Nature* **399**, 700–704.

Matsui, E., Kawasaki, S., Ishida, H., Ishikawa, K., Kosugi, Y., Kikuchi, H., Kawarabayashi, Y., and Matsui, I. (1999). *J. Biol. Chem.* **274**, 18297–18309.

Mazin, A. V., Bornarth, C. J., Solinger, J. A., Heyer, W. D., and Kowalczykowski, S. C. (2000a). *Mol. Cell* **6**, 583–592.

Mazin, A. V., Zaitseva, E., Sung, P., and Kowalczykowski, S. C. (2000b). *EMBO J.* **19**, 1148–1156.

McCullough, A. K., Dodson, M. L., and Lloyd, R. S. (1999). *Annu. Rev. Biochem.* **68**, 255–285.

McGlynn, P., and Lloyd, R. G. (2000). *Cell* **101**, 35–45.

McKim, K. S., and Hayashi-Hagihara, A. (1998). *Genes Dev.* **12**, 2932–2942.

Mendonca, V. M., Klepin, H. D., and Matson, S. W. (1995). *J. Bacteriol.* **177**, 1326–1335.

Meyer, R. R., and Laine, P. S. (1990). *Microbiol. Rev.* **54**, 342–380.

Michel, B. (2000). *Trends Biochem. Sci.* **25**, 173–178.

Michel, B., Ehrlich, S. D., and Uzest, M. (1997). *EMBO J.* **16**, 430–438.

Modrich, P. (1991). *Annu. Rev. Genet.* **25**, 229–253.

Mortensen, U. H., Bendixen, C., Sunjevaric, I., and Rothstein, R. (1996). *Proc. Natl. Acad. Sci. USA* **93**, 10729–10734.

Nakayama, H., Nakayama, K., Nakayama, R., Irino, N., Nakayama, Y., and Hanawalt, P. C. (1984). *Mol. Gen. Genet.* **195**, 474–480.

Nakayama, K., Irino, N., and Nakayama, H. (1985). *Mol. Gen. Genet.* **200**, 266–271.

Neddermann, P., Gallinari, P., Lettieri, T., Schmid, D., Truong, O., Hsuan, J. J., Wiebauer, K., and Jiricny, J. (1996). *J. Biol. Chem.* **271**, 12767–12774.

New, J. H., Sugiyama, T., Zaitseva, E., and Kowalczykowski, S. C. (1998). *Nature* **391**, 407–410.

Nicolas, A., Treco, D., Schultes, N. P., and Szostak, J. W. (1989). *Nature* **338**, 35–39.

Ogawa, H., Johzuka, K., Nakagawa, T., Leem, S. H., and Hagihara, A. H. (1995). *Adv. Biophys.* **31**, 67–76.

Ogawa, T., Yu, X., Shinohara, A., and Egelman, E. H. (1993). *Science* **259**, 1896–1899.

Ogrunc, M., Becker, D. F., Ragsdale, S. W., and Sancar, A. (1998). *J. Bacteriol.* **180**, 5796–5798.

Olsen, L. C., Aasland, R., Wittwer, C. U., Krokan, H. E., and Helland, D. E. (1989). *EMBO J.* **8**, 3121–3125.

Olsson, M., and Lindahl, T. (1980). *J. Biol. Chem.* **255**, 10569–10571.

Pâques, F., and Haber, J. E. (1999). *Microbiol. Mol. Biol. Rev.* **63**, 349–404.

Paull, T. T., and Gellert, M. (1998). *Mol. Cell* **1**, 969–979.

Paull, T. T., and Gellert, M. (1999). *Genes Dev.* **13**, 1276–1288.

Petrini, J. H. (1999). *Am. J. Hum. Genet.* **64**, 1264–1269.

Petronzelli, F., Riccio, A., Markham, G. D., Seeholzer, S. H., Stoerker, J., Genuardi, M., Yeung, A. T., Matsumoto, Y., and Bellacosa, A. (2000). *J. Biol. Chem.* **275**, 32422–32429.

Petukhova, G., Stratton, S., and Sung, P. (1998). *Nature* **393**, 91–94.

Petukhova, G., Van Komen, S., Vergano, S., Klein, H., and Sung, P. (1999). *J. Biol. Chem.* **274**, 29453–29462.

Philipova, D., Mullen, J. R., Maniar, H. S., Lu, J., Gu, C., and Brill, S. J. (1996). *Genes Dev.* **10**, 2222–2233.

Pierce, A. J., Johnson, R. D., Thompson, L. H., and Jasin, M. (1999). *Genes Dev.* **13**, 2633–2638.

Pochart, P., Woltering, D., and Hollingsworth, N. M. (1997). *J. Biol. Chem.* **272**, 30345–30349.

Postow, L., Ullsperger, C., Keller, R. W., Bustamante, C., Vologodskii, A. V., and Cozzarelli, N. R. (2001). *J. Biol. Chem.* **276**, 2790–2796.

Prakash, S., and Prakash, L. (2000). *Mutat. Res.* **451**, 13–24.

Puranam, K. L., and Blackshear, P. J. (1994). *J. Biol. Chem.* **269**, 29838–29845.

Radding, C. M. (1989). *Biochim. Biophys. Acta* **1008**, 131–145.

Rao, H. G., Rosenfeld, A., and Wetmur, J. G. (1998). *J. Bacteriol.* **180**, 5406–5412.

Rattray, A. J., and Symington, L. S. (1994). *Genetics* **138**, 587–595.

Raymond, W. E., and Kleckner, N. (1993). *Nucleic Acids Res.* **21**, 3851–3856.

Resnick, M. A. (1976). *J. Theor. Biol.* **59**, 97–106.

Rice, M. C., Smith, S. T., Bullrich, F., Havre, P., and Kmiec, E. B. (1997). *Proc. Natl. Acad. Sci. USA* **94**, 7417–7422.

Romanienko, P. J., and Camerini-Otero, R. D. (2000). *Mol. Cell* **6**, 975–987.

Rothstein, R., Michel, B., and Gangloff, S. (2000). *Genes Dev.* **14**, 1–10.

Sack, S. Z., Liu, Y., German, J., and Green, N. S. (1998). *Clin. Exp. Immunol.* **112**, 248–254.

Saffi, J., Pereira, V. R., and Henriques, J. A. (2000). *Curr. Genet.* **37**, 75–78.

Sancar, A. (1996). *Annu. Rev. Biochem.* **65**, 43–81.

Sancar, G. B., Smith, F. W., and Heelis, P. F. (1987). *J. Biol. Chem.* **262**, 15457–15465.

Sandigursky, M., and Franklin, W. A. (1999). *Curr. Biol.* **9**, 531–534.

Sandigursky, M., and Franklin, W. A. (2000). *J. Biol. Chem.* **275**, 19146–19149.

Sandler, S. J., and Marians, K. J. (2000). *J. Bacteriol.* **182,** 9–13.
Sandler, S. J., Satin, L. H., Samra, H. S., and Clark, A. J. (1996). *Nucleic Acids Res.* **24,** 2125–2132.
Schild, D., Lio, Y., Collins, D. W., Tsomondo, T., and Chen, D. J. (2000). *J. Biol. Chem.* **275,** 16443–16449.
Schmidt, K. J., Beck, K. E., and Grogan, D. W. (1999). *Genetics* **152,** 1407–1415.
Schwacha, A., and Kleckner, N. (1995). *Cell* **83,** 783–791.
Seigneur, M., Bidnenko, V., Ehrlich, S. D., and Michel, B. (1998). *Cell* **95,** 419–430.
Seitz, E. M., and Kowalczykowski, S. C. (2000). *Mol. Microbiol.* **37,** 555–560.
Seitz, E. M., Brockman, J. P., Sandler, S. J., Clark, A. J., and Kowalczykowski, S. C. (1998). *Genes Dev.* **12,** 1248–1253.
Seki, M., Miyazawa, H., Tada, S., *et al.* (1994). *Nucleic Acids Res.* **22,** 4566–4573.
Shah, R., Cosstick, R., and West, S. C. (1997). *EMBO J.* **16,** 1464–1472.
Sharples, G. J., and Leach, D. R. (1995). *Mol. Microbiol.* **17,** 1215–1217.
Shen, J. C., and Loeb, L. A. (2000). *Trends Genet.* **16,** 213–220.
Shinohara, A., and Ogawa, T. (1998). *Nature* **391,** 404–407.
Shinohara, A., Ogawa, H., and Ogawa, T. (1992). *Cell* **69,** 457–470.
Shinohara, A., Shinohara, M., Ohta, T., Matsuda, S., and Ogawa, T. (1998). *Genes Cells* **3,** 145–156.
Sinclair, D. A., and Guarente, L. (1997). *Cell* **91,** 1033–1042.
Skorvaga, M., Raven, N. D., and Margison, G. P. (1998). *Proc. Natl. Acad. Sci. USA* **95,** 6711–6715.
Smith, D. R., Doucette-Stamm, L. A., Deloughery, C., *et al.* (1997). *J. Bacteriol.* **179,** 7135–7155.
Smith, G. R., Kunes, S. M., Schultz, D. W., Taylor, A., and Triman, K. L. (1980). *In* "Mechanistic Studies of DNA Replication and Genetic Recombination" B. Alberts, ed. pp. 927–931. Academic Press, New York.
Spies, M., Kil, Y., Masui, R., Kato, R., Kujo, C., Ohshima, T., Kuramitsu, S., and Lanzov, V. (2000). *Eur. J. Biochem.* **267,** 1125–1137.
Stasiak, A., and Egelman, E. H. (1986). *Biochem. Soc. Trans.* **14,** 218–220.
Stasiak, A., and Egelman, E. H. (1994). *Experientia* **50,** 192–203.
Stasiak, A., Di Capua, E., and Koller, T. (1981). *J. Mol. Biol.* **151,** 557–564.
Stasiak, A., Stasiak, A. Z., and Koller, T. (1984). *Cold Spring Harbor Symp. Quant. Biol.* **49,** 561–570.
Sugiyama, T., Zaitseva, E. M., and Kowalczykowski, S. C. (1997). *J. Biol. Chem.* **272,** 7940–7945.
Sugiyama, T., New, J. H., and Kowalczykowski, S. C. (1998). *Proc. Natl. Acad. Sci. USA* **95,** 6049–6054.
Sun, H., Treco, D., Schultes, N. P., and Szostak, J. W. (1989). *Nature* **338,** 87–90.
Sung, P. (1994). *Science* **265,** 1241–1243.
Sung, P. (1997a). *Genes Dev.* **11,** 1111–1121.
Sung, P. (1997b). *J. Biol. Chem.* **272,** 28194–28197.
Sung, P., and Robberson, D. L. (1995). *Cell* **82,** 453–461.
Sung, P., Trujillo, K. M., and Van Komen, S. (2000). *Mutat. Res.* **451,** 257–275.
Szostak, J. W., Orr-Weaver, T. L., Rothstein, R. J., and Stahl, F. W. (1983). *Cell* **33,** 25–35.
Tan, T. L., Essers, J., Citterio, E., Swagemakers, S. M., de Wit, J., Benson, F. E., Hoeijmakers, J. H., and Kanaar, R. (1999). *Curr. Biol.* **9,** 325–328.
Tang, M., Shen, X., Frank, E. G., O'Donnell, M., Woodgate, R., and Goodman, M. F. (1999). *Proc. Natl. Acad. Sci. USA* **96,** 8919–8924.
Tracy, R. B., and Kowalczykowski, S. C. (1996). *Genes Dev.* **10,** 1890–1903.

Tucker, J. D., Jones, N. J., Allen, N. A., Minkler, J. L., Thompson, L. H., and Carrano, A. V. (1991). *Mutat. Res.* **254**, 143–152.

Umezu, K., Chi, N. W., and Kolodner, R. D. (1993). *Proc. Natl. Acad. Sci. USA* **90**, 3875–3879.

Usui, T., Ohta, T., Oshiumi, H., Tomizawa, J., Ogawa, H., and Ogawa, T. (1998). *Cell* **95**, 705–716.

Van Dyck, E., Stasiak, A. Z., Stasiak, A., and West, S. C. (1999). *Nature* **398**, 728–731.

Van Komen, S., Petukhova, G., Sigurdsson, S., Stratton, S., and Sung, P. (2000). *Mol. Cell* **6**, 563–572.

Wadsworth, R. I., and White, M. F. (2001). *Nucleic Acids Res.* **29**, 914–920.

Wagner, J., Gruz, P., Kim, S. R., Yamada, M., Matsui, K., Fuchs, R. P., and Nohmi, T. (1999). *Mol. Cell* **4**, 281–286.

Watrin, L., and Prieur, D. (1996). *Curr. Microbiol.* **33**, 377–382.

Watt, P. M., Louis, E. J., Borts, R. H., and Hickson, I. D. (1995). *Cell* **81**, 253–260.

Watt, P. M., Hickson, I. D., Borts, R. H., and Louis, E. J. (1996). *Genetics* **144**, 935–945.

Webb, B. L., Cox, M. M., and Inman, R. B. (1997). *Cell* **91**, 347–356.

West, S. C. (1992). *Annu. Rev. Biochem.* **61**, 603–640.

West, S. C. (1994a). *Cell* **76**, 9–15.

West, S. C. (1994b). *Ann. N.Y. Acad. Sci.* **726**, 156–163; discussion, 163–154.

West, S. C. (1997). *Annu. Rev. Genet.* **31**, 213–244.

Whitby, M. C., and Lloyd, R. G. (1998). *J. Biol. Chem.* **273**, 19729–19739.

White, M. F., Giraud-Panis, M. J., Pohler, J. R., and Lilley, D. M. (1997). *J. Mol. Biol.* **269**, 647–664.

Wold, M. S. (1997). *Annu. Rev. Biochem.* **66**, 61–92.

Wood, R. D. (1996). *Annu. Rev. Biochem.* **65**, 135–167.

Woodgate, R. (1999). *Genes Dev.* **13**, 2191–2195.

Woods, W. G., and Dyall-Smith, M. L. (1997). *Mol. Microbiol.* **23**, 791–797.

Xiao, W., and Samson, L. (1992). *Nucleic Acids Res.* **20**, 3599–3606.

Yang, H., Fitz-Gibbon, S., Marcotte, E. M., Tai, J. H., Hyman, E. C., and Miller, J. H. (2000). *J. Bacteriol.* **182**, 1272–1279.

Yang, W. (2000). *Mutat. Res.* **460**, 245–256.

Yasui, A., Takao, M., Oikawa, A., Kiener, A., Walsh, C. T., and Eker, A. P. (1988). *Nucleic Acids Res.* **16**, 4447–4463.

Yu, C. E., Oshima, J., Fu, Y. H., Wijsman, E. M., Hisama, F., Alisch, R., Matthews, S., Nakura, J., Miki, T., Ouais, S., Martin, G. M., Mulligan, J., and Schellenberg, G. D. (1996). *Science* **272**, 258–262.

Basal and Regulated Transcription in Archaea

Jörg Soppa

Institute for Microbiology
Biocentre Niederursel
J. W. Goethe University Frankfurt
D-60439 Frankfurt, Germany

I. Introduction

About 20 years ago it was recognized that the Archaea form a third domain of life, apart from the Bacteria and Eukarya (former names: archaebacteria, eubacteria, and eukaryotes). Many archaeal species

171

were isolated from environments extreme in temperature, pH value, salt concentration, or pressure, therefore the Archaea were often thought to be "ancient extremophiles" which today cannot compete with the more modern Bacteria and Eukarya in "normal" habitats. However, during recent years newly developed methods of molecular ecology have opened the possibility of monitoring the microbial composition of ecosystems without the requirement of a cultivation method. It turned out that the Archaea are widespread and can be discovered in oceans, freshwater lakes, sediments, and soils around the world and have also been found as endosymbionts (Pace, 1997). In addition, these studies showed that the diversity is much higher than anticipated on the basis of the isolated species, and a new taxon on the kingdom level could even be defined (Barns et al., 1996). Unfortunately these mesophilic species could not be cultivated until now and thus today cannot be used for biochemical, physiological, or molecular genetic characterization. Phylogenetic analyses based on 16S rDNA sequences indicated that the Archaea form a monophyletic group separated from the Bacteria and Eukarya (Woese et al., 1990). Rooting of the universal tree of life using families of duplicated paralogous genes indicated that the Archaea and Eukarya are sister groups and share a common ancestor to the exclusion of the Bacteria. In congruence with this view, archaeal and eukaryotic information processing systems were found to be more closely related to one another than to the bacterial counterpart, and the basal transcription machinery is a prominent example. On the other hand, the Archaea share a variety of metabolic pathways with the Bacteria to the exclusion of the Eukarya, challenging the view of solely vertical inheritance. During the last few years a large number of microbial genome sequences have become available and indicated that genomes, at least in the past, have been very flexible and that horizontal gene transfer has been or still is much more frequent than previously anticipated (Doolittle, 1999). This would make the conservative species concept, which includes a common history of organisms with (most of) their genes, obsolet for the early history of life and possibly even for today's microorganisms.

A further problem is that it might be impossible to reconstruct the correct phylogenetic relationships of domains based on sequence data recovered from recent organisms because long branch attraction hinders recovering the correct topology and because many sites might be saturated with mutations and therefore be uninformative concerning early evolution (Phillipe and Forterre, 1999). These considerations could lead to the view that it might not be meaningful to delineate features of the domain of the Archaea, e.g., the basal transcription apparatus, and compare them with the counterparts of the other two domains.

However, the following (and other) evidence indicates that the Archaea are a monophyletic group united by a variety of common features: (1) analysis of the fraction of shared orthologues using 13 whole-genome sequences showed that the Archaea, Bacteria, and Eukarya are three separated groups (Snel *et al.*, 1999); (2) the membranes of Archaea are composed of ether lipids of glycerol and isoprenoid alcohols, while the other two domains have ester lipids of glycerol and fatty acids (de Rosa *et al.*, 1986); (3) the glycerol component of the lipids has a different stereochemistry in the Archaea compared to the other two domains (Koga *et al.*, 1998); and (4) as detailed in this article, the Archaea have a common basal transcription apparatus, deviating from the bacterial counterpart and being similar, but not identical, to the eukaryotic model. In general, it has been shown that "informational genes" encoding products which take part in cellular information processing, e.g., replication, transcription, and translation, are not easily being laterally transferred between species, in contrast to "operational genes" encoding elements of metabolic pathways (Jain *et al.*, 1999). An explanation for this finding could be that the elements of information processing pathways have many interactions and are even involved in multicomponent complexes such as the ribosome. The transfer of a single element would be prohibited because it would not function in the network of specific physical interactions in the new host, e.g., a bacterial σ factor cannot functionally interact with eukaryotic polymerase subunits or with archaeal promoter elements. In this view, informational genes including the genes for the basal transcriptions apparatus would be vital for the definition of species and phylogenetic comparisons. In agreement with this view it has been hypothesized that the last common ancestor was not a single cell but the gene pool of a population and that early in evolution genes were freely exchangable (Woese, 1998). Only the development of higher fidelity and higher specificity in replication, transcription, and translation, meaning the invention of complexes of proteins of increasing size, would have been the prerequisite for the formation of "modern" organisms with an increasing fraction of vertically transmitted genes. The genes encoding elements of cellular information processing would have been the first to lose their lateral interchangability and, again, would be most typical for phylogenetic groups (Woese, 1998).

These considerations should underscore that the basal transcription apparatus is central to cellular life. It is required for the transcription of all genes of a cell and thus is at the start of expressing the genotype of a cell into a phenotype. Furthermore, regulation of expression often takes place at the level of transcription initiation, therefore the basal transcription apparatus must interact not only with basal promoter elements, but also with global and gene-specific regulators.

Soon after it was discovered that the Archaea form a third group of organims distinct from the Bacteria, it became clear that they do not follow the bacterial paradigm but have a multicomponent RNA polymerase similar to that of the Eukarya. Knowledge about the archaeal transcription apparatus has increased tremendously since then and different aspects have regularly been summarized (recent reviews: Soppa, 1999a; Bell and Jackson, 1998; Soppa *et al.*, 1998; Reeve *et al.*, 1997; Thomm, 1996; Langer *et al.*, 1995; Brown *et al.*, 1989). Important progress has been made recently which has not been summarized before, e.g., on the polarity determinants of transcription, transcription elongation, *in vitro* systems for transcriptional regulation, RNA polymerase structure, nucleosomes, and bioinformatic processing of genome sequences.

The basal transcription apparatus is comprised of cis-acting promoter elements, transcription initiation factors which recognize these elements, a multicomponent polymerase, and factors required for elongation and termination. *In vivo*, the basal transcription apparatus has to interact with nucleosome-bound instead of naked DNA. Transcriptional regulators have to interact with elements of the basal transcription apparatus to inhibit or enhance transcription. This article gives an overview of the current understanding of the basal transcription apparatus and transcriptional regulation in the Archaea, with an emphasis on new developments.

II. Promoter Elements

Early experiments on archaeal transcription initiation resting on a limited set of stable RNA genes from methanogens revealed that the Archaea do not follow the bacterial paradigm and that the canonical −10 and −35 regions do not exist (Wich *et al.*, 1986). In 1988 the idea was put forward that archaeal promoters might be similar to eukaryotic polymerase II promoters and might include a TATA box (Reiter *et al.*, 1988; Thomm and Wich, 1988). As at that time no archaeal transformation system was available which would have allowed promoter characterization *in vivo*, efforts were concentrated on establishing *in vitro* transcription systems to open the possibility of identifying and analyzing the components necessary for specific transcription initation (Frey *et al.*, 1990; Hüdepohl *et al.*, 1990; Knaub and Klein, 1990). Today, *in vitro* transcription systems for several species of methanogens and sulfolobales are available, including a defined homologous system composed of purified RNA polymerase and transcription factors from *Sulfolobus acidocaldarius* (Bell and Jackson, 2000b), and a system for the hyperthermophilic species *Pyrococcus furiosus* was even established working at 80°C (Hethke *et al.*, 1996). No such system has yet been established for a haloarchaeal species. However, until very recently haloarchaea were the only group which could be transformed, making it possible to complement the *in vitro* approaches with promoter characterizations *in vivo* using reporter genes (Palmer and Daniels, 1995; Danner and Soppa, 1996). In a study aiming at the identification of basal promoter elements shared by the Archaea or selected archaeal subgoups, all cases of biochemically determined 5' ends of transcripts were collected and nucleotide frequencies in this databank of putative promoter regions were analyzed statistically (Soppa, 1999b). The existence of three basal promoter elements, discussed below, was verified through nonrandom nucleotide distributions: (1) a TATA box, (2) a transcription factor B recognition element, and (3) an initiator element. Figure 1 gives an overview of archaeal, bacterial, and eukaryotic promoter structures.

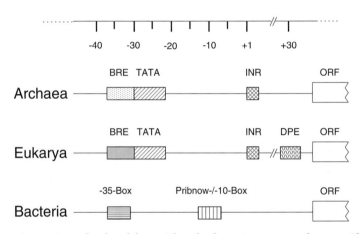

FIG. 1. Comparison of archaeal, bacterial, and eukaryotic promoter elements. The promoter elements are shown schematically and are aligned to the scale at top. The numbering refers to the transcriptional start site (+1). BRE, transcription factor (II) B recognition element; TATA, TATA box; INR, initiator element; DPE, downstream promoter element; ORF, open reading frame.

A. THE TATA BOX

The TATA box was discovered very early as a major archaeal promoter element, and it was called "box A" or the "distal promoter element" before its homology with the eukaryotic TATA box was recognized. A variety of studies using linker scanning or site-specific mutagenesis underscored that the TATA box is essential for promoter activity in all archaea investigated, that its sequence can modulate promoter strength over at least two orders of magnitude, and that certain point mutations within the TATA box are incompatible with its activity (e.g., Reiter *et al.*, 1990;

Hausner *et al.,* 1991; Hain *et al.,* 1992; Palmer and Daniels, 1995; Danner and Soppa, 1996). The TATA box is recognized by an archaeal TATA box binding protein (TBP) which is homologous to the eukaryotic transcription factor, and crystallization of protein DNA complexes has shown that TBP contacts 8 bp (base pairs) of DNA, defining the length of this element (see Section IV). In agreement with the fact that the Archaea have only one RNA polymerase instead of the three eukaryotic enzymes, the TATA box is required for transcription of all types of genes, protein-coding genes as well as rRNA and other stable RNA genes. The TATA box is optimally centered around $-26/-27$, and several lines of evidence suggest that its spacing to the transcription initation site is somewhat flexible and can vary by one or two positions around the ideal value: (1) the introduction of small deletions or insertions 3' of the TATA box did not alter the start site selection in an *in vitro Sulfolobus* transcription system, while variation of three nucleotides or more shifted the start site (Reiter *et al.,* 1990); (2) a considerable fraction of natural archaeal promoter sequences fits the TATA box consensus only if they are shifted by one or two nucleotides; and (3) the same is true for haloarchaeal promoter sequences selected after random mutagenesis (Danner and Soppa, 1996).

Three systematic studies were performed to characterize the sequence requirements of an archaeal TATA box experimentally. In the first study all single-nucleotide changes at seven positions of a rRNA TATA box of *Sulfolobus shibatae* were constructed and the resulting promoter strengths of the 21 mutants and the wild type were quantitated using an *in vitro* transcription system (Hain *et al.,* 1992). In the second study all single-nucleotide changes at 11 positions of a tRNA promoter box from *Haloferax volcanii* were constructed and the promoter strengths of the 33 mutants were compared to the wild-type sequence *in vivo* using a transcriptional reporter gene (Palmer and Daniels, 1995). Both studies showed that at several positions two or even three different nucleotides are compatible with a high promoter strength, whereas other nucleotides at the same positions diminish promoter activity. In the third study 14 nucleotides around the TATA box of a protein-encoding gene of *H. volcanii* were randomized, resulting in a library of about 10^8 plasmids, and sequences with a high promoter activity *in vivo* were selected making use of a marker gene. The TATA box sequences recovered after the selection were A/T-rich, but a variety of sequences with a high promoter activity *in vivo* was found (Danner and Soppa, 1996). The experimentally observed high variability of the TATA box sequence is in excellent agreement with the mode of interaction between the TATA box and the TATA box binding protein described below (Section III.A).

Another approach is the analysis of natural promoters with known transcriptional start sites (Soppa, 1999b). It underscored the variability of TATA box sequences but also revealed statistical preferences for certain nucleotides at different positions and indicated that the optimal TATA box sequence is not identical for different archaeal groups, i.e., TTTTTAAA was found for Crenarchaeota (mostly sulfolobales), TTTATATA was found for methanogens, and NTTTWWWN was found for halophiles (W = A or T; N = A or T or G or C). The consensus sequences derived from natural promoters fit well the results of the experimental studies on haloarchaea and sulfolobales (Fig. 1). No comparison is possible for methanogens, because no systematic TATA box characterization has been reported until now.

B. THE TRANSCRIPTION FACTOR B RECOGNITION ELEMENT

The existence of the transcription factor B element (BRE) upstream of the TATA box was discovered in the Archaea and Eukarya simultaneously. Qureshi and Jackson (1998) chose a strong *Sulfolobus* promoter for a detailed characterization and found that the region upstream of the TATA box is important for its promoter strength. The region increases the strength of a weaker promoter if it is fused to its TATA box. A mutant analysis confirmed that the region three to six nucleotides upstream of the TATA box is important for promoter function in an *in vitro* transcription system, and several nucleotides upstream and two nucleotides downstream of the TATA box gave signals in *in vitro* protection assays using the ternary complex of DNA, TBP, and transcription factor B (TFB). SELEX experiments were used to verify that the archaeal TFB selects specific sequences from a randomized pool of DNA fragments, proving a sequence-specific interaction between TFB and several nucleotides on both sides of the TATA box. Most prominent were an A three nucleotides and a G six nucleotides upstream of the TATA box, but unequal distributions after selection were also obvious for additional positions, albeit to a lesser extent. An important observation was that TBP alone does not select for DNA fragments with a high promoter strength from a random library, while the combination of TBP and TFB does, underscoring that BRE is indeed an important basal transcription element for *Sulfolobus* and that it has an important role in transcription initiation. Until now the BRE has been investigated systematically only using the *Sulfolobus* system, but the data are in agreement with earlier circumstantial observations, e.g., that a purine-rich sequence upstream of the TATA box is important for the strength of a haloarchaeal promoter *in vivo* (Palmer and Daniels, 1995) or that replacement of the region upstream of the TATA box of the GDH gene diminished transcription to 6% in an *in vivo* transcription system of *Pyrococcus furiosus* (Hethke et al., 1996). Analysis of natural promoters revealed the preference for two A's at positions $-33/-34$ and indicated that a BRE is important for transcription initiation in all the archaea *in vivo* in the context of the whole transcription apparatus, the presence of general DNA binding proteins, and at the physiological conformeric state of chromosomal DNA (Soppa, 1999b). However, it still remains to be determined whether different archaeal groups have different BRE consensus sequences and how many and which nucleotides upstream and downstream of the TATA box influcence transcription initiation in the different systems.

The BRE was also found to be a basal promoter element in eukaryotes. Lagrange et al. (1998) used the human TFIIB to select for DNA fragments with an affinity for this transcription factor from a random fragment library. A nonrandom sequence distribution in the selected subpopulation was found at eight positions upstream of the TATA box, and the consensus sequence -38 SSRCGCC -32 was derived (S = G or C; R = A or G). Mutations were introduced at positions -37 and -34 and it was verified that they decreased TFIIB binding as well as *in vitro* transcription efficiency, proving that the BRE is a bona fide basal promoter element in the eukaryotic system. In the binary system of TFIIB and DNA, cross-linking of TFIIB was observed only at positions upstream of the TATA box, while in the ternary system of DNA, TFIIB and TBP nucleotides both upstream and downstream of the TATA box could be cross-linked to TFIIB, showing (1) that the primary sequence-specific binding site of eukaryotic TFIIB is also upstream of the TATA box and (2) that TBP-induced bending of the TATA box DNA is a prerequisite to allow TFIIB to contact nucleotides on both sides of the TATA box (see Section IV).

In summary, the newly discovered BRE is a basal promoter element in the Archaea and Eukarya, the consensus sequences are different in the only two experimentally analyzed systems, i.e., human and *Sulfolobus,* and the importance of the BRE upstream of different genes in different species remains to be clarified.

C. The Initiator Element

The third archaeal basal promoter element is the initiator element around the transcription initiation site itself. Using an *in vitro* transcription system of *Sulfolobus,* it was shown that mutations at the initiator element are less severe than TATA box mutations, that upon changing the spacing between the two elements the TATA box overides the initiator element control, and upon insertions or deletions of three or more nucleotides transcription starts at a new artificial "initiator element" around the optimal distance to the TATA box (Reiter *et al.,* 1990).

The analysis of natural promoters indicates that an initiator element is also present throughout the Archaea but that it might be of different importance for transcription initiation for different groups, i.e., it is hardly detectable in haloarchaeal sequences and very pronounced at the initiation sites of methanogens and sulfolobales (Soppa, 1999b). No systematic studies to characterize the initiator element have been performed until now, the sequence requirements have not been determined experimentally, and it is not known which protein recognizes it and at which stage of the initiation process it is read out. Recently specifically labeled DNA fragments were cross-linked to the proteins in the initiation complex of *Pyrococcus,* and it was shown that the large subunits A′ and B are in the vicinity of the initiator element, while A″ is not (Bartlett *et al.,* 2000).

D. ADDITIONAL BASAL PROMOTER ELEMENTS

Apart from the three elements mentioned, no further basal promoter element has been found in archaeal systems. It should be explicitly noted that there is no indication for the presence of a "downstream promoter element" (DPE), which is found in a variety of eukaryotic promoters and in *Drosophila* is as widely used as the TATA box (Kutach and Kadonaga, 2000). This is in agreement with the model that the eukaryotic DPE is contacted by a TBP-associated factor (TAF) and that TAFs are not present in the Archaea.

III. Transcription Initiation Factors

In the Bacteria, promoter recognition is accomplished by a complex of core RNA polymerase with a tightly bound σ factor which is lost during elongation. In the Eukarya, promoter elements are first recognized by transcription factors, which subsequently recruit the polymerase. Factors involved in initiation at polymerase II promoters upstream of protein-encoding genes are TFIIA, TFIIB, TFIID, TFIIE, TFIIF, and TFIIH, most of them complexes of several different polypeptides. In the Archaea, it was discovered about 10 years ago that purified RNA polymerase cannot initiate transcription *in vitro* accurately by itself, but transcription factors must be added (Frey *et al.,* 1990; Hüdepohl *et al.,* 1990), and thus archeal transcription initiation is similar to the eukaryotic model. Using a variety of approaches including biochemical isolation, immunoprecipitation with specific antibodies, and genome sequence analysis, it became clear that the main and probably the only transcription initiation factors in the Archaea are the TATA box binding protein (TBP) and TFB, a homologue of the eukaryotic factor TFIIB (as archaea have only one RNA polymerase, the "II" is omitted when archaeal proteins are named in accordance with eukaryotic polymerase II-related proteins).

A. THE TATA BOX BINDING PROTEIN

In the Eukarya, TBP is the most general transcription factor involved in transcription initiation with all three RNA polymerases, acting in complexes with different polypeptides at the three types of promoters. At polymerase II promoters, TBP binding to the TATA box is the initial step in promoter recognition. *In vitro,* TBP alone is sufficient for transcription initiation in concert with TFIIB and polymerase; *in vivo* it acts in a complex called TFIID, in which it is bound to TBP-associated factors (TAFs).

After the first TBP sequences were discovered in the Archaea (e.g., Marsh *et al.,* 1994; Rowlands *et al.,* 1994), it soon became clear that TBP is widely distributed and is a basal archaeal transcription factor. Today, more than 20 sequences are known. In general, each archaeal genome contains a single TBP gene (exceptions are discussed in Section IX). Archaeal TBPs consist of about 180 amino acids and are highly conserved; the similarities between different archaeal TBPs range between 40 and 55% (throughout this article similarity means the fraction of identical amino acids). TBPs are comprised of two repeats of about 90 amino acids, thus it seems that the recent *tbp* gene arose by an ancient gene duplication. The similarities of the first to the second repeats are slightly higher than 40%, indicating that archaeal TBPs are highly symmetrical proteins and can be regarded as "internal dimers" (Fig. 2).

Eukaryotic TBPs share with their archaeal homologues the two repeat domains of about 90 amino acids and, in addition, contain an N-terminal extension which is very variable both in length and in sequence throughout the Eukarya. The function of this N-terminal domain is not known. Remarkably, the similarity of the first and the second repeats of eukaryotic TBPs is only 26%, about 15% lower than for archaeal TBPs. Probably eukaryotic TBPs have evolved further from the ancestral symmetric state because they have to interact with a larger variety of different transcription factors, e.g., with TAFs to form TFIID. The similarities between archaeal TBPs and the homologous parts of eukaryotic TBPs range between 20 and 40%. Within the Eukarya, the degree of conservation is very different; it is extremely high in higher Eukarya, but between higher and lower Eukarya the degree of conservation is about the same as between the Archaea and Eukarya, e.g., 27% between *Homo sapiens* and *Plasmodium falciparum.*

A. The TBP of *Pyrococcus woesei*

B. Comparison of TBPs

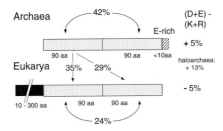

FIG. 2. The TATA box binding protein. (A) Structure of the TATA box binding protein of *Pyrococcus woesei* [PDB entry number 1PCZ (DeDecker *et al.*, 1996)]. To visualize the internal symmetry of the protein, the first and second repeat domains (compare text) are shown in different shades of gray. (B) Comparison of archaeal and eukaryotic TBPs. The domain structure is shown schematically. Average values of the fraction of identical amino acids are indicated. The acidic C-terminal extension is present only in a subgroup of archaea (see text). The excess of negatively charged amino acids (positive value) or positively charged amino acids (negative value) is depicted on the right. aa, amino acids.

The alignment of archaeal and eukaryotic TBP sequences and construction of a phylogenetic tree revealed that they form two distinct monophyletic groups. The similarities of the first repeats of archaeal TBPs are only slightly greater to the first than to the second repeats of eukaryotic TBPs. Therefore it seems that the gene duplication had already occurred in the last common ancestor of the Archaea and Eukarya but that the fusion gene had not evolved far at the time that the two lineages split and that the protein at that time was still nearly absolutely symmetrical. It seems safe to assume that the cognate DNA element, the precursor of the TATA box, had perfect dyad symmetry at that time.

Severeal structures have been determined for eukaryotic or archaeal TBPs alone or in complex with DNA, TF(II)B, or TFIIA (e.g., Tsai and Sigler, 2000; Littefield *et al.*, 1999; Kosa *et al.*, 1997; DeDecker *et al.*,

1996; Kim and Burley, 1994; Nikolov *et al.*, 1995; Chasman *et al.*, 1993). Not unexpectedly, it turned out that the structure of the archaeal protein from *Pyrococcus woesei* is very similar to the structures of TBPs (C-terminal domains) from yeast, *Arabidopsis thalinana,* and *Homo sapiens.* TBP from all sources is a saddle-shaped protein of α/β structure with a high twofold symmetry also for the eukaryotic proteins (Fig. 2). The concave underside of the protein binds 8 bp of DNA, thereby creating a high distortion of the TATA box. Two pair of phenylalanines, which are absolutely conserved in archaeal and eukaryotic TBPs, intercalate between the first and the last 2 bp of the TATA box. The central 6 bp adopt a conformation which was unprecedented and is today called "TATA-DNA" (Guzikevich-Guerstein and Shakked, 1996). It is somewhat related to A-DNA, and a molecular dynamic simulation suggests that TATA-DNA has the intrinsic property of adopting an A-DNA-like conformation, which might be an intermediate in complex formation (Flatters *et al.,* 1997). The TATA-DNA conformation is characterized by very wide and shallow minor grooves, which interact with the concave underside of TBP. Most of the protein–DNA interactions are nonpolar and hydrophobic. TBP binding to the minor groove and the absence of electrostatic interactions with bases indicate that TBP recognizes the deformability of the TATA box more than its specific sequence. The minor groove is chemically more monotonous than the major groove, therefore many DNA binding proteins with a broad sequence specificity bind to the minor groove. This is in good agreement with the sequence variability of archaeal TATA boxes described above. It also fits to a study in which one eukaryotic TBP was cocrystallized with 10 sequence modifications of a TATA box fragment, and the structure of the protein, the path of the DNA in the complexes, and the protein–DNA interfaces were found to be identical in all cases (Patikoglou *et al.,* 1999). On both sides of the TATA box the DNA is in the canonical B-DNA conformation. Nearly all of the amino acids at the DNA binding surface of TBP are absolutely conserved in the Archaea and Eukarya. In eukaryotic TBPs, the amino acids which are conserved in the first and second repeat domain (26%) are mostly confined to the DNA binding domain, while the sequence identity at equivalent positions of the convex upper surface is low. In contrast, the convex surface of archaeal TBPs contains many identical amino acids at equivalent positions of the two repeat domains, in accordance with the higher primary sequence conservation of the two domains in archaeal TBPs (42%). This unequal distribution of conserved residues at different surfaces of the protein structure strengthens the above-mentioned conclusion that the reason for the different degree of conservation is the higher number of interaction partners of eukaryotic TBPs. Therefore it cannot be expected that archaeal TBPs can functionally replace eukaryotic TBPs in factor-dependent transcription. However, it has been reported that human and yeast TBPs can functionally replace the archaeal protein in an *in vitro* transcription system (Wettach *et al.*, 1995), underscoring the high structural similarity of archaeal and eukaryotic TBPs. Two differences between archaeal and eukaryotic TBPs have already been mentioned, i.e., the presence of an additional N-terminal domain in the eukaryotic proteins and the higher degree of internal symmetry of the archaeal proteins. A third difference is that archaeal TBPs have an excess of acidic amino acids, whereas eukaryotic TBPs have an excess of basic amino acids. It will be interesting to unravel what the consequences of these differences are. The TBPs of three groups of Archaea, i.e., *Pyrococcus, Thermococcus,* and *Sulfolobus,* contain a short C-terminal extension that is not present other archaeal sequences. The extension contains a run of five or six glutamic acid resisues (sometimes interrupted by a single glycine or leucine residue) and thus is very acidic. Its conservation in several species indicates that it is important for function, either for binding another transcription factor or as a target for a posttranslational modification regulating function. This is the first example of the specialization of a highly conserved basal transcription factor in some archaeal species; further examples are discussed below (Section IX).

B. Transcription Factor B

At eukaryotic polymerase II promoters the next factor to bind after TBP is TFIIB, before additional transcription factors and the polymerase are recruited. Archaea harbor a homologue to TFIIB, which is called TFB. Sequence analysis of a prepublished sequence led to the first discovery of an archaeal TFB in 1992 (Ouzounis and Sander, 1992); today more than 20 TFB sequences are known. In general, archaeal genomes contain a single *tfb* gene; some exceptions are discussed in Section IX.

Archaeal and eukaryotic TF(II)B's consist of three domains: an N-terminal region—TF(II)B$_N$—comprised of about 100–120 amino acids; and a C-terminal region—TF(II)B$_C$—comprised of two repeats of about 90 amino acids (see Fig. 3). The degree of conservation is substantially higher for archaeal TFBs than for archaeal TBPs, i.e., the similarities are 50–60%. Similarly to the TBPs, the similarities of the two repeat domains are higher for the archaeal TFBs (25–37%) than for the eukaryotic TFIIBs (11–25%). In addition, the similarities of the

A. TFB$_C$ of *Pyrococcus woesei*

B. Comparison of TFBs/TFIIBs

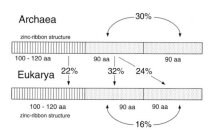

FIG. 3. Transcription factor B. (A) Structure of the C-terminal domain of transcription factor B (TFB$_C$) of *Pyrococcus woesei*. The structure of TFB$_C$ was determined in a ternary complex together with TBP and a BRE- and TATA box-containing DNA fragment [PDB entry number 1D3U (Littlefield *et al.*, 1999)]. The N-terminal domain of TFB was not included in structure determination. To visualize the internal symmetry of TFB$_C$, the two repeats are shown in different shades of gray. (B) Comparison of archaeal TFBs with eukaryotic TFIIBs. The domain structure of TF(II)B is shown schematically. Average values of the fraction of identical amino acids are indicated. aa, amino acids.

N-terminal domains are much higher within the Archaea (50–60%) than within the Eukarya (30–40%). Again, the interaction of the eukaryotic protein with a higher number of global and gene-specific transcription factors compared to the archaeal TFB can easily explain these observations.

The similarities between archaeal TFBs and eukaryotic TFIIBs are between 20 and 30% and thus slightly lower than between archaeal and eukaryotic TBPs. The first repeat of the archaeal TFBs is on average only 8% more similar to the first than to the second repeat of the eukaryotic TFIIBs, which indicates that the two repeats of the TF(II)B precursor were also nearly identical at the time that the two lineages separated.

It was discovered simulatiously in archaea and eukaryotes that TF(II)B is a sequence-specific DNA binding protein and recognizes a transcription factor B recognition element (BRE) upstream of the TATA box (see above; Lagragne *et al.*, 1998; Qureshi and Jackson, 1998). The ability to recognize the BRE is localized in the C-terminal domain of TF(II)B, because the binding affinities of TF(II)B and TF(II)B$_C$ to several wild-type and mutant BREs are virtually identical (Bell and Jackson, 2000b; Lagrange *et al.*, 1998), and TFB$_C$ is able to select specific sequences from a random DNA fragment library (Qureshi and Jackson, 1998). At that time these findings were very surprising since the structure determination of TBP/TF(II)B/DNA ternary complexes from both an archaeon (Kosa *et al.*, 1997) and a eukaryote (Nikolov *et al.*, 1995) had not revealed any base-specific DNA contacts of TF(II)B$_C$. However, the DNA used for crystallization was rather short and did not contain a BRE, and it turned out that this had inhibited the formation of base-specific contacts. The ternary complexes of the archaeal and the eukaryotic systems were recrystallized using DNA fragments containing BREs derived from biochemical data, and in both cases sequence-specific DNA contacts of TF(II)B were obvious in the new structures (see Fig. 4).

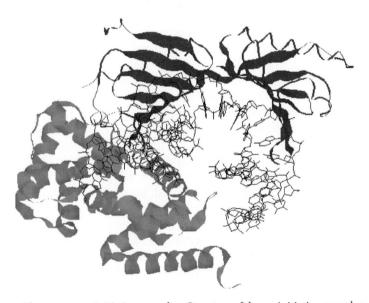

Fig. 4. The ternary preinitiation complex. Structure of the preinitiation complex, comprised of TBP and TFB from *Pyrococcus wosei* and a BRE- and TATA box-containing DNA fragment [PDB entry number 1D3U (Littlefield *et al.*, 1999)]. TBP is shown at the top, in darker gray. TFB is shown at the left, in lighter gray. The DNA is shown in wire-frame representation. Note the asymmetric binding of TFB to the C-terminal (left) repeat of TBP, the highly distorted path of the DNA at the concave underside of TBP, and the TFB interactions with DNA on both sides of the TATA box, which are more intense and base specific at the BRE upstream (left) of the TATA box.

The TFB of *Pyrococcus woeseii* makes contacts several nucleotides upstream of the TATA box, with positions −3 and −6 being the strongest
specificity determinants, and it makes contacts with phosphates up to
nine positions upstream and six positions downstream of the TATA box
(Littlefield *et al.,* 1999). The strong bending of the DNA at the TATA
box, brought about by the TBP–DNA interactions (see above), is a prerequisite for the ability of TFB to contact both sides of the TATA box. A
prominent example is a single amino acid, arginine 188 of *Pyrococcus*
TFB, which contacts phosphates on both sids of the TATA box in the
ternary complex. These TFB contacts stabilize the TBP-induced conformational deformation of the DNA. Essentially identical results were
obtained using human TFIIB$_C$, human TBP$_C$, and the adenovirus major
late promoter (Tsai and Sigler, 2000). In the human system, an additional
base-specific TFIIB$_C$ contact was detected five positions downstream of
the TATA box, which had not been anticipated before. Both the archaeal
and the eukaryotic TF(II)B$_C$'s make the base-specific contacts upstream
of the TATA box with a helix–turn–helix (HTH) motif localized in the
second repeat domain (helix H5'), which binds to the major groove of
the DNA. This corrobarates an earlier prediction based on a primary sequence analysis and is in excellent aggrement with the observation that
mutations in this helix influence the recognition of a BRE base pair three
positions upstream of the TATA box (Lagrange *et al.,* 1998). The DNA interactions of the TF(II)B$_C$ HTH motif are reminiscent of the mode of binding of previously analyzed bacterial and eukaryotic HTH DNA binding
proteins. The binding ot TF(II)B$_C$ to the major groove upstream and the
minor groove downstream of the TATA box agrees very well with SELEX
experiments which had shown that the major determinants for TFB
recognition are located upstream of the TATA box (Qureshi and Jackson,
1998). The exclusive binding to the major groove in the upstream region leaves the minor groove untouched, and thus additional proteins,
e.g., regulatory factors, could bind there simultaneously with TF(II)B
(Tsai and Sigler, 2000).

The structure of TFIIB$_C$ was first resolved by NMR spectroscopy and it
was revealed that the two repeats have a structure similar to the cyclin A
fold (Bagby *et al.,* 1995). The X-ray structures of archaeal and eukaryotic
TF(II)B$_C$'s in the ternary complex established that the structure of the
two repeats is virtually identical in contact with TBP and DNA in the
crystallized form compared to the solution structure (Tsai and Sigler,
2000; Littlefield *et al.,* 1999; Kosa *et al.,* 1997; Nikolov *et al.,* 1995).
However, the two repeats are arranged differently from one another in
the solution structure than in the ternary complex. The comparison
indicated that upon TF(II)B binding to the TBP–DNA binary complex,
an en bloc rotation of the two cyclin repeats of about 100° is necessary

to allow the formation of the contacts with the major groove upstream of the TATA box and the minor groove downstream of the TATA box (Tsai and Sigler, 2000).

While the C-terminal domain of TF(II)B mediates DNA binding, the N-terminal domain is involved in polymerase recruitment and in late stages of the initiation process (see Section IV). The structure of part of the the N-terminal TF(II)B domain was first solved for the archaeal protein from *Pyrococcus furiosus* (Zhu *et al.,* 1996) and later for the human protein (Chen *et al.,* 2000). It contains a zinc ribbon motif of three antiparallel β strands, a motif that has also been found in other proteins involved in transcription, e.g., the elongation factor TF(II)S and the polymerase subunit RPB9. A model containing the structure of the resolved part of the N-terminal domain, the structure of the C-terminal domain, and a simulation of the remaining part of the protein indicated that the N-terminal domain could reach position -1 and is in agreement with the functional studies described below that TF(II)B influences open complex formation and promoter clearance.

C. FURTHER TRANSCRIPTION INITIATION FACTORS

Faithful transcription initiation has been accomplished using TBP, TFB, and purified RNA polymerase of several archaea, which could be taken as an indication that archaea contain no further transcription initiation factors. However, in some eukaryotic *in vitro* transcription systems TBP and TFB are also sufficient for initiation, while that is clearly not the case *in vivo*. Therefore the archaeal *in vitro* systems do not prove the lack of further basal transcription initiation factors. While from a philosophical point of view the absence of a factor cannot be proven for obvious reasons, the following paragraphs list a variety of arguments that the Archaea indeed use a much smaller set of basal transcription initiation factors than the Eukarya.

In an elegant set of experiments the factor requirement for transcription initiation of *Sulfolobus shibatae* was analyzed (Qureshi *et al.,* 1997). An *in vitro* transcription system using crude cell extract was established. Immunodepletion and TBP, TFB, and polymerase verified that all three components are essential for transcription and, because in each case the other two components were not coprecipitated, that they are not stably associated in the cell extract. Supplementation of the depleted extracts with recombinant TBP or TFB or with a highly purified polymerase restored the activity in the respective depleted extracts. Gel filtration gave no indication that either of the two factors was associated with another protein, and the purified polymerase did not contain any nonpolymerase subunits. Thus there was no indication for the presence or the need for a transcription factor apart from TBP and TFB on several promoters used.

Today about 10 archaeal genome sequences are available (the genome sequences can be addressed and analyzed using the program PEDANT at http://pedant.mips.biochem.mpg.de). None contains recognizable homologues of TAFs or of subunits of the eukaryotic factors TFIIA, TFIIE, TFIIF, and TFIIH. However, all archaeal genomes contain a gene encoding a protein of 200 amino acids with similarity to the N-terminal half of the α-subunit of TFIIE. The eukaryotic TFIIE is heterotetramer of $\alpha_2\beta_2$ composition. Two of its functions are to recruit TFIIH to the promoter and to promote the polymerase-phosphorylating activity of TFIIH. Both functions seem to be unnecessary in the Archaea (see below). But TFIIE is not well characterized in the Eukarya, and recently it has been shown that TFIIEα alone can influence TBP binding to DNA (Yokomori et al., 1998). It contains a zinc ribbon motif that is required for that function and that is conserved in the known eukaryotic TFIIEα (CXXCX$_{21}$CXXC). The presence of the zinc ribbon is also conserved in the archaeal proteins, but its sequence is more variable (CX$_{2-4}$CX$_{14-15}$CXXC). While the sequence similarity and the existence of the zinc ribbon motif open the possibility that the archaeal protein is a homologue of TFIIEα and a third basal transcription factor, experimental evidence for its expression, its function, or its involvement in transcription initiation is still lacking.

There are more arguments for the absence of an archaeal TFIIH homologue than the lack of recognizable sequence similarity in genome sequences. The function of TFIIH in the Eukarya is to phosphorylate the largest RNA polymerase subunit at its C-terminal domain (CTD). The underphosphorylated polymerase is recruited to the promoter, and CTD phosphorylation by TFIIH or other kinases is a prerequisite for the transition from the initiation to the elongation phase of transcription. Archaeal RNA polymerases do not contain this CTD and thus a phosphorylation switch does not seem to operate. This could be underscored by the experimental evidence that the $\beta-\gamma$ phosphoanhydride bond of ATP, which is required during eukaryotic CTD phosphorylation, is not required for transcription in an *in vitro* system of *Sulfolobus* (Bell et al., 1998). Therefore the Archaea apparently do not require a TFIIH homologue and, in turn, also do not need the TFIIH recruiting and stimulating activities of TFIIE.

IV. Preinitiation Complex Formation and Polarity of Transcription

The available data indicate that the archaeal preinitiation complex consists of a DNA fragment containing the BRE, TATA box, and initiation element, the transcription factors TFB and TBP, and RNA polymerase. The involvement of the TFEα homologue, for which no experimental data exist, cannot be ruled out, but no additional factors seem to be involved. For some time the factors which determine the polarity of transcription were thoroughly discussed. TBPs from yeast and from *Arabidobsis thaliana* were found to bind in an oriented manner to the TATA box in crystallized binary complexes (J. L. Kim *et al.,* 1993; Y. Kim *et al.,* 1993), and it was thought that deviations from perfect symmetry in the repeat domains would cause a preference for one orientation. However, it was shown that in solution yeast TBP binds to two TATA boxes in both orientations and that the preference for one of them is only modest (Cox *et al.,* 1997). Therefore even in the Eukarya the TBP–TATA box interaction cannot determine polarity, and this can safely be assumed to be even more true for the Archaea, which harbor a TBP of much higher symmetry. Indeed, in an *in vitro* transcription system the switch of TATA box orientation had no influence on the direction of transcription (Bell *et al.,* 1999). The discovery of the BRE in the Archaea and Eukarya explained how asymmetical TFB binding to the convex upper side of only the C-terminal TBP repeat in combination with sequence-specific binding upstream of the TATA box could orient the two transcription factors faithfully on the DNA in the ternary complex and, through asymmetrical binding of RNA polymerase, could lead to the observed high fidelity in the direction of transcription. It was indeed verified in an *in vitro* transcription system from *Sulfolobus* that switching the DNA sequences surrounding the TATA box, changing exclusively the BRE orientation, and leaving the orientation of the TATA box and the initiation element untouched resulted in a complete reversal of the direction of transcription (Bell *et al.,* 1999). Similarly, yeast TBP, which in solution was shown to bind to the TATA box of the adenovirus major late promoter in both orientations, became more than 90% unidirectional in the quaternary complex together with TFIIB and TFIIA (Kays and Schepartz, 2000).

The question of polarity was complicated for some time by the observation that TBP and TFB in the archaeal ternary complex (Kosa *et al.,* 1997) had an opposite orientation compared to several ternary structures determined with eukaryotic proteins (e.g., Nikolov *et al.,* 1995) and that, in none of the complexes, could base-specific DNA contacts of TF(II)B could be observed. As mentioned above, the DNA fragments used for crystallization did not contain a BRE, which had not yet been discovered at that time, and recrystallization (Littlefield *et al.,* 1999; Tsai and Sigler, 2000) nicely verified the biochemical evidence, i.e., that ternary complex orientation is identical in the Archaea and Eukarya and that a much larger surface of TF(II)B than observed previously interacts with

DNA, including several contacts in the major groove with bases of the BRE.

The structure of the archaeal ternary complex indicates that binding of TBP and TFB to promoter DNA is cooperative (compare Fig. 4). While to my knowledge this has not been quantitated, it is in agreement with many observations that the affinity of TBP and TFB for DNA is rather low and that ternary complex formation is observed at protein and DNA concentrations which would not allow binding of the individual proteins (e.g., Qureshi and Jackson, 1998).

In all available structures of archaeal and eukaryotic ternary complexes TF(II)B$_C$ was used for crystallization, while the N-terminal domain was omitted. The affinities of binding to the ternary complexes are about the same for *Sulfolobus solfataricus* TFB$_C$ and TFB, indicating that the N-terminal domain does not play an important role in DNA binding. However, in contrast to the C-terminal domain, only the full-length protein can recruit RNA polymerase into a quaternary complex, giving evidence for a direct interaction of the N-terminal domain with the polymerase (Bell and Jackson, 2000b). An interaction of TFB with the polymerase could also be shown for *Methanobacterium thermoautotrophicum,* and, in contrast to the *Sulfolobus* system, in this case it was detected in the absence of TBP and DNA (Darcy *et al.,* 1999). Size fractionation of a crude cell extract revealed that TFB and polymerase cosedimented in a sucrose gradient, and furthermore, polymerase could be coimmunoprecipitated with anti-TFB antibodies, indicating a stable interaction in solution. Several experiments indicate that eukaryotic TFIIB is also involved in polymerase recruitment, e.g., only full-length TFIIB, in contrast to TFIIB$_C$, was shown to be active in an *in vitro* transcription system (Malik *et al.,* 1993) and a mutant analysis showed that the N-terminal region of yeast TFIIB is important for polymerase interaction (Bangur *et al.,* 1997). Using the *Sulfolobus* system it was shown that archaeal TFB has additional functions in the process of transcription initiation apart from polarity determination and polymerase recruitment (Bell and Jackson, 2000). A point mutation at a highly conserved position in the N-terminal domain immediately downstream of the zinc ribbon motif was constructed and found to be undistinguishable from the wild-type protein in ternary complex formation, polymerase recruitment, and even open complex formation. Nevertheless, the wild type and mutant differed considerably in the amounts of transcripts formed in an *in vitro* system. The differences were sensitive to the NTP concentrations as well as to the sequence of the initiator element at the transcriptional start site. These data indicate that TFB also plays a role in late stages of transcription initiation, e.g., during promoter clearence.

V. The Different Phases of Transcription: From Initiation to Termination

A. PREINITIATION AND BEYOND

Knowledge about the different steps in the initiation process from archaeal RNA polymerase binding to the TBP/TFB/DNA ternary complex until a stable elongation complex is reached is very scarce. It has been shown that, in contrast to eukaryotic systems, in *Sulfolobus* TBP, TFB, and polymerase are sufficient for open complex formation and that neither additional transcription factors nor ATP hydrolysis is necessary (Bell *et al.*, 1998; Bell and Jackson, 2000b). The region from −12 to −1 becomes sensitive to the single strand-specific reagent potassium permanganate, and upon the addition of three ribonucleotide triphosphates, this sensitive region can be chased several positions into the transcribed region until the lack of the fourth ribonucleotide triphosphate stalls the complex (Bell *et al.*, 1998). These data agree well with the length of the transcription bubble within the eukaryotic polymerase.

In the Eukarya, after open complex formation there is a phase of abortive transcription, in which small RNAs of 2 to 10 nucleotides are synthesized and released from the complex, until productive promoter clearence occurs and polymerase II enters the elongation phase (Keene and Luse, 1999). Whether this takes place during archaeal transcription initiation as well has not been determined. The similarities of archaeal and eukaryotic RNA polymerases and the advances in polymerase structure determination, described below, make it tempting to speculate that this is a shared feature. As pointed out above, the analysis of an archaeal TFB mutant has made clear that at least one additional step has to be accomplished between open complex formation and elongation. The existence of different stages during the initiation process was also observed in the mammalian system, where it was shown that the DNA region around +40 is required transiently, after open complex formation and synthesis of the first <10 phosphodiester bonds and before the transcripts have reached a length of more than about 14 nucleotides. It was concluded that "promoter escape requires that the RNA polymerase II transcription complex undergoes a critical structural transition" (Dvir *et al.*, 1997).

B. ELONGATION

The presence of an archaeal homologue to the eukaryotic transcription elongation factor TFIIS was first predicted based upon an open reading frame of *Sulfolobus acidocaldarius* discovered accidentally downstream of a gene for a RNA polymerase subunit (Langer and Zillig, 1993). This view was challenged after the discovery of an orthologue in

Thermococcus celer and the hypothesis was put forward that it might instead be a small polymerase subunit (Kaine *et al.,* 1994). The determination of several archaeal genome sequences revealed that this gene is conserved in the Archaea and enabled a sequence analysis of the emerging protein family (Soppa, 1999a; Hausner *et al.,* 2000). Several lines of evidence indicated that it is more likely a polymerase subunit than an elongation factor: (1) the archaeal proteins have a size of about 100–130 amino acids, comparable a eukaryotic polymerase subunit conserved in polymerases I, II, and III (called RPA12.2, RPB9, and RPC11), while eukaryotic TFIIS is about 300 amino acids long; (2) the degree of similarity is on average greater to the polymerase subunits than to TFIIS; (3) the archaeal protein and the polymerase subunits contain two zinc binding motifs, while eukaryotic TFIIS contains only one; and (4) the gene is localized near polymerase subunit genes in *S. acidocaldarius.*

Despite this circumstantial evidence, biochemical characterization of the *Methanococcus thermolithotrophiscus* protein revealed that it is a bona fide archaeal elongation factor and not a polymerase subunit (Hausner *et al.,* 2000). The archaeal protein will be called aTFS to point out its limited similarity to eukaryotic TFIIS. aTFS is not present in the purified polymerase, and size fractionation of a crude cell extract showed that it has an elution profile distinct from that of the polymerase and thus was not separated from the polymerase in the course of purification. It was shown that aTFS exhibits some activities which are similar to the activities of eukaryotic TFIIS: (1) aTFS induces a cleavage activity of the RNA polymerase at stalled elongation complexes, (2) this activity removes predominantly dinucleotides from the 3′ end of the nascent transcript as the polymerase retracts, and (3) the amount of arrested elongation complex is reduced and the amount of full-length runoff transcript is enhanced by aTFS in an *in vitro* transcription system. These data verify that the conserved archaeal aTFS is an elongation factor, which enables the polymerase to resume transcription at sites where it was stalled or arrested.

The sequence similarities indicate a common ancestry of aTFS, the three polymerase subunits RPA12.2, RPB9, and RPC11, and TFIIS, which had aquired an additional N-terminal domain. It seems that the four eukaryotic proteins have specialized during evolution, e.g., polymerase III has a factor-independent cleavage activity, which depends on the subunit RPC11. A RPC11-deficient polymerase III is unable to release dinucleotides at stalled elongation complexes, but the activity can be restored by the subsequent addition of RPC11, which, however, is not able to integrate into the polymerase and thus, in this artificial situation, has retained the ability to act as an elongation factor (Chedin *et al.,* 1998). As Hausner *et al.* (2000) point out, the polymerase II sub-unit RPB9 deviates from the other four proteins in the region of the zinc ribbon motif retained in TFIIS and, specifically, has lost a glutamic acid residue which has been shown to be important for function in TFIIS (Jeon *et al.,* 1994) and RPC11 (Chedin *et al.,* 1998). In agreement with this, a polymerase II deficient in this subunit is still competent of TFIIS-induced transcript cleavage, indicating that RPB9 is not involved in this reaction. It is tempting to speculate that the precursor protein evolved as an elongation factor in the common ancestor of the Archaea and Eukarya, increasing the efficiency of transcription. Archaeal aTFS would have retained this function, while the eukaryotic protein became integrated into the polymerase before its specialization into the three derivatives. In this model polymerase II would have reintroduced the use of the elongation factor, but in a larger version, thereby allowing subunit RPB9 to acquire a new function.

C. Termination

The mechanism of termination of transcription in the Archaea is hardly understood. Early results of transcriptional termination studies were collected by Brown and colleages (1989), and not much has been added to that picture in the last decade. The determination of the 3′ ends of a variety of mRNAs or stable RNAs from different archaeal species has often located the termination site within or very close to an oligo(dT) tract. In general, "staggered termination" was observed, i.e., transcrip-tion terminated not on a single nucleotide, but in a closely defined region on a few up to more than 10 nucleotides. However, no common struc-ture for an "archaeal terminator" could be defined until now, because the oligo(dT) tract was located either downstream of a putative stem–loop structure, similar to bacterial ρ-independent terminators, or upstream of a putative stem–loop structure directly following the 3′ end of a gene, or without a putative stem–loop structure in its vicinity. The construc-tion of mutants verified for the oligo(dT) tract of the *tRNA*^{Val} gene of *Methanococcus thermolitothrophicus* that it is essential for termination (Thomm *et al.,* 1994). Deletion of the last two thymidine residues of the wild-type sequence 5′ TTTTAATTTT 3′ severely decreased the ter-mination efficiency, and deletion of two additional thymidine residues abolished termination at this site completely. However, a deletion up-stream of the oligo(dT) tract, within the tRNA gene, also resulted in a reduced termination efficiency and enhanced readthrough, indicating that the oligo(dT) tract is a required but not a sufficient determinant for termination. In a second study a reporter gene was used to verify that

the T-rich element 5′ TTATTCTTT 3′ acts *in vivo* as an efficient terminator of transcription in *Haloferax volcanii* and that both the inversion
of the element and the mutation of the last two thymidines to guanosines
totally abolish its activity (Thompson *et al.*, 1999). In this case 88 bp
was transferred into the nonnative surrounding of a reporter gene from
yeast and thus this region appeared to to be the sole determinant for
termination. The presence of inverted repeats close to the T-rich element was not reported, and a nearby direct repeat was shown not to be
important for termination. At the moment it appears as if no uniform
archaeal termination signal exists, therefore further studies are needed
to characterize the complete termination structures for different classes
of genes in different archaea.

VI. RNA Polymerase

A. RNA POLYMERASE COMPOSITION

Very soon after it was proposed that the Archaea are a third domain of
organisms distinct from the Bacteria and Eukarya, it became clear that archaeal polymerases do not follow the bacterial paradigm. Bacterial RNA
polymerases contain only four different subunits and have the compo-
sition $\alpha_2\beta\beta'\omega$. In contrast to bacteria, eukaryotes harbor three different
polymerases with specialized functions, which are multicomponent en-
zymes of 12 to 17 subunits. By 1983 it was clear that archaea contain
only one type of polymerase, that RNA polymerases of several archaea of
diverse subgroups are comprised of at least 10 subunits, and that, on the
basis of not only the composition but also the immunological crossreac-
tivity, archaeal polymerases are more closely related to their eukaryotic
than to their bacterial counterparts (Huet *et al.*, 1983). Hence RNA polymerases were the first component of the basal transcription apparatus
for which the archaeal–eukaryotic relationship was discovered, and in
fact this observation triggered much of the subsequent work described
above. Zillig's group had a huge impact on the field in the following
decade, and the reader is referred to earlier reviews summarizing the
results obtained until the mid-1990s, which included isolation of the
genes for the larger subunits of a variety of archaea, cloning of the genes
for 8 subunits (of about 12 detected) of the *Sulfolobus* enzyme, phylogenetic reconstruction of polymerase subunit relationships, and introduction of a factor-dependent *in vitro* transcription system (Langer *et al.*,
1995; Zillig *et al.*, 1992, 1993). The results clearly underscored the relationship of the archaeal polymerase and the eukaryotic polymerase II.

The best-characterized archaeal RNA polymerases are the enzymes
from *Sulfolobus acidocaldarius* (Langer *et al.*, 1995) and the enzyme

recently isolated from *Methanobacterium thermoautotrophicum* (Darcy *et al.,* 1999). Table I shows a comparison of the subunit compositions of the two archaeal enzymes with the eukaryotic polymerase from yeast and the bacterial enzyme from *E. coli.* While it has been known for a long time that the three largest subunits of the polymerase—called A, B, and D—are conserved in the Archaea and have homologues in the three eukaryotic polymerases and in the bacterial enzyme, the following examples illustrate the difficulties in determining the exact number and identities of the small subunits. The enzyme isolated from *M. thermoautothrophicum* contains two subunits—F and P—which had not been detected in the genome sequence: one had not been recognized as a polymerase subunit, and the other had not been annotated as an open reading frame (ORF). The reasons were the only modest sequence similarity of F to the eukaryotic RPB4 subunit and the fact that P is smaller than the homologous RPB12 and consists of fewer than 50 amino acids, thus escaping ORF-finding routines. F had been isolated with the polymerase from *Sulfolobus,* but its being a real subunit was questioned because it was easily lost during purification and *in vitro* transcription was possible without it (Langer *et al.,* 1995), and P had not been found as

TABLE I

SUBUNIT COMPOSITION OF ARCHAEAL, BACTERIAL, AND EUKARYOTIC POLYMERASES

Yeast polymerase II	MW (kDa)	*Methanobacterium thermoautotrophicum*[a]	Complex *in vitro*[b]	*Sulfolobus acidocaldarius*[c]	*E. coli*
Rpb1	191.6	A′, A″		A′, A″	β′
Rpb2	138.8	B′, B″		B	β
Rpb3	35.3	D	X	D	α
Rpb4	25.4	F		F[d]	
Rpb5	25.1	H		H	
Rpb6	17.9	K		K	
Rpb7	19.1	E′, E″[e]		E	
Rpb8	16.5	—		—	
Rpb9	14.3	—		—	
Rpb10	8.3	N[e]	X	N	
Rpb11	13.6	L	X	L	α
Rpb12	7.7	P	X	—	

[a] The polymerase of *M. thermoautotrophicum* was isolated by Darcy *et al.* (1999).

[b] Subunits of the *Methanococcus jannaschii* polymerase which form a complex *in vitro* (Werner *et al.,* 2000).

[c] The polymerase of *Sulfolobus acidocaldarius* was isolated by Langer *et al.* (1995).

[d] Subunit F was present in the isolated polymerase, but the gene was not cloned by Langer *et al.* (1995).

[e] Subunits E″ and N were not found in the isolated polymerase but are predicted from the genome sequence.

part of the *Sulfolobus* enzyme. Nevertheless, it is clear that F and P are archaeal polymerase subunits, because (1) they are isolated consistently in stochiometric amounts with the *Methanobacterium* enzyme, (2) the genes as well as their genomic localization are conserved in the Archaea, (3) homologous subunits of eukaryotic polymerase II are known, (4) heterologously produced P of *Methanococcus jannaschii* can form a quaternary complex with three other subunits (Eloranta *et al.*, 1998), and (5) the yeast homologue of F—RPB4—is also less tightly bound to the polymerase than other subunits and an RPB4-deficient enzyme is functional (see below). Contrary to the above example, two subunits have not been identified in the purified *Methanobacterium* enzyme which had been predicted to be present based on the genome sequence. But these subunits—E″ and N—are very small and could well have been included in the isolated complex but overlooked due to technical limitations. N can safely be assumed to be an archaeal polymerase subunit because it is conserved in the Archaea, it has a eukaryotic orthologue, and N from *M. jannaschii* is part of the above-mentioned quaternary complex. The situation is somewhat more difficult with E″, a polypeptide of about 60 amino acids. It is conserved in the Archaea but not in the Eukarya, and the only reason for its designation as a polymerase subunit is that the *Sulfolobus* polymerase contains a fusion protein which combines E′ and E″ into a single polypeptide.

In summary, of the 12 subunits of the yeast polymerase II, 10 seem to be present in archaeal polymerases. The large subunit A is generally represented by 2 polypeptides (A′ and A″), and the same is true in many cases for the large subunit B and for the small subunit E; therefore the archaeal enzyme consists of 11–13 polypeptides. The large subunits are shared by all archaeal, eukaryotic, and bacterial enzymes, whereas archaeal and eukaryotic polymerases share six small subunits which do not have counterparts in the bacterial enzyme. With the exception of E″, no archaeal polymerase subunit has been discovered which does not have a eukaryotic homologue. Archaeal polymerases do not seem to contain counterparts for eukaryotic subunits RPB8 and RPB9. There is no indication of a RPB8 homologue in the Archaea, and in the structure of the eukaryotic polymerase, RPB8 is located at the periphery of the enzyme complex (Cramer *et al.*, 2000), therefore it might well be missing in the archaeal polymerase. RPB9 is the subunit with similarity to aTFS and TFIIS, and since aTFS has been proven not to be a polymerase subunit in *Methanoscoccus* (Hausner *et al.*, 2000) and *Methanobacterium* (Darcy *et al.*, 1999), the RPB9 homologue seems to be present but a transcription factor in Archaea (see above), which, however, would be associated with the polymerase during elongation phase. The structure of the C-terminal domain of aTFS, containing one of the two zinc ribbon motifs, has been determined for the protein from *Thermococcus celer* (Wang *et al.*, 1998). The structure could be used to determine the position of RPB9 in the eukaryotic polymerase II structure (Cramer *et al.*, 2000), underscoring the similarity of the two proteins.

It remains to be determined whether variations of this general view exist in different archaeal or eukaryotic species or in one species under different conditions. In yeast, it has been shown that subunit RPB4 is essential for growth at high and low temperatures, but that at intermediate temperatures an RPB4 deletion strain is viable, albeit with a reduced growth rate (Woychik and Young, 1989). Isolated yeast polymerase II contained only substoichiometric amounts of subunits RPB4 and RPB7. The polymerase isolated from a RPB4 deletion strain contained neither of them and was compromised *in vitro* only in initiation, and not in elongation of transcription (Fu *et al.*, 1999). The RPB4 homologous archaeal subunit F was found not to bind as tightly to the polymerase as other subunits (Langer *et al.*, 1995), therefore it seems at least possible that the participation of one or a few of the smaller subunits might be variable.

B. STRUCTURES AND INTERACTIONS OF SUBUNITS

The structures of two archaeal polymerase subunits and of a TFS have been resolved recently. Two structures were obtained for subunit H, which is homologous to RPB5 (Thiru *et al.*, 1999; Yee *et al.*, 2000). Structure determination of RPB5 from yeast (Todone *et al.*, 2000) proved that the result of Yee *et al.* is correct. Subunit H consists of a four-stranded β sheet that is flanked on both sides by an α-helix, giving the overall structure an unprecedented mushroom shape. Archaeal H subunits are about 70 amino acids long, while eukaryotic RBP5's contain an additional N-terminal extension of about 140 amino acids. It turned out that RPB5 has a bipartite structure; the C-terminal part forms a separate structural domain, which is very similar to the archaeal H-subunit structure. The eukaryote-specific N-terminal sequences fold into a second distinct protein domain which makes relatively few direct contacts with the C-terminal conserved structural domain (Todone *et al.*, 2000). A yeast mutant with a single amino acid change in the N-terminal part is unaffected in basal transcription but is defective in transcriptional activation (Miayo and Woychik, 1998), and the interaction of the N-terminal domain with a viral transcriptional activator has been reported (Cheong *et al.*, 1995). Thus it seems that the main function of the additional N-terminal domain in eukaryotic RPB5 is the interaction with transcriptional activators which are not present in the Archaea. Determination of the polymerase II structure has revealed that the C-terminal domain of eukaryotic RPB5, which corresponds to H, is in contact with DNA, while the N-terminal domain is not (Cramer *et al.*, 2000).

Structure determination of the RPB10 homologue subunit N from *M. thermoautothrophicum* revealed that it consists of an unprecedented zinc bundle, i.e., three α-helices stabilized by a zinc ion, which is coordinated to four cysteine residues contained in an atypical motif of the conseneus sequence $CXXCX_nCC$ (Mackereth *et al.*, 2000). The structure of the archaeal subunit was helpful in a recent reconstruction of yeast polymerase II structure, which included the use of high-resolution structures of different subunits of different species of archaea, bacteria, and eukaryotes (Cramer *et al.*, 2000). It can be taken as reinforcement for this approach that a major part of the archaeal and the eukaryotic protein is interchangeable. Hybrid proteins were constructed from the *Sulfolobus* and the yeast protein, and a version that contained a majority of archaeal-derived amino acids was found to function in yeast *in vivo*, while an additional replacement of a eukaryotic motif led to a deficiency in polymerase I (but not II) assembly (Gadal *et al.*, 1999).

The functional equivalence of archaeal and eukaryotic subunits was corrobarated by two further studies. Using isolated eukaryotic subunits *in vitro*, it was found that a ternary RPBs3–11–10 complex and a quaternary RPBs3–11–10–12 complex can be formed. First, it was shown that the equivalent D–L–N subunit ternary complex forms *in vitro* from heterologously produced *Methanococcus jannaschii* proteins (Eloranta *et al.*, 1998); later this was also verified for the quaternary D–L–N–P complex (Werner *et al.*, 2000). Furthermore, a specific interaction that had been observed between the human homologues RPB4 and RPB7 also occurs in an equivalent manner between the archaeal subunits F and E (Werner *et al.*, 2000). In addition, in this case the binding partners could even be swapped across the domain boundaries, i.e., archaeal F can form a specific complex with human RPB7. Taken together, these data indicate that not only the subunits but also their contacts are highly conserved in archaea and eukaryotes, which is of special importance for the D–L–N–P complex because it represents the initial structural framework onto which the large subunits assemble to form a functional polymerase.

C. RNA POLYMERASE STRUCTURE AND MECHANISM

In the last 2 years major breakthroughs have been accomplished in understanding the structures of the bacterial polymerase and eukaryotic polymerase II. For details the reader is referred to recent publications (Cramer *et al.*, 2000; Naryshkin *et al.*, 2000; Opalka *et al.*, 2000; Mooney and Landick, 1999; Poglitsch *et al.*, 1999; Fu *et al.*, 1999; Zhang *et al.*, 1999). Due to its great similarity to the eukaryotic enzyme in terms of subunit composition as well as subunit interactions, the archaeal polymerase can safely be assumed to be of a similar structure; therefore the major conclusions of these studies are summarized here. Special emphasis is placed on the publication by Cramer *et al.* (2000), in which a 20-year-long attempt to understand eukaryotic RNA polymerase II culminates. A backbone model of yeast polymerase at a 3-Å resolution is presented, and high-resolution structures of small subunits from the Archaea, Bacteria, and Eukarya are used to localize the yeast homologues in the polymerase model (see Fig. 5). The model includes 10 of the 12 subunits, because RPB4 and RPB7 were found to be underrepresented in enzyme preparations, inhibited crystal formation, and were therefore omitted. The two large subunits form distinct masses with a cleft between them. At the base the large subunits are anchored by a subassembly of RPB3, RPB10, RPB11, and RPB12. As described above, these four subunits form a quaternary complex *in vitro,* which is also built by the archaeal orthologues. RPB3 and RPB11 include regions which are similar to the bacterial α-subunit, and the elucidation of the structure of the bacterial polymerase from *Thermus aquaticus* at a 3.3-Å resolution (Zhang *et al.,* 1999) has placed the α_2 dimer at a position equivalent to that of the eukaryotic quaternary subassembly. The large subunits of the bacterial enzyme, β and β', are assembled on the α_2 platform. In general, the bacterial and eukaryotic (and probably the archaeal) polymerases share a DNA- and RNA-contacting core, which is surrounded by eukaryotic (and archaeal-)-specific elements. RBP8, which is found only in the eukaryotic enzyme, has a very peripheral position.

The eukaryotic enzyme possesses two jaws, composed of domains of the large subunits and eukaryotic-/archaeal-specific subunits. About 20 bp of downstream DNA (that has not yet been transcribed) can be placed between the jaws in the canonical B-DNA conformation. The synopsis of several approaches indicates that parts of the polymerase are mobile and the polymerase can exist in an open conformation necessary for initation and termination and a conformation important for elongation, in which the polymerase closes on the DNA, explaining the high processivity of the enzyme. The DNA is melted near the active site, which contains a tightly bound Mg ion. The linear extension of the path of the downstream DNA into the enzyme is inhibited by a "wall" made of part of RPB2; the DNA is not only melted but also tilted, and the upstream DNA leaves the enzyme at an approximately 90° angle to the downstream DNA. Near the active site about 8 bp of the DNA is melted and forms the "transcription bubble," in which the template strand is base-paired with the 3' end of the nascent RNA. Two funnels

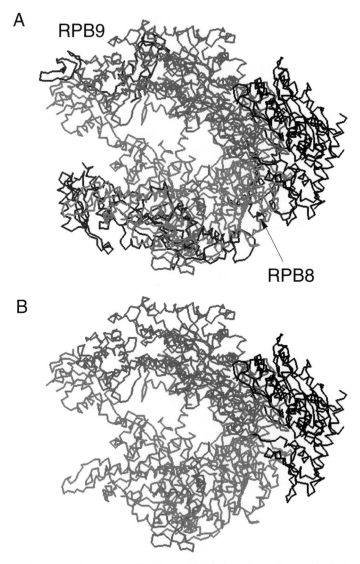

FIG. 5. The RNA polymerase. A backbone model of 10 of 12 subunits of eukaryotic RNA polymerase II was constructed by Cramer *et al.* (2000) (PDB entry number 1EN0) with the inclusion of predetermined high-resolution structures of subunits from the Archaea, Bacteria, and Eukarya. RBP4 and RBP7, which are conserved in the Archaea and Eukarya but not present in the Bacteria, were not part of the crystallized complex. (A) The whole structure is shown. The subunits RBP3, RBP10, RBP11, and RBP12 are localized at the right end of the complex and are colored black (the four eukaryotic subunits as well as their archaeal homologues form a quaternary complex *in vitro*). The two large subunits are shaded in light gray, RBP2 (upper part) a little bit darker than RBP1 (lower part).

in the polymerase were observed which could accommodate a single-stranded nucleic acid, and it seems that the nascent RNA (about nucleotides 9 to 18 from the 3′ end) leaves the enzyme at a place near the DNA entry site and helps to lock the polymerase in the closed conformation. A hole leads to the active site from the "underside" of the polymerase, and it was concluded that the substrates (rNTP's) reach the active site through this opening. A second role has been ascribed to this hole. The polymerase is able to backtrack on the template to overcome obstacles encountered during the process of trancription. While the transcription bubble moves backward, the 3′ end of the RNA becomes liberated, and it could be accommodated in the substrate hole before it is liberated by TFIIS.

The similarity of the subunit compositions and their interactions indicates that most of the conclusions based on the polymerase II structure can be generalized to the archaeal enzyme. This view is reinforced by a recent study in which base-specifically labeled DNA fragments were cross-linked to the archaeal polymerase of *P. fursiosus* (Bartlett *et al.*, 2000). This approach enabled determination of the contacts of the large subunits (A′, A″, B) with specific DNA positions in the initiation complex. The results were in full agreement with the structural data discussed above, again underscoring the similarity of the archaeal polymerase and the eukaryotic polymerase II.

VII. Archaeal Histones and Nucleosomes

Until now the interaction of the basal transcription apparatus with naked DNA has been discussed, but at least in Euryarchaeota it is confronted with nucleosome-bound DNA *in vivo*. Knowledge about archaeal histones and nucleosomes has increased dramatically during recent years, due mainly to the work of Reeve and colleages, who concentrated on the *Methanothermus fervidus* proteins. Several reviews and exciting new findings have been published recently (e.g., Sandman and Reeve, 2000; Bailey *et al.*, 2000; Decanniere *et al.*, 2000; Soares *et al.*, 2000; Pereira and Reeve, 1999) and an up-to-date picture is given in another article in this book (by Sandman and Reeve). In short, *M. fervidus* has the coding capacity for two histone orthologues, called HmfA and HmfB, which can form homodimers as well as heterodimers.

The subunits RBP5, RBP6, RBP8, and RBP9 are shaded in dark gray, and the localization of RBP8 and RBP9 is indicated. (B) Same as A, but the archaeal-/eukaryotic-specific subunits RBP5, RBP6, RBP8, and RBP9 are omitted. The universal subunits RPB1 and RBP2 are shown as well as the four small complex-forming subunits, which build a platform for the large subunits similar to the α_2 dimer in the bacterial polymerase (Zhang *et al.*, 1999).

Archaeal nucleosomes are comprised of histone tetramers instead of the octamers found in eukaryotic nucleosomes, but eukaroytic nucleosomes contain a tetrameric core of histones H3 and H4, which, in a variety of aspects, resembles the archaeal nucleosome. The structure of archaeal histones contains the histone fold known from eukaryotic histones, but the extension of the eukaryotic histones required for reversible acetylation is not present. As first proposed by Ouzounis and Kyrpides (1996) based on sequence comparisons, the histone fold is also found in some eukaryotic transcription factors, linking DNA packaging and regulation of transcription. About 60 nucleotides of DNA are wrapped around archaeal nucleosomes and are protected from nuclease digestion. The nucleosomes recognize positioning signals, which include T-rich oligomers spaced at regular intervals of about 10 bp. Whole-genome analysis has revealed long-range oscillations (sequence periodicities) with a phase of 10 bp for eukaryotes and histone-containing euryarchaea and 11 bp for bacteria and Crenarchaeota (Herzel *et al.*, 1998; Bailey *et al.*, 2000), which can be taken as an argument that nucleosomes are regularly spaced in archaea, that nucleosomal packaging of DNA influenced whole-genome evolution, and that the data on *M. fervidus* histones can be generalized to other euryarchaeal species.

Studies on eukaryotic nucleosomes have revealed great differences between DNA packaged into nuclesomes with histone octamers and DNA packaged into (H3–H4)$_2$-tetramers; e.g., the latter is much more accessible to transcription factor binding and as a template for transcription (Tse *et al.*, 1998; Hansen and Wolffe, 1994). Presently it is not known how the archaeal basal transcription apparatus can cope with DNA packaged into nucleosomes, e.g., (1) whether both initiaton and elongation, neither, or only one of the processes is influenced and (2) how transcription efficiency, velocity, and processivity are effected. The use of nucleosome-bound DNA in an *in vitro* transcription system is badly needed to tackle these and related questions, and with the availability of *in vitro* transcription systems for a growing number of species, the recombinant large-scale production of archaeal histones and the definition of nucleosome positioning signals, such a system has come into reach.

The Crenarchaeota analyzed to date do not contain histones and the analyses of whole-genome sequence periodicities indicate that DNA is not packaged in a comparable way to euryarchaeal nucleosomes. Nevertheless, Crenarchaeota contain small basic DNA binding proteins, which have been shown to colocalize with DNA inside the cell (Bohrmann *et al.*, 1990, 1994), therefore in Crenarchaeota the basal transcription apparatus also has to cope with protein-bound DNA *in vivo*.

VIII. Regulation of Gene Expression

The expression of genetic information can be regulated at many levels, from the initiation of transcription to the reversible covalent modification, the allosteric regulation of enzymatic activity, and the differential subcellular localisation of proteins. In bacteria, gene expression is regulated mostly at the level of transcription, while many mechanisms are realized in eukaryotes. In contrast to the major advances in the description of the basal transcription apparatus, the understanding of gene regulation in archaea is only in its beginning. As archaea share with bacteria a lifestyle as free-living single-cellular organisms with a relatively short generation time, and thus constantly have to grow and multiply, it can be envisaged that transcriptional regulation will be a prominent mechanism of regulation of gene expression. Transcriptional regulation means that the activity of the basal transcriptional apparatus has to be influenced—either inhibited or induced—at single genes, at regulons comprised of several or many coordinately regulated genes, or on the whole-genome level. Sections IX and XI therefore give an overview of the current understanding of global transcriptional regulation and of the examples of gene-specific transcriptional regulation currently under investigation.

IX. Global Regulation of Gene Expression

A. Multiplication of Initiation Factors in Haloarchaea

While most archaeal genomes harbor single genes encoding TBP and TFB, haloarchaea were found to posses multiple—up to eight—alleles. This was first discovered by Daniels' group, working with *Haloferax volcanii* (Thompson *et al.*, 1999; Reeve *et al.*, 1997), and whole-genome sequencing verified that it is also true for *Halobacterium salinarum* (Baliga *et al.*, 2000; Oesterhelt *et al.*, unpuslished). Theoretically more than 40 combinations of TBPs and TFBs are possible in *H. salinarum*, and it is tempting to speculate that transcription from various classes of promoters could be regulated differentially by the expression level of individual initiation factor isoforms. Thus it seems that haloarchaea would have transformed the basal transcription apparatus into a variant which is simultaneously capable of initiation and global regulation of transcription. Two observations corroborate this interpretation: (1) transcription of one of six *tfb* alleles of *H. volcanii* was found to be six- to eightfold induced in response to a heat shock (Thompson *et al.*, 1999), and (2) it has been shown that the heat-shock control elements in the promoters of three chaperonin genes of *H. volcanii* are located in a small region directly upstream and downstream of the TATA box (Thompson and Daniels, 1998), i.e., exactly where TFB binds during transcription initiation (see above). Taken together, it seems that *H. volcanii* would possess a dedicated "heat-shock TFB" and use it to control expression of the heat-shock regulon. It will be interesting to study whether and how the genomes of haloarchaea are organized into different regulons by initiation factor isoforms and how differential promoter recognition is realized at the molecular level.

B. Modified Versions of Initiation Factors

Comparative analysis of primary sequences of initiation factors indicates that there may exist modified versions of initiation factors with specialized functions (Soppa, 1999a). One example is *Pyrococcus horikoshii*, which contains two *tfb* genes. The first codes for a bona fide TFB, while the second deduced TFB deviates considerably from standard TFBs. It consists of only the two repeat domains, while the N-terminal domain is lacking totally. As it has been shown that the N-terminal domain is inportant for polymerase recruitment (see above), this TFB version cannot fulfill its normal function in transcription initiation. It could well be that it is important for transcription at a special subclass of promoters, possibly in concert with a specific regulator replacing the function of the N-terminal domain. The second example was found in *H. salinarum*. One of its deduced TBPs deviates from a bona fide TBP in several respects: (1) its overall similarity to other haloarchaeal TBPs is much lower than normal (40 instead of 80%), (2) the internal symmetry is lower than in other TBPs (27% instead of about 40%), (3) two of the four phenylalanines which intercalate between the first and the last TATA box base pairs are replaced by tyrosines, and (4) some additional residues are changed which are conserved in all other TBPs. Therefore, it might well be that this TBP functions only at a small subset of promoters. This would be analogous to the situation in higher eukaryotes, where TBP-related factors (TRFs) have been described. In *Drosophila*, TBP and TRF are 56% identical, but TRF is important for development and functions only at a small subset of genes *in vivo* (Hansen *et al.*, 1997).

C. TBP Interacting Proteins

Circumstantial evidence indicates that some archaea may contain TBP interacting proteins which could be involved in regulation or other transcription-coupled processes. The genomes of *A. fulgidus, Aeropyrum pernix, P. abyssei, P. horikoshii,* and *S. shibatae* contain a gene coding for a putative protein with similarity to a eukaryotic protein called TIP49 (aTBP interacting protein of 49-kDa molecular mass). The rat protein was cloned based on its affinity to TBP, and the rat and the human proteins were shown to interact with TBP in nuclear extracts (Kanemaki *et al.,* 1997; Makino *et al.,* 1998; Kurokawa *et al.,* 1999). Archaeal and eukaryotic proteins are well conserved, and in fact the similarities of the archaeal proteins to the human protein is even greater than the similarity of the human protein to the yeast protein. The function of this emerging protein family is not understood. The proteins are about 450 amino acids long, contain sequence motifs which indicate their ability to bind ATP/GTP (Walker box A and Walker box B), and show a limited similarity to the bacterial helicase RuvB.

A TBP interacting protein was isolated from cell extract of *Pyrococcus kodakaraensis* using affinity chromatography (Matsuda *et al.,* 1999). The gene was cloned; the deduced protein has a molecular mass of 25.5 kDa and therefore is called TIP26. Heterologously produced TIP 26 was shown to bind to TBP *in vitro* with a K_D of about 1.3 μM. TBP binding to a TATA box was inhibited by TIP26, indicating that it may be a negative regulator of transcription initiation. TIP26-encoding genes are also present in the genomes of *P. abyssei* and *P. horikoshii,* but not in the other archaeal genomes, therefore this putative regulator seems to be limited to the genus of *Pyrococcus.*

D. DNA STRUCTURE

It might well be that DNA topology is involved in global regulation of gene expression in the Archaea, similar to the situation in the Bacteria, where it was found that the superhelical density of the chromosome can influence transcription. Using an *in vitro* transcription system of the hyperthermophilic archaeon *Sulfolobus shibatae,* it has been shown that negatively supercoiled, relaxed, and positively supercoiled templates can be transcribed at the physiological temperature of 75°C, while only negatively supercoiled DNA can be transcribed at lower temperatures (Bell *et al.,* 1998). Transcription with the latter templates at low temperatures was inhibited at the initiation step, therefore it was concluded that thermal energy is required for promoter melting by RNA polymerase. The natural state of DNA in hyperthermophilic archaea is relaxed or positively supercoiled, therefore temperature changes in the environment should influence the transcription initiation frequency and efficiency. In agreement with this idea, it has indeed been found that hyperthermophilic archaea change the positive superhelical density of their DNA in response to a cold shock or heat shock (Lopez-Garcia and Forterre, 1999, 2000), which in itself is an indication that DNA topology and transcription are not independent.

In haloarchaea, more than 20 Z-DNA-containing genomic fragments were isolated making use of an anti-Z-DNA antibody (Kim and DasSarma, 1996). These regions could well influence transcription if they were localized in promoter regions. For the bacterioopsin gene it was verified that its transcription is influenced by the superhelicity of the DNA and that a non-B-DNA conformation, probably Z-DNA, is involved in transcriptional regulation (Yang *et al.*, 1996).

E. Differential Nucleosome Formation

As mentioned, euryarchaeota contain histones which package the DNA into nucleosomes. There are several lines of evidence which indicate that differential packaging of chromosomal DNA occurs, and it can be assumed that this would influence transcription on a global scale. (1) It was shown that the two histones of *M. fervidus*—HmfA and HmfB—are differentially produced, HmfA is the major histone during exponential growth, while both are present equally in the stationary phase (Sandmann *et al.*, 1994). The structures of the two histones are very similar but not identical (Decanniere *et al.*, 2000), therefore it could well be that the two homodimers and the heterodimer, which have been shown to form *in vivo,* package DNA slightly differently. (2) A more pronounced difference between the exponential and the stationary phase has been observed with *H. salinarum.* The DNA of early exponential cells was found to be protein-free, in contrast to the DNA of stationary phase cells, which was packaged into regular nucleosome-like structures (Takayanagi *et al.*, 1992). (3) The discovery that a predicted heat-shock control element is present in most promoters of archaeal histone genes (Gelfand *et al.*, 2000) indicates that chromosomal packaging changes in response to a heat shock, which can be envisaged to result in global changes in transcription. (4) In *M. fervidus* it was shown that nucleosomes are not positioned arbitrarily but that nucleosome positioning signals exist, and the analysis of genome periodicities made it likely that nucleosome positioning is a common feature of Euryarchaeota (Bailey *et al.*, 2000). Sequence-specific positioning of nucleosomes in promoter regions could influence transcription and make it responsive to differential packaging and/or the activity of additional transcription factors.

F. Global Regulators

The analysis of archaeal genomes reveals the presence of many putative DNA binding proteins of several families (see Section X), some of which will most likely turn out to be global regulators of transcription.

In recent years work on archaeal members of the Lrp family of DNA binding proteins has started, especially using proteins from *Pyrococcus* and *Sulfolobus*. The Lrp family is widely distributed in bacteria, and the best-studied example, Lrp from *E. coli*, is a global regulator which induces or represses the expression of nearly 80 genes. *E. coli* Lrp—the leucine-responsive protein—controls primarily genes encoding products involved in amino acid metabolism and transport, and Lrp activity at many, but not all, promoters can be influenced by leucine availability. *E. coli* Lrp represses the transcription of its coding gene in a leucine-independent manner. *Pyrococcus* Lrp was shown to bind to the promoter of its own gene at a region overlapping the TATA box and thereby to regulate its transcription negatively (Brinkman *et al.*, 2000). Similarly, *Sulfolobus* Lrp was shown to bind to its own promoter in the TATA box region, a mutant analysis revealed that a helix–turn–helix (HTH) motif is important for its function, and the abundance of the *lrp* transcript and its accumulation in the stationary phase suggest that Lrp may indeed be a global regulator (Enoru-Eta *et al.*, 2000; Napoli *et al.*, 1999; Charlier *et al.*, 1997). However, none of the target genes of these archaeal Lrp's (apart from the feedback control) have been identified yet, and thus further studies are required to verify that they are (global) regulators of transcription and identify the metabolic pathways under their control.

A negative selection procedure was used to isolate mutants of *H. volcanii* which have lost the ability to grow anaerobically via nitrate respiration. A variety of genes was identified to be essential for nitrate-respirative growth, among them one gene encoding a regulator of the Lrp family (Wanner and Soppa, 1999, unpublished). This is the first example of the identification of an archaeal regulator based on its function *in vivo*. This haloarchaeal Lrp family member seems to control an anaerobic respiration regulon and thus could well be a global regulator, but in this case the protein is not yet available and its DNA binding capacity is unknown. Taken together, experimental data are emerging that archaeal members of the Lrp family are regulators of transcription, and a more detailed characterization can be expected soon.

X. A Bioinformatic View on Archaeal Transcription

The advent of whole-genome sequencing opened the possibility of generating new information by data mining, and this bioinformatic approach is rapidly gaining power with the increasing number of genome sequences available for comparative analyses. Three recent studies are mentioned which focus on the analysis of archaeal transcription. Kyrpides and Ouzounis (1999) analyzed the four archaeal genome sequences completed at that time (from *Archaeoglobus fulgidus, M. jannaschii, M. thermoautotrophicum,* and *Pyrococcus horikoshii*) for the presence of transcription-related genes and compared the deduced gene products with proteins from the Bacteria and Eukarya. They found only 12 transcription-related proteins which are shared by the Archaea and Eukarya to the exclusion of the Bacteria: TBP, TFB, aTFS, TFEα, seven RNA polymerase subunits, and histones. In general, each of the four genomes contained one orthologue of each gene; in some cases several paralogues were present. These data indicate the absence of typical eukaryotic-like transcriptional regulators in the Archaea. Fourteen genes were identified which are present in the Archaea, Bacteria, and Eukarya and were called universal genes, including, of course, the two large polymerase subunits. In this automatic approach the similarity of the bacterial α-subunit to the archaeal and eukaryotic subunits was not considered high enough to classify the gene as universal. Surprisingly, "histone deacetylases" were included in this group, which clearly have to have another substrate in archaea, which contain only the core histone fold, and bacteria, which lack histones totally. It was hypothesized that the reversible (de-)acetylation of a DNA binding protein might be an ancient mechanism of regulating gene expression, and it will be interesting to see whether this turns out to be true and to identify the archaeal and bacterial substrates. A total of 168 proteins falling into 24 families was found to be shared by the Archaea and Bacteria to the exclusion of the Eukarya. Contrary to the above-mentioned classes, in this case the distribution in the four archaeal genomes was quite variable. Many of the proteins were bacterial transcriptional regulators, indicating that in archaea a "eukaryotic-like" basal transcription apparatus acts in concert with "bacterial-like" regulatory proteins.

Aravind and Koonin (1999) analyzed the same four archaeal genomes for the presence of likely DNA binding proteins and classified them into families. They also found that archaea contain bacterial-like helix–turn–helix (HTH) proteins and revealed that the concentration of HTH proteins in archaeal and bacterial genomes (fraction corrected for genome size) is about the same. Interestingly, it was found that in archaea the HTH domains are fused to a great variety of other domains, from replication system components to metabolic enzymes, and it will be interesting to investigate whether these are really DNA binding proteins which integrate various cell cycle and metabolic signals into a gene-regulatory network. It was also found that archaea harbor in their genomes genes coding for two other putative DNA binding domains, i.e., the Arc/MetJ domain and the zinc ribbon. The concentration of theses genes was found to be considerably higher in archaeal than in bacterial genomes. Again, these results should induce experimental examination.

Gelfand *et al.* (2000) aimed at the prediction of transcription regulatory sites in the Archaea. They found that an iterative profile search using only one genome was unsuccessful but that a comparative genomic analysis using initial profile "seeds" derived from experimental data allowed the prediction of the regulatory signals and members of several regulons. One example was the heat-shock regulon. Starting from some characterized archaeal heat-shock genes, a search profile was derived which, in combination with the the phylogenetic knowledge about conserved orthologue groups, allowed the prediction of several families of archaeal heat shock-induced proteins. Among them were two families of chaperonins, which could have been expected, but the identification of the archaeal histones as putative heat shock-induced proteins was a surprise. The prediction corroborates the hypothesis that histones might be involved in transcriptional regulation (see above) and proposes a specific and unanticipated condition under which they might be differentially produced and influence DNA packaging.

These examples illustrate, on one hand, the power of bioinformatic approaches and, on the other hand, that experimental studies are needed to test their hypotheses or to generate starting points for analysis. Clearly, the integration of bioinformatics with genetics, biochemistry, and the emerging techniques of functional genomics would, and hopefully will, deepen our understanding of regulatory hierachies and networks in the Archaea.

XI. Gene-Specific Transcriptional Regulation

The investigation of gene-specific regulation in the Archaea is still in its beginning, but it is a rapidly developing field. A variety of model systems for regulatory studies has been established in recent years, which are summarized in Table II (only one reference is listed for each system and research group). A survey of these studies exceeds the scope of this review, but a few new developments will be mentioned.

In two cases it has been possible to unravel the mechanism of action of an archaeal regulatory protein *in vitro,* and two principles were found. Bell *et al.* (1999a) characterized MDR1 from *Archaeoglobus fulgidus,* a homologue of the bacterial metal-dependent transcriptional regulator DtxR. The *mdr1* gene is the first gene of an operon also containing three genes encoding subunits of an ABC transporter, which probably is involved in metal import. All four genes are cotranscribed, and transcription initiates at the *mdr1* promoter. It was shown that transcription is inhibited by the presence of several metal ions *in vivo.* Heterologously produced MDR1 binds to three operator sites of its own promoter *in vitro* and thereby inhibits *in vitro* transcription. Interestingly, MDR1

TABLE II

RECENT REPORTS ON TRANSCRIPTIONAL REGULATION IN ARCHAEA

Factor	Species	Regulated genes/pathway	Signal(s)	Reference
DtxR	*Archaeoglobus fulgidus*	ABC transporter operon	Metal ions	Bell et al. (2000a)
Brp, Bat	*Halobacterium salinarum*	*bop*	O_2, light	Baliga and DasSarma (2000)
GvpE	*Halobacterium salinarum*	*gvp* genes	Growth phase	Krüger et al. (1998)
ArcR	*Halobacterium salinarum*	*arcACB*	Arginine	Ruepp and Soppa (1996)
Brp, Bat	*Halobacterium salinarum*	*bop*	O_2, light	Gropp et al. (1995)
?	*Haloferax mediterranei*	Halocin H4	Growth phase	Cheung et al. (1997)
?	*Haloferax volcanii*	TFB isoform	Heat shock	Thompson et al. (1999)
TFB?	*Haloferax volcanii*	Chaperonin	Heat shock	Thompson and Daniels (1998)
Lrp-fam.	*Haloferax volcanii*	Nitrate respiration	?	Wanner and Soppa (1999)
Tfx	*Methanobacterium thermoautotrophicum*	*fmd*ECB	W	Hochheimer et al. (1999)
?	*Methanobacterium thermoautotrophicum*	*mrc* and *mrt* genes, e.g.	H_2	Morgan et al. (1997)
?	*Methanococcus maripaludis*	*nif* genes	N-source	Kessler et al. (1998)
?	*Methanococcus maripaludis*	*glnA*	N-source	Cohen-Kupiec et al. (1999)
?	*Methanococcus voltae*	*frc* and *vhc* genes	Se	Sorgenfrei et al. (1997)
?	*Methanosarcina barceri*	*nif*HDK2	N-source	Chien et al. (1998)
Lrp	*Pyrococcus furiosus*	Lrp	?	Brinkman et al. (2000)
?	*Pyrococcus furiosus*	*lamA* operon	Cellobiose	Voorhorst et al. (1999)
Lrp	*Sulfolobus acidocaldarius*	Lrp	?	Enoru-Eta (2000)
Lrp	*Sulfolobus acidocaldarius*	Lrp	?	Napoli (1999)
Lrp	*Sulfolobus solfataricus*	Lrp	?	Bell and Jackson (2000)
Lrp	*Sulfolobus solfataricus*	Lrp	?	Charlier et al. (1997)
?	*Sulfolobus solfataricus*	*malA*, *lacS*	Carbon source	Haseltine et al. (1999)

binds downstream of the BRE and TATA box in the region of -18 to $+67$ and does not prevent the binding ot TFB and TBP but inhibits the recruitment of RNA polymerase and thereby transcription inititation. By *in vivo* cross-linking of proteins and DNA, the *in vitro* data were verified, i.e., that MDR1 binds its own promoter in the presence, but not in the absence, of free metal ions. Bell and Jackson (2000) also investigated the mechanism of action of Lrs14, a *Sulfolobus* protein of the Lrp protein family (see above). They showed that Lrs14 is a dimer in solution and binds to multiple sites in its own promoter. In contrast to the example descibed above, the Lrs14 binding sites overlap with the BRE and TATA box, and it was found that the binding of Lrs14 and that of TFB/TBP are mutually exclusive and therefore Lrs14 inhibits transcription by preventing preinitiation complex formation.

In the majority of cases listed in Table II transcriptional regulation is mediated by a "bacterial-type" regulator with a HTH DNA binding motif. An exception is the GvpE protein, which was found to be an activator of transcription of gas vesicle gene clusters of several haloarchaeal species (Krüger *et al.,* 1998). It contains a coiled-coil (leucine zipper) motif, which is very similar to an equivalent motif in the eukaryotic GCN4 protein. The coiled-coil motif is probably necessary for dimerization and it was found to be essential for GvpE function *in vivo*. As an activator of transcription, GvpE should interact with a component of the basal transcription apparatus, which remains to be identified.

Additional examples emerge which indicate that the presence of an activator is necessary for transcription to occur. The *in vitro* transcription system of *M. thermoautotrophicum* faithfully produces transcripts from a histone-gene promoter but not from promoters of several genes encoding enzymes involved in methanogenesis (Darcy *et al.,* 1999). As mentioned, a *H. volcanii* mutant was isolated which had lost the ability to grow via nitrate respiration, and the mutation was tracked to a member of the Lrp family, which accordingly should act as an activator *in vivo* (Wanner and Soppa, unpublished).

Two examples have been reported indicating that transcription initiation is possible in the absence of one or both of the basal initiation factors described above. (1) It was found that a BRE is not involved in transcription of the bacterioopsin gene of *H. salinarum in vivo* and thus initiation should occur without TFB participation (Baliga and Das Sarma, 2000). In this case, the presence and location of an upstream activator sequence are essential for transcription, thus it seems that the TFB function in initiation has been replaced by a gene-specific activator. (2) Using an *in vitro* transcription system of *Sulfolobus* it was found that a minor promoter of the *tfb* gene is recognized by the polymerase alone and transcription is not stimulated further by the addition of TBP and TFB

(Qureshi *et al.,* 1997), while the major promoter is factor dependent. One explanation seems to be that this specialized promoter ensures that a basal level of TFB is constantly produced and a possible negative feedback loop—i.e., that a decrease in TFB level would result in an even further decrease in TFB production—is circumvented.

XII. Concluding Remarks and Outlook

This article hopefully has shown that the areas exploring basal transcription and transcriptional regulation in the Archaea are developing rapidly and major advances have been made recently. A variety of important questions which were under controversial discussion 2 years ago, e.g., the polarity determinants of transcription in the Archaea and Eukarya (see Soppa, 1999a), has now been clearly resolved. Exciting new findings have been reported on different steps in the initiation phase of transcription and on the factor requirements during elongation. A major breakthrough has been reached in understanding the structure of RNA polymerase, which is based on the combined effort of many groups working with single subunits, subunit complexes, whole enzymes, and mutant derivatives *in vitro* and *in vivo.* Characterization of the regulation of transcription in the Archaea is still not very evolved, but it can be expected to take off soon based on numerous new developments in recent years, e.g., the increasing availability of genome sequences, the establishment of new and defined *in vitro* transcription systems, a highly efficient transformation system for methanogens, the introduction of many model systems for regulatory studies (see Table II), and genomewide mutagenesis approaches.

In my opinion, important challenges for archaeal transcription research in the coming years concern the following questions and areas.

1. Until now attention has been focused mainly on the preinitiation phase of transcription. The subsequent phases of transcription, from open complex formation to termination, and their factor requirement should be characterized.
2. The currently available structures of the bacterial polymerase and eukaryotic polymerase II will hopefully be refined. This will allow us to reveal more clearly the similarities and differences of the archaeal polymerase versus the enzymes of the other two domains and will help to establish the high-resolution structure of the archaeal polymerase. Together, successful efforts in these areas would help us to understand the molecular mechanisms of different activities of the polymerase, from DNA melting to proofreading and antitermination.

3. Until now archaeal transcription *in vitro* has been characterized exclusively using naked DNA. It will be interesting to study how the basal transcription apparatus can cope with nucleosome-bound DNA and how constitutive and regulated transcription will be affected.

4. The interaction of global and gene-specific regulators with the basal transcription apparatus should be studied in a variety of cases. This is more trivial in the case of negative regulators, which just inhibit different components of the basal transcription apparatus from recognizing DNA elements, but will be very interesting in the case of positive regulators such as GvpE.

5. The methods of functional genomics should and will be established, i.e., transcriptome and proteome analyses, but also genome-wide molecular genetic approaches. This will allow us to identify the targets of global and gene-specific regulators and thereby to gain insight into regulatory hierachies and networks in archaeal transcriptional regulation.

6. It will be important to study several species of diverse archaeal groups to be able to advance from single examples to generalizable rules.

7. It would be especially fruitful if research on archaeal model species advanced to allow the integration of the power of different approaches; i.e., ideally an *in vitro* transcription system allowing the study of regulation with a nucleosome-bound template should be available for a species with a highly efficient transformation system and advanced molecular genetics, for which—of course—the genome sequence should be known, functional genomic methods should be introduced, cell biological methods allowing the real-time analysis of the intracellular localization of diverse constituents should be applicable, and proteins and protein complexes should hasten to form highly ordered crystals. Even though this will not come into being tomorrow, the integration of different approaches to study a single species *in silico, in vitro,* and, last but also first, *in vivo* will certainly deepen our understanding of basal and regulated transcription.

REFERENCES

Aravind, L., and Koonin, E. V. (1999). *Nucleic Acids Res.* **27,** 4658–4670.
Bagby, S., Kim, S., Maldonada, E., Tong, K. I., Reinberg, D., and Ikura, M. (1995). *Cell* **82,** 857–867.
Bailey, K. A., Pereira, S. L., Widom, J., and Reeve, J. N. (2000). *J. Mol. Biol.* **303,** 25–34.
Baliga, N. S., and DasSarma, S. (2000). *Mol. Microbiol.* **36,** 1175–1183.

Baliga, N., Goo, Y., Ng, W., Hood, L., Daniels, C., and DasSarma, S. (2000). *Mol. Microbiol.* **36**, 1184–1185.

Bangur, C. S., Pardee, T., and Ponticelli, A. S. (1997). *Mol. Cell. Biol.* **17**, 6784–6793.

Barns, S. M., Delwiche, C. F., Palmer, J. D., and Pace, N. R. (1996). *Proc. Natl. Acad. Sci. USA* **93**, 9188–9193.

Bartlett, M. S., Thomm, M., and Geiduschek, E. P. (2000). *Nature Struct. Biol.* **7**, 782–785.

Bell, S. D., and Jackson, S. P. (1998). *Trends Microbiol.* **6**, 222–228.

Bell, S. D., and Jackson, S. P. (2000a). *J. Biol. Chem.* **275**, 31624–31629.

Bell, S. D., and Jackson, S. P. (2000b). *J. Biol. Chem.* **275**, 12934–12940.

Bell, S. D., Jaxel, C., Nadal, M., Kosa, P. F., and Jackson, S. P. (1998). *Proc. Natl. Acad. Sci. USA* **95**, 15218–15222.

Bell, S. D., Kosa, P. L., Sigler, P. B., and Jackson, S. P. (1999). *Proc. Natl. Acad. Sci. USA* **96**, 13662–13667.

Bell, S. D., Cairns, S. S., Robson, R. L., and Jackson, S. P. (2000). *Mol. Cell.* **4**, 971–982.

Bohrmann, B., Arnold-Schulz-Gahmen, B., and Kellenberger, E. (1990). *J. Struct. Biol.* **104**, 112–119.

Bohrmann, B., Kellenberger, E., Arnold-Schulz-Gahmen, B., Sreenivas, K., Suryanarayana, T., Stroup, D., and Reeve, J. N. (1994). *J. Struct. Biol.* **112**, 70–78.

Brinkman, A. B., Dahlke, I., Tuininga, J. E., Lammers, T., Dumay, V., de Heus, E., Lebbink, J. H., Thomm, M., de Vos, W. M., and van Der Oost. J. (2000). *J. Biol. Chem.* **275**, 38160–38169.

Brown, J. W., Daniels, C. J., and Reeve, J. N. (1989). *Crit. Rev. Microbiol.* **16**, 287–338.

Charlier, D., Roovers, M., Thia-Toong, T.-L., Durbecq, V., and Glansdorff, N. (1997). *Gene* **201**, 63–68.

Chasman, D. I., Flaherty, K. M., Sharp, P. A., and Kornberg, R. D. (1993). *Proc. Natl. Acad. Sci. USA* **90**, 8174–7478.

Chedin, S., Riva, M., Schultz, P., Sentenac, A., and Carles, C. (1998). *Genes Dev.* **12**, 3857–3871.

Chen, H. T., Legault, P., Glushka, J., Omichinski, J. G., and Scott, R. A. (2000). *Protein Sci.* **9**, 1743–1752.

Cheong, J. H., Yi, M., Lin, Y., and Murakami, S. (1995). *EMBO J.* **14**, 143–150.

Cheung, J., Danna, K. J., O'Connor, E. M., Price, L. B., and Shand, R. F. (1997). *J. Bacteriol.* **179**, 548–551.

Chien, Y., Helmann, J. D., and Zinder, S. H. (1998). *J. Bacteriol.* **180**, 2723–2728.

Cohen-Kupiec, R., Marx, C. J., and Leigh, J. A. (1999). *J. Bacteriol.* **181**, 256–261.

Cox, J. M., Hayward, M. M., Sanchez, J. F., Gegnas, L. D., van der Zee, S., Dennis, J. H., Sigler, P. B., and Schepartz, A. (1997). *Proc. Natl. Acad. Sci. USA* **94**, 13475–13480.

Cramer, P., Bushnell, D. A., Fu, J., Gratt, A., Maier-Davis, B., Thompson, N. E., Burgess, R. R., Edwards, A. M., David, P. R., and Kornberg, R. D. (2000). *Science* **288**, 640–649.

Danner, S., and Soppa, J. (1996). *Mol. Microbiol.* **19**, 1265–1276.

Darcy, T. J., Hausner, W., Awery, D. E., Edwards, A. M., Thomm, M., and Reeve, J. N. (1999). *J. Bacteriol.* **181**, 4424–4429.

Decanniere, K., Babu, A. M., Sandman, K., Reeve, J. N., and Heinemann, U. (2000). *J. Mol. Biol.* **303**, 35–47.

DeDecker, B. S., O'Brien, R., Fleming, P. J., Geiger, J. H., Jackson, S. P., and Sigler, P. (1996). *J. Mol. Biol.* **264**, 1072–1084.

de Rosa, M., Gambacorta, A., and Gkiozzi, A. (1986). *Microbiol. Rev.* **50**, 70–80.

Doolittle, W. F. (1999). *Science* **284**, 2124–2128.

Dvir, A., Tan, S., Conaway, J. W., and Conaway, R. C. (1997). *J. Biol. Chem.* **272**, 28175–28178.

Eloranta, J. J., Kato, A., Teng, M. S., and Weinzierl, R. O. (1998). *Nucleic Acids Res.* **26**, 5562–5567.

Enoru-Eta, J., Gigot, D., Thia-Toong, T. L., Glansdorff, N., and Charlier, D. (2000). *J. Bacteriol.* **182**, 3661–3672.

Flatters, D., Young, M., Beveridge, D., and Lavery, R. (1997). *J. Biomol. Struct.* **14**, 757–765.

Frey, G., Thomm, M., Brüdigam, B., Gohl, H. P., and Hausner, W. (1990). *Nucleic Acids Res.* **18**, 1361–1367.

Fu, J., Gnatt, A. L., Bushnell, D. A., Jensen, G. J., Thompson, N. E., Burgess, R. R., David, P. R., and Kornberg, R. D. (1999). *Cell* **98**, 799–810.

Gadal, O., Shpakovski, G. V., and Thuriaux, P. (1999). *J. Biol. Chem.* **274**, 8421–8427.

Gelfand, M. S., Koonin, E. V., and Mironov, A. A. (2000). *Nucleic Acids Res.* **28**, 695–705.

Gropp, F., Gropp, R., and Betlach, M. C. (1995). *Mol. Microbiol.* **16**, 357–364.

Guzikevich-Guerstein, G., and Shakked, Z. (1996). *Nature Struct. Biol.* **3**, 32–37.

Hain, J., Reiter, W.-D., Hüdepohl, U., and Zillig, W. (1992). *Nucleic Acids Res.* **20**, 5423–5428.

Hampsey, M. (1998). *Microbiol. Mol. Biol. Rev.* **62**, 465–503.

Hansen, J. C., and Wolffe, A. P. (1994). *Proc. Natl. Acad. Sci. USA* **91**, 2339–2343.

Hansen, S. K., Takada, S., Jacobson, R. H., Lis, J. T., and Tijan, R. (1997). *Cell* **91**, 71–83.

Haseltine, C., Montalvo-Rodriguez, R., Bini, E., Carl, A., and Blum, P. (1999). *J. Bacteriol.* **181**, 3920–3927.

Hausner, W., Frey, G., and Thomm, M. (1991). *J. Mol. Biol.* **222**, 495–508.

Hausner, W., Lange, U., and Musfeldt, M. (2000). *J. Biol. Chem.* **275**, 12393–12399.

Herzel, H., Weiss, O., and Trifonov, E. (1998). *J. Biomol. Struct. Dyn.* **16**, 1–5.

Hethke, C., Geerling A. C. M./Hausner, W., de Vos, W. M., and Thomm, M. (1996). *Nucleic Acids Res.* **24**, 2369–2376.

Hochheimer, A., Hedderich, R., and Thauer, R. K. (1999). *Mol. Microbiol.* **31**, 641–650.

Hüdephol, U., Reiter, W. D., and Zillig, W. (1990). *Proc. Natl. Acad. Sci. USA* **87**, 5851–5855.

Huet, J., Schnabel, R., Sentenac, A., and Zillig, W. (1983). *EMBO J.* **2**, 1291–1294.

Jain, R., Rivera, M. C., and Lake, J. A. (1999). *Proc. Natl. Acad. Sci. USA* **96**, 3801–3806.

Jeon, C., Yoon, H., and Agarwal, K. (1994). *Proc. Natl. Acad. Sci. USA* **91**, 9106–9110.

Kadonaga, J. T. (1998). *Cell* **92**, 307–313.

Kaine, B. P., Mehr, I. J., and Woese, C. R. (1994). *Proc. Natl. Acad. Sci. USA* **91**, 3854–3856.

Kanemaki, M., Makino, Y., Yoshida, T., Kishimoto, T., Koga, A., Yamamoto, K., Yamamoto, M., Moncollin, V., Egly, J. M., Muramatsu, M., and Tamura, T. (1997). *Biochem. Biophys. Res. Commun.* **235**, 64–68.

Kays, A. R., and Scheparts, A. (2000). *Chem. Biol.* **7**, 601–610.

Keene, R. G., and Luse, D. (1999). *J. Biol. Chem.* **274**, 11526–11534.

Kessler, P. S., Blank, C., and Leigh, J. A. (1998). *J. Bacteriol.* **180**, 1504–1511.

Kim, J. M., and Dassarma, S. (1996). *J. Biol. Chem.* **271**, 19724–19731.

Kim, J. L., and Burley, S. K. (1994). *Struct. Biol.* **1**, 638–652.

Kim, J. L., Nikolov, D. B., and Burley, S. K. (1993). *Nature* **365**, 520–527.

Kim, Y., Geiger, J. H., Hahn, S., and Sigler, P. B. (1993). *Nature* **365**, 512–520.

Knaub, S., and Klein, A. (1990). *Nucleic Acids Res.* **18**, 1441–1446.

Koga, Y., Kyuragi, T., Nishihara, M., and Sone, N. (1998). *J. Mol. Evol.* **46**, 54–63.

Kosa, P. F., Ghosh, G., DeDecker, B. S., and Sigler, P. B. (1997). *Proc. Natl. Acad. Sci. USA* **94**, 6042–6047.

Krüger, K., Hermann, T., Armbruster, V., and Pfeifer, F. (1998). *J. Mol. Biol.* **279**, 761–771.

Kurokawa, Y., Kanemaki, M., Makino, Y., and Tamura, T. A. (1999). *DNA Seq.* **10**, 37–42.

Kutach, A. K., and Kadonaga, J. T. (2000). *Mol. Cell. Biol.* **20**, 4754–4764.

Kyrpides, N. C., and Ouzounis, C. A. (1999). *Proc. Natl. Acad. Sci. USA* **96**, 8545–8550.

Lagrange, T., Kapanidis, A. N., Tang, H., Reinberg, D., and Ebright, R. H. (1998). *Genes Dev.* **12**, 34–44.

Langer, D., and Zillig, W. (1993). *Nucleic Acids Res.* **21**, 2251–2252.

Langer, D., Hain, J., Thuriaux, P., and Zillig, W. (1995). *Proc. Natl. Acad. Sci. USA* **92**, 5768–5772.

Littlefield, O., Korkhin, Y., and Sigler, P. B. (1999). *Proc. Natl. Acad. Sci. USA* **96**, 13668–13673.

Lopez-Garcia, P., and Forterre, P. (1999). *Mol. Microbiol.* **33**, 766–777.

Lopez-Garcia, P., and Forterre, P. (2000). *Bioessays* **22**, 738–746.

Machereth, C. D., Arrowsmith, C. H., Edwards, A. M., and McIntosh, L. P. (2000). *Proc. Natl. Acad. Sci. USA* **97**, 6316–6321.

Makino, Y., Mimori, T., Koike, C., Kanemaki, M., Kurokawa, Y., Inoue, S., Kishimoto, T., and Tamura, T. (1998). *Biochem. Biophys. Res. Commun.* **245**, 819–823.

Malik, S., Lee, D. K., and Roeder, R. G. (1993). *Mol. Cell. Biol.* **13**, 6253–6259.

Marsh, T. L., Reich, C. I., Whitelock, R. B., and Olsen, G. J. (1994). *Proc. Natl. Acad. Sci. USA* **91**, 4180–4184.

Matsuda, T., Morikawa, M., Haruki, M., Higashibata, H., Imanaka, T., and Kanaya, S. (1999). *FEBS Lett.* **457**, 38–42.

Miayo, T., and Woychik, N. A. (1998). *Proc. Natl. Acad. Sci. USA* **95**, 15281–15286.

Mooney, R. A., and Landick, R. (1999). *Cell* **98**, 687–690.

Morgan, R. M., Pihl, T. D., Nölling, J., and Reeve, J. N. (1997). *J. Bacteriol.* **179**, 889–898.

Napoli, A., van der Oost, J., Sensen, C., Charlebois, R. L., Rossl, M., and Ciaramella, M. (1999). *J. Bacteriol.* **181**, 1474–1480.

Naryshkin, N., Revyakin, A., Kim, Y., Mekler, V., and Ebright, R. H. (2000). *Cell* **101**, 601–611.

Naryshkina, T., Rogulja, D., Golub, L., and Severinov, K. (2000). *J. Biol. Chem.* **275**, 31183–31190.

Nikolov, D. B., Chen, H., Halay, E. D., Usheva, A. A., Hisatake, K., Lee, D. K., Roeder, R. G., and Burley, S. K. (1995). *Nature* **377**, 119–128.

Opalka, N., Mooney, R. A., Richter, C., Severinov, K., Landick, R., and Darst, S. A. (2000). *Proc. Natl. Acad. Sci. USA* **97**, 617–622.

Ouzounis, C. A., and Kurpides, N. C. (1996). *J. Mol. Evol.* **42**, 234–239.

Ouzounis, C., and Sander, C. (1992). *Cell* **71**, 189–190.

Pace, N. R. (1997). *Science* **276**, 734–740.

Palmer, J. R., and Daniels, C. J. (1995). *J. Bacteriol.* **177**, 1844–1849.

Patikoglou, G. A., Kim, J. L., Sun, L., Yang, S. H., Kodadek, T., and Burley, S. (1999). *Genes Dev.* **13**, 3217–3230.

Pereira, S. L., and Reeve, J. N. (1999). *J. Mol. Biol.* **289**, 675–681.

Philippe, H., and Forterre, P. (1999). *J. Mol. Evol.* **49**, 509–523.

Poglitsch, C. L., Meredith, G. D., Gnatt, A. L., Jensen, G. J., Chang, W. H., Fu, J., and Kornberg, R. D. (1999). *Cell* **98**, 791–798.

Qureshi, S. A., and Jackson, S. (1998). *Mol. Cell* **1**, 389–400.

Qureshi, S. A., Bell, S. D., and Jackson, S. P. (1997). *EMBO J.* **16**, 2927–2936.

Reeve, J. N., Sandman, K., and Daniels, C. J. (1997). *Cell* **89**, 999–1002.

Reiter, W. D., Palm, P., and Zillig, W. (1988). *Nucleic Acids Res.* **16**, 1–19.

Reiter, W. D., Hüdepohl, U., and Zillig, W. (1990). *Proc. Natl. Acad. Sci. USA* **87**, 9509–9513.

Rowlands, T., Baumann, P., and Jackson, P. (1994). *Science* **264**, 1326–1329.

Ruepp, A., and Soppa, J. (1996). *J. Bacteriol.* **178**, 4942–4947.

Sandman, K., and Reeve, J. N. (2000). *Arch. Microbiol.* **173**, 165–169.

Sandman, K., Grayling, R. A., Dobrinski, B., Lurz, R., and Reeve, J. N. (1994). *Proc. Natl. Acad. Sci. USA* **91**, 12624–12628.

Snel, B., Bork, P., and Huynen, M. A. (1999). *Nature Genet.* **21**, 108–110.

Soares, D. J., Sandmann, K., and Reeve, J. (2000). *J. Mol. Biol.* **297**, 39–47.

Soppa, J. (1999a). *Mol. Microbiol.* **31**, 1295–1305.

Soppa, J. (1999b). *Mol. Microbiol.* **31**, 1589–1601.

Soppa, J., and Link, T. A. (1997). *Eur. J. Biochem.* **249**, 318–324.

Soppa, J., Link T. A., z. Mühlen, A., Ruepp, A., and Vatter, P. (1998). *In* "Microbiology and Biogeochemistry of Hypersaline Enviroments" (A. Ohren, ed.), pp. 249–263. CRC Press, Boca Raton, FL.

Sorgenfrei, O., Müller, S., Pfeiffer, M., Sniezko, I., and Klein, A. (1997). *Arch. Microbiol.* **167**, 189–195.

Takayanagi, S., Morimura, S., Kusaoke, H., Yokoyama, Y., Kano, K., and Shioda, M. (1992). *J. Bacteriol.* **174**, 7207–7216.

Thiru, A., Hodach, M., Eloranta, J. J., Kostourou, V., Weinzierl, R. O., and Matthews, S. (1999). *J. Mol. Biol.* **287**, 753–760.

Thomm, M. (1996). *FEMS Microbiol. Rev.* **18**, 159–171.

Thomm, M., and Wich, G. (1988). *Nucleic Acids Res.* **16**, 151–163.

Thomm, M., Hausner, W., and Hethke, C. (1994). *Syst. Appl. Microbiol.* **16**, 648–655.

Thompson, D. K., and Daniels, C. J. (1998). *Mol. Microbiol.* **27**, 541–551.

Thompson, D. K., Palmer, J. R., and Daniels, C. J. (1999). *Mol. Microbiol.* **33**, 1081–1092.

Todone, F., Weinzierl, R. O., Brick, P., and Onesti, S. (2000). *Proc. Natl. Acad. Sci. USA* **97**, 6306–6310.

Tsai, F. T., and Sigler, P. B. (2000). *EMBO J.* **19**, 25–36.

Tse, C., Fletcher, T. M., and Hansen, J. C. (1998). *Proc. Natl. Acad. Sci. USA* **95**, 12169–12173.

Voorhorst, W. G., Gueguen, Y., Geerling, A. C., Schut, G., Dahlke, I., Thomm, M., van der Oost, J., and de Vos, W. (1999). *J. Bacteriol.* **181**, 3777–3783.

Wang, B., Jones, D. N. M., Kaine, B. P., and Weiss, M. A. (1998). *Structure* **6**, 555–569.

Wanner, C., and Soppa, J. (1999). *Genetics* **152**, 1417–1428.

Werner, F., Eloranta, J. J., and Weinzierl, O. J. (2000). *Nucleic Acids Res.* **28**, 4299–4305.

Wettach, J., Gohl, H. P., Tschochner, H., and Thomm, M. (1995). *Proc. Natl. Acad. Sci. USA* **92**, 472–476.

Wich, G., Hummel, H., Jarsch, M., Bär, U., and Böck, A. (1986). *Nucleic Acids Res.* **14**, 2459–2479.

Woese, C. (1998). *Proc. Natl. Acad. Sci. USA* **95**, 6854–6859.

Woese, C., Kandler, O., and Wheelis, M. L. (1990). *Proc. Natl. Acad. Sci. USA* **87**, 4576–4579.

Woychik, N. A., and Young, R. A. (1989). *Mol. Cell. Biol.* **9**, 2854–2859.

Yang, C. F., Kim, J. M., Molinari, E., and Dassarma, S. (1996). *J. Bacteriol.* **178**, 840–845.

Yee, A., Booth, V., Dharamsi, A., Engel, A., Edwards, A. M., and Arrowsmith, C. H. (2000). *Proc. Natl. Acad. Sci. USA* **97**, 6311–6315.

Yokomori, K., Verrijzer, C. P., and Tijan, R. (1998). *Proc. Natl. Acad. Sci. USA* **95**, 6722–6727.

Zhang, G., Campbell, E. A., Minakhin, L., Richter, C., Severinov, K., and Darst, S. A. (1999). *Cell* **98**, 811–824.

Zhu, W., Zeng, Q., Colangelo, C. M., Lewis, M., Summers, M. F., and Scott, R. A. (1996). *Nature Struct. Biol.* **3**, 122–124.

Zillig, W., Palm, P., Langer, D., Klenk, H. P., Lanzendörfer, M., Hüdepohl, U., and Hain, J. (1992). *Biochem. Soc. Symp.* **1992**, 79–88.

Zillig, W., Palm, P., Klenk, H. P., Langer, D., Hüdepohl, U., Hain, J., Lanzendörfer, M., and Holz, I. (1993). *In* "The Biochemistry of Archaea" (M. Kates, D. J. Kushner, and A. T. Matheson, eds.), pp. 367–391. Elsevier, New York.

Protein Folding and Molecular Chaperones in Archaea

MICHEL R. LEROUX

Department of Molecular Biology and Biochemistry
Simon Fraser University
Burnaby, British Columbia, Canada V5A 1S6

ADVANCES IN APPLIED MICROBIOLOGY, VOLUME 50

I. Introduction

The central dogma of molecular biology, DNA → RNA → Protein, underscores two biological processes (transcription and translation) that are fundamental to life. Simply stated, an organism's genomic blueprint (DNA) is transcribed into messenger RNA molecules, which are in turn translated into the proteins required for performing the bulk of a cell's functions. This chapter presents a comparative view of how archaea carry out the final step in protein formation, which is protein folding and assembly. Related aspects of cellular information processing employed by archaea are addressed in other chapters in this text. These include transcription, DNA repair and recombination and, protein turnover.

The biogenesis of proteins comprises folding and assembly, events that were originally presumed to be spontaneous but are now known in many cases to be assisted by families of proteins termed molecular chaperones. The transport of proteins to and across biological membranes is also an essential step in protein biogenesis that can be assisted by chaperones. Following the useful lifetime of a protein, degradation is an obligatory event which ensures that the proteome functions at peak efficiency; it is notable that a conserved family of chaperones has recently emerged as important constituents, or cohorts, of proteolytic degradation machineries. Perhaps not surprisingly, chaperones are also involved in a number of housekeeping cellular processes involving the regulation of protein function. Molecular chaperones thus play important roles in the birth, life, and death of a significant fraction of an organism's protein repertoire.

In Sections II and III, a broad overview of well-characterized molecular chaperone families and their cellular functions is presented, covering the three domains of life (Bacteria, Archaea, and Eukarya). In subsequent sections, in-depth descriptions of the structures and functions of a variety of representative molecular chaperones are given, with special emphasis placed on eukaryotic-like chaperone families that exist in archaea.

II. Molecular Chaperones

A. DEFINITION AND PRINCIPLE OF ACTION

Historically, the term "molecular chaperone" was first coined to describe a function ascribed to nucleoplasmin, a protein rich in basic charges that prevents unproductive associations between histones and DNA (Laskey *et al.*, 1978). This basic functional characteristic of nucleoplamin, which parallels that of a human chaperone ensuring proper interactions between different people, was later recognized by John Ellis to comprise a diverse family of unrelated proteins (Ellis, 1987; Ellis and Hemmingsen, 1989). Presently, well over 50 types of molecular chaperones and chaperone cofactors are known. Their cellular functions are as varied as their primary sequences and structures and include assisting *de novo* protein folding and assembly, preventing protein aggregation and renaturing proteins under stress conditions, facilitating the transport of proteins to different cell locations and across membranes, undoing protein aggregates, promoting protease-dependent protein degradation by unfolding, and modulating the conformation of a variety of signalling and other proteins (Beissinger and Buchner, 1998; Bukau and Horwich, 1998; Ellis and Hartl, 1999; Feldman and Frydman, 2000; Hartl, 1996; Horwich *et al.*, 1999; Leroux and Hartl, 2000a; Saibil, 2000).

Remarkably, the wide breadth of functions demonstrated by chaperones stems from variations on one basic, common mechanistic principle: the ability to interact with and stabilize surfaces that require proper intra- or intermolecular self-assembly. Intramolecular self-assembly refers to the folding of the protein to the native state, whereas intermolecular self-assembly denotes protein interactions with like or unlike proteins to form homo- or hetero-oligomeric complexes, respectively (Ellis and van der Vies, 1991). Most often, although not always, the transiently exposed surfaces arise from the exposure of hydrophobic residues that are normally buried within the folded, native proteins. As chaperones are by definition not part of the final protein structure or assembly, a variety of chaperone mechanisms for disengaging from the protein substrates is utilized. In a large number of cases (e.g., for Hsp70, Hsp90 and chaperonins), the hydrolysis of nucleotides plays an important role, but interactions with cofactors or transfers to other chaperones for further processing may also be required. For example, both prefoldin and small heat-shock protein chaperones stabilize only nonnative proteins and require ATP-dependent chaperones for folding (Leroux and Hartl, 2000a). Some cofactors not only enhance the folding efficiency of the chaperone but also may possess chaperone (substrate-binding) functions on their own [e.g., the Hsp40/DnaJ family of Hsp70 cofactors (Cyr *et al.*, 1994; Kelley, 1998)].

Peptidyl-prolyl isomerase (PPI) and protein disulfide isomerase (PDI) proteins catalyze the slow cis–trans isomerization of prolyl bonds and formation of correct disulfide bonds in nonnative polypeptides, respectively (Ferrari and Soling, 1999; Gothel and Marahiel, 1999; Maruyama and Furutani, 2000; Schiene and Fischer, 2000). The PPI and PDI families therefore assist protein folding but are typically categorized separately from molecular chaperones under the label "protein fold-ing catalysts" (Gething and Sambrook, 1992). This designation can be somewhat arbitrary, however, since a growing number of family mem-bers (e.g., bacterial trigger factor and DsbG) exhibit both catalytic ac-tivity (PPI and PDI activities, respectively) and a general ability to bind and stabilize nonnative structures in proteins (Wang and Tsou, 1998).

Aside from their general mechanism of action, perhaps the most im-portant consideration regarding molecular chaperones is their scope of action: Do they bind and stabilize a large spectrum of nonnative polypeptides (as does Hsp70, which interacts with nascent chains, and small heat-shock proteins, which bind a variety of proteins unfolded because of cellular stress, for example)? Or are they specific for only one or a few types of substrates, as is the case with the collagen-specific chaperone Hsp47 and the bacterial PapD/FimC periplasmic chaperones, which assist the folding and assembly of pili structures? For many differ-ent chaperones, shedding light into the question of substrate specificity remains a top priority.

B. MOLECULAR CHAPERONE FAMILIES

The exposure of cells to stress conditions such as nonphysiolog-ical temperatures and proteotoxic agents (heavy metals, for exam-ple) causes protein denaturation and a concomitant upregulation in the expression of the so-called stress or heat-shock proteins (Hsps) (Hightower, 1980; Lindquist and Craig, 1988; Morimoto et al., 1997). This phenomenon, termed the stress or heat-shock response, is ob-served in all organisms, including archaea (Cavicchioli et al., 2000; Cannio et al., 2000; Conway de Macario and Macario, 2000; Macario et al., 1999b). Most of the prominent Hsps, named according to their molecular weights, have been identified and found to be bona fide molecular chaperones. Indeed, the recent discovery of a novel and un-usual redox-activated chaperone, Hsp33, was made possible by the char-acterization of heat-inducible Escherichia coli genes previously identi-fied in a genomic expression screen (Chuang and Blattner, 1993; Jakob et al., 1999). The best-studied stress-inducible chaperone families in-clude small Hsps (12–42 kDa), Hsp60 (chaperonin) and the Hsp10 co-factor, Hsp70 and its chaperone cofactor Hsp40, Hsp90 and associ-ated cofactors, and members of the Hsp100/Clp/AAA ATPase family (Beissinger and Buchner, 1998; Leroux and Hartl, 2000a; Saibil, 2000). As would be expected for a diverse group of proteins that perform numerous housekeeping functions, some chaperones are present dur-ing normal growth conditions and their expression may or may not be upregulated during stress conditions; those that are strictly stress-inducible are likely to play critical functions in organism survival under adverse situations. A list of the major molecular chaperone families and brief descriptions of their functions is presented in Table I. The role of various chaperones in *de novo* protein folding is summarized schematically in Fig. 1.

The number of molecular chaperones that exist within cells is likely to be significantly larger than we are presently aware of, as judged by the latest unearthing of prefoldin, a chaperone first identified in *Saccharomyces cerevisiae* but also present in archaea (Geissler *et al.,* 1998; Leroux *et al.,* 1999), and clusterin, a eukaryotic protein with significant sequence similarity to small Hsps (Wilson and Easterbrook-Smith, 2000). Perhaps most remarkably, various related members of a protein family (AAA ATPases), found in all three domains and known to be associated with protein complexes involved in proteolytic degradation, are also capable of functioning as molecular chaperones. The latest series of discoveries that have brought this to our attention—that chaperones can be involved in both degradation and folding, two essentially opposite cellular processes—would be viewed as paradoxical were it not for the increasing realization that chaperones play essential *quality control* roles in cells (Brodsky and McCracken, 1999; Ellgaard *et al.,* 1999; Gottesman *et al.,* 1997; Horwich *et al.,* 1999; Suzuki *et al.,* 1997).

C. Distribution of Chaperones across the Three Domains

Given that the processes of protein folding and assembly, transport, protection from stress, and degradation play essential roles in all organisms, it might be expected that the molecular chaperones which carry out these functions would be highly conserved in all three domains of life. With the materialization of complete genomic DNA sequences for several eukaryotes and countless bacteria and archaea, it has now become possible to catalog the full complement of known molecular chaperones in whole organisms. Table I reveals one important aspect of molecular chaperones: few molecular chaperones are as widely represented as some housekeeping proteins (e.g., ribosomal subunits and various proteins involved in translation). Some of the functions carried out by chaperones either may be highly specialized (and thus not needed in all organisms) or are functionally redundant with members from different or related chaperone families. Alternatively, uncharacterized proteins that display a weak sequence similarity to, or which are not homologous to known chaperone families, may possess important and perhaps complementary chaperone functions (Leroux and Hartl, 2000a).

TABLE I

Distribution of Molecular Chaperones in the Three Domains of Life

Protein family	Proposed function(s)	Archaea	Bacteria		Eukarya			
		Cytosol	Cytosol	Periplasm	Cytosol	Mitochon.	Chloropl.	ER
Chaperones								
Trigger factor (TF)	Ribosome-bound chaperone and peptidyl-prolyl isomerase. Cooperates with DnaK during *de novo* protein folding	—	TF	—	—	—	—	—
NAC	Ribosome-bound factor that interacts with nascent chains and regulates the targeting of signal sequence-bearing proteins to the ER	α-NAC	—	—	$\alpha\beta$-NAC	—	—	—
Hsp70/DnaK	Binds linear hydrophobic segments in nonnative proteins and assists folding in an ATP- and cofactor-dependent manner	DnaK (only in some archaea)	DnaK	—	Hsp70	Mitochon. Hsp70	Chloropl. Hsp70	BiP
Hsp40/DnaJ	Binds nonnative proteins and stimulates ATP hydrolysis of Hsp70/DnaK	DnaJ (only in some archaea)	DnaJ	—	Hsp40	Mitochon. Hsp40	Chloropl. Hsp40	ER Hsp40
GrpE	ATP–ADP nucleotide exchange factor for Hsp70/DnaK	GrpE (only in some archaea)	GrpE	—	GrpE	Mitochon. GrpE	Chloropl. GrpE	GrpE
Prefoldin	Octopus-shaped hexameric complex stabilizes nonnative proteins with its coiled-coil "tentacles" for subsequent folding by chaperonin	Prefoldin/GimC	—	—	Prefoldin/ GimC	—	—	—
Chaperonin	Double-ring structures with central cavities mediate the ATP-dependent refolding of nonnative polypeptides	Thermosome (Group II chaperonin)	GroEL (Group I chaperonin)	—	TRiC/CCT (Group II chaperonin)	Hsp60 (Group I chaperonin)	RuBP (Group I chaperonin)	—
Hsp10	Group I chaperonin cofactor. Bowl-shaped heptamer caps chaperonin cylinder during folding	—	GroES	—	—	Hsp10	Cpn20	—
Cofactors A–E	Specifically involved in the postchaperonin folding/assembly of tubulins	—	—	—	Cofactors A–E	—	—	—

Small Hsps	Large spherical assemblies prevent protein aggregation. Cooperate with other chaperones (Hsp70/chaperonins) for protein folding	Small Hsp	Ibp	—	Small Hsps, α-crystallin	Mitochon. small Hsp	Chloropl. small Hsp	—
Hsp90	Abundant chaperone plays various roles, including refolding under stress conditions, modulation of protein conformation. Functions with a variety of cofactors in eukaryotes	—	HtpG	—	Hsp90	—	—	Grp94/Gp96
AAA ATPases	Hexameric chaperone complexes that display both folding and unfolding activities. Often coupled to protease for protein degradation	VAT, Lon, PAN, others	ClpA/B/X, Lon, FtsH, HslU, etc.	—	Various families (see text)	Yta, Yme in *S. cerevisiae*	FtsH related	—
Hsp33	Chaperone activated by "redox switch" under conditions of oxidative stress	—	Hsp33	—	—	—	—	—
Calnexin	Integral membrane chaperone promotes folding and oligomerization of glycoproteins	—	—	—	—	—	—	Calnexin
Calreticulin	ER lumen chaperone is involved in glycoprotein folding	—	—	—	—	—	—	Calreticulin
PapD/FimC	Periplasmic chaperones stabilizes and directs the assembly of pili subunits to the membrane	—	—	PapD/FimC	—	—	—	—
SecB	Protein export chaperone	—	SecB	—	—	—	—	—
Folding Catalysts								
Peptidyl-prolyl isomerase (PPI)	PPIases catalyze the slow cis-trans isomerization of prolyl bonds	FKBP members mostly	FKBP, cyclophilin, parvulin	SurA	Various families	Cyclophilin D	Cyclophilin related	Cyclophilin related
Protein disulfide isomerase (PDI)	PDIases assist correct disulfide bond formation in proteins	TrxA; others with a thioredoxin domain	Thioredoxins	Dsb members	Thioredoxins	—	—	Various members

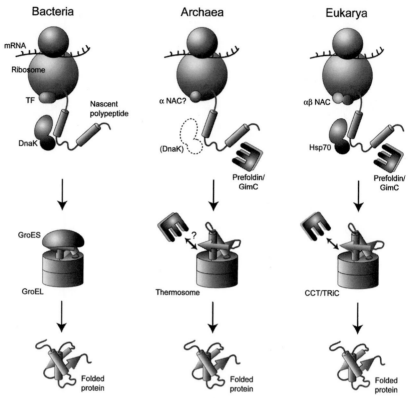

FIG. 1. Molecular chaperones are involved in *de novo* protein folding in the three domains of life. In bacteria, two cooperating molecular chaperones interact with a significant proportion of nascent polypeptide chains: trigger factor (TF), a peptidyl-prolyl isomerase with the ability to interact with unfolded proteins, and the Hsp70 system (DnaK, including its cofactors DnaJ and GrpE). The nascent chains may fold upon release from the ribosome or interact with the chaperonin system consisting of GroEL and its cofactor GroES. In eukaryotes, nascent chains first bind the ribosome-bound nascent polypeptide-associated complex (NAC), which mediates the targeting of preproteins to the endoplasmic reticulum but may also have a general ability to stabilize nascent chains. The Hsp70 chaperone system (including Hsp40, the eukaryotic homologue of DnaJ, and various other Hsp70 cofactors) binds to a wide spectrum of nascent chains. Prefoldin/GimC is a chaperone involved in the biogenesis of actins, tubulins, and perhaps other proteins. Incompletely folded actins and tubulins, as well as a variety of other proteins, are folded by CCT/TRiC, presumably with the assistance of prefoldin. In archaea, a NAC-like protein subunit also exists, but its function and ability to bind the ribosome have not been demonstrated. Many archaea lack an Hsp70 system (outlined in the figure), but all possess prefoldin, which may play an Hsp70-like function during *de novo* folding by transferring nascent polypeptides to the chaperonin (thermosome) for folding to the native state.

1. Archaeal Chaperones

Archaea *universally* harbor only a limited number of chaperone families, namely, small Hsp(s), chaperonin, AAA ATPases, prefoldin, NAC (nascent polypeptide-associated complex), and folding catalysts (FKBP-type peptidyl-prolyl isomerase and disulfide isomerases). The most intriguing omission from this list is the Hsp70 family of chaperones, which is absent from many archaea but is found in the cytosol of all bacteria and eukaryotes, often in multiple copies, and in eukaryotic mitochondria (mt-Hsp70), chloroplasts (cp-Hsp70), and endoplasmic reticulum (BiP) (Table 1). Interestingly, some archaeal chaperones—such as the chaperonin termed "thermosome," prefoldin, NAC, and some AAA ATPases—are distinctly eukaryal-like and either have no representative members in bacteria or have a more distant homologue, as in the case of the chaperonin (Horwich and Willison, 1993; Gutsche *et al.*, 1999; Leroux *et al.*, 1999; Zwickl *et al.*, 1999). In contrast to eukaryotes (see below), however, archaea appear to encode a rather restricted set of members from each chaperone family (Macario and Conway de Macario, 2001). One notable exception is that the euryarchaeote *Halobacterium* sp. NRC-1 encodes four small Hsps (most have one or two) (Ng *et al.*, 2000). Many chaperone complexes are homooligomeric or heterooligomeric, however; an example of this is the double-ring chaperonin, which is assembled from one type of subunit in *Methanococcus jannaschii,* two subunits in *Methanobacterium thermoautotrophicum,* and probably three subunits in *Sulfolobus* (Archibald *et al.*, 1999; Gutsche *et al.*, 1999).

Nearly all sequenced archaeal genomes possess a complete, or common, core of molecular chaperones. A previously-believed exception, discussed later, is that the genome of *Aeropyrum pernix* (Kawarabayasi *et al.*, 1999) encodes only one of the two known subunits of archaeal prefoldin; it does, however, lack a homologue of the Lon chaperone-protease. In several cases, one or more (but not all) archaeal species have apparently acquired chaperone(s) via lateral gene transfers (e.g., bacterial ClpA, an Hsp100 family member, Hsp70 and its cofactors, the bacterial protein export chaperone SecB, and various folding catalysts).

2. Bacterial Chaperones

Bacteria are, like archaea, prokaryotes—and both might be expected to boast very similar intracellular processes and thus an equally similar ensemble of molecular chaperone families. Interestingly, bacteria typically have all known archaeal chaperones except prefoldin and encode some chaperones that are ubiquitous in eukaryotes, such as an Hsp70 system (including DnaJ and GrpE cofactors) and HtpG, a homologue of the Hsp90 chaperone. Trigger factor, a PPI and chaperone, is common to all bacteria and has no counterpart in archaea or eukaryotes (Hesterkamp and Bukau, 1996a). In addition, several other bacterial-specific chaperones include the periplasmic PapD/FimC chaperone involved in pilus biogenesis in gram-negative pathogens, Hsp33, various folding catalysts, and SecB (Driessen *et al.*, 1998; Jakob *et al.*, 1999; Sauer *et al.*, 2000). Bacteria thus appear to have a richer selection of chaperones than archaea, but it should be noted that there is more variability in the families present in bacteria compared to those found in archaea. For instance, many bacteria do not encode PapD/FimC, Hsp33, small Hsp, HtpG, SecB, or certain protein folding catalysts; strikingly, at least one lacks a chaperonin system which is essential in *E. coli* (Glass *et al.*, 2000).

3. Eukaryotic Chaperones

As a whole, eukaryotes not only possess the largest number of different molecular chaperone families, but also have the highest count of orthologues and cofactors (Pahl *et al.*, 1997). Among the highly conserved eukaryotic chaperones are cytosolic and organellar Hsp70 and chaperonin proteins, Hsp90 and numerous associated chaperone cofactors, small Hsps, prefoldin, AAA ATPases, NAC, and various protein folding catalysts (Table I). The size of certain chaperone families is remarkable: for example, *S. cerevisiae* has 14 Hsp70 proteins, plants, and the nematode *Caenorhabditis elegans* have 15 or more small Hsps. Similarly, eukaryotes have experienced a tremendous expansion in the number of chaperone cofactors. The relative abundance of molecular chaperone families and cochaperones is likely to support the more diversified cellular processes existing in eukaryotes (Kelley, 1998; Rassow *et al.*, 1995). Eukaryotes also have organellar chaperones that either are specific to the domain (e.g., calnexin and calreticulin in the endoplasmic reticulum, which are involved in the folding and assembly of glycoproteins) or are of endosymbiotic origin (e.g., Hsp70 and cofactors, the Hsp60/Hsp10 chaperonin system and AAA ATPases) (Leroux and Hartl, 2000a).

The subunit complexity of many oligomeric chaperones is also strikingly greater in eukaryotes than in prokaryotes. For example, prefoldin contains six different subunits and the cytosolic chaperonin, termed CCT or TRiC, contains eight related proteins (the archaeal counterparts typically have two and one or two subunits, respectively). In the case of the latter two chaperones, the increase in complexity is likely to be necessary for accomodating specialized functions, namely, the folding of actins and tubulins, which are not present in prokaryotes (reviewed by Leroux and Hartl, 2000b; see also Section VI.B.4). The presence of multiple members of a chaperone family, especially as often occurs in eukaryotes, is not necessarily an indication of functional redundancy, however. Differences in gene expression (whether spatial, temporal, or conditional), in combination with variable substrate specificities can make seemingly redundant chaperones perform different functions (Leroux and Hartl, 2000a). A noteworthy example is that some *C. elegans* small Hsps are expressed only under heat-shock conditions, some are expressed only during certain developmental stages, and the tissue distribution of the various members varies significantly (Ding and Candido, 2000; Leroux *et al.*, 1997a; Linder *et al.*, 1996; Stringham *et al.*, 1992).

4. Molecular Chaperone Requirements and Genomic Complexity

From comparing the genomes of eukarya, archaea, and bacteria, it would appear as though there is not one single molecular chaperone family that is *strictly* conserved across all domains, except perhaps for a member of the PDI family (see Section XII.B). While Hsp70, small Hsps, and various related members of the AAA ATPase family are highly conserved, not all sequenced genomes encode these chaperones. Chaperonins, which assist *de novo* protein folding, are found in all eukaryotes and archaea. Up until now, GroEL and GroES chaperonin systems were also believed to be absolutely conserved in bacteria. Its absence in the recently sequenced genome of the mucosal bacterial pathogen *Ureaplasma urealyticum* poses a mystery that will need to be clarified (Glass *et al.*, 2000). Given the abundance of different molecular chaperone families in yeast and multicellular eukaryotes, it will be extremely interesting to scrutinize the sequence of the smallest known eukaryotic genome (2.9 million base pairs), that of the parasitic microsporidian *Encephalitozoon cuniculi* (Vivares and Metenier, 2000)—in search of a minimal set of chaperones in a compact eukaryotic genome. It should be noted, however, that the chaperone requirements of organisms does not appear to correlate with the complexity, or size, of the genome. For example, the number of different chaperones encoded by the relatively small *Thermoplasma acidophilum* archaeal genome [1.56 Mb (Ruepp *et al.*, 2000)] is essentially identical to that of *Halobacterium* sp. [2.57 Mb (Ng *et al.*, 2000)]. Similarly, the smallest genome for a prokaryote, that of the bacterium *Mycoplasma genitalium* [0.58 Mb (Fraser *et al.*, 1995)], encodes the same chaperones as those of *Chlamydia trachomatis* [almost twice the genome size at 1.05 Mb (Stephens *et al.*, 1998)]. On the other hand, *Rickettsia prowazekii* [1.10 Mb (Andersson *et al.*, 1998)] also includes HtpG and Hsp22 (a small Hsp). In conclusion, it appears likely that there is sufficient functional overlap in the known (and presently unknown) chaperones to make it unnecessary for organisms to have an all-encompassing set of molecular chaperone families.

III. Protein Folding *in Vitro* versus *in Vivo*

The term protein folding refers to the process a polypeptide undergoes to acquire a native conformation and thus biological activity. From a theoretical perspective, folding within a reasonable time span can be understood only by the fact that a limited number of common structural motifs, mainly β-sheets, α-helices, and loops, form and self-assemble cooperatively and hierarchically to make up the final product (Dinner *et al.*, 2000). These concepts, combined with emerging views of protein folding occurring along "folding landscapes," adequately resolve Levinthal's paradox, which puts forth that it would take an astronomical amount of time for proteins to sample all possible conformations to reach the native conformation (Levinthal, 1968; Karplus, 1997). Because of the technical challenges involved in studying protein folding *in vivo,* most of our knowledge regarding protein folding is derived from *in vitro* biochemical and biophysical experiments on model proteins (Radford, 2000).

A. Folding in the Test Tube

A seminal discovery, made in the early 1970s by Christian Anfinsen, was instrumental in propulsing the field of protein folding forward. By showing that ribonuclease A could be unfolded in a denaturant and subsequently refolded to a fully functional state by dilution in buffer, he demonstrated that the information contained within a polypeptide chain is sufficient to specify its native conformation (Anfinsen, 1973). In other words, protein folding and assembly could be assumed to occur without any external factors. Such spontaneous refolding of a protein in the test tube is a remarkable tool which biochemists and biophysicists use to understand folding principles, and such studies have led to the concept of protein folding intermediates and molten globules (nonnative proteins with significant tertiary structure), for example (Kuwajima and Arai, 2000; Rumbley *et al.*, 2001). Although numerous proteins can be refolded under appropriate conditions *in vitro* (e.g., at low protein concentrations and temperatures), unfolded proteins expose hydrophobic amino acids that make them highly prone to associate incorrectly both inter- and intra-molecularly, resulting in misfolding and aggregation (Gething and Sambrook, 1992). The importance of two critical events that facilitate the folding of newly made polypeptides in the cell started to be fully appreciated a mere decade before the turn of the century: the folding at the cotranslational level, and the participation of molecular chaperones (Ellis and van der Vies, 1991).

B. COTRANSLATIONAL AND MOLECULAR CHAPERONE-ASSISTED PROTEIN FOLDING

Proteins synthesized on the ribosome (or extruded through a membrane) must, because of physical constraints, fold in a gradual and sequential manner (i.e., cotranslationally). Secondary structure elements likely form spontaneously and self-assemble to create the subdomains and larger domains of fully synthesized proteins (Hartl, 1996; Kolb *et al.*, 1995). Because such secondary and supersecondary structures expose hydrophobic residues that are normally shielded when present in the native tertiary (and quaternary) conformation of the protein, off-pathway misfolding and aggregation events are possible, or perhaps even likely, in the crowded cellular milieu (Bukau *et al.*, 1996; Ellis and Hartl, 1999; Leroux and Hartl, 2000a). Indeed, the intracellular concentration of macromolecules (preexisting and nascent proteins, as well as nucleic acids) is estimated as 340 mg/ml in *E. coli*, and such crowding is predicted to increase the propensity for nonnative proteins to undergo off-pathway folding events (Ellis and Hartl, 1996; Minton, 2000). The ability of molecular chaperones transiently to bind and shield nonnative structures in proteins is ideally suited to assist the folding process: it prevents premature misfolding and aggregation of the nascent protein until it is entirely made and can fold to the native state (Fig. 1). Similarly, native proteins undergoing denaturation will have a strong tendency to aggregate unless they can be stabilized by chaperones and be permitted to enter a productive folding pathway (e.g., see Schröder *et al.*, 1993). There is now also evidence that chaperones of the Hsp100/Clp family can rescue proteins that have aggregated as a consequence of cellular stress (Glover and Lindquist, 1998; Motohashi *et al.*, 1999; Zolkiewski, 1999; Goloubinoff *et al.*, 1999; see also Section IX.A). Evidently, characterizing the structures and *in vitro* functions of different families of molecular chaperones can assist us in determining how protein folding occurs *in vivo*.

IV. Ribosome-Bound Molecular Chaperones: Trigger Factor and NAC

Some chaperones are bound to the ribosome in close proximity to the polypeptide exit tunnel and are thus perfectly positioned to stabilize nascent chains and assist their folding in a timely manner (Bukau *et al.*, 1996; Hartl, 1996; Rassow and Pfanner, 1996). Two such

factors associated with ribosomes are trigger factor and the nascent
polypeptide-associated complex (NAC).

A. TRIGGER FACTOR

All bacterial genomes sequenced thus far encode a ribosome-
associated protein, known as trigger factor (TF), that is not found in
archaea or eukaryotes (Hesterkamp and Bukau, 1996a; Stoller et al.,
1995). Being an FKBP-like peptidyl prolyl isomerase with an ability
to bind nonnative proteins, TF is both a classical protein folding cata-
lyst and a molecular chaperone (Hesterkamp and Bukau, 1996b; Scholz
et al., 1997; Stoller et al., 1996). Although the importance of TF in
catalyzing the cis–trans isomerization of proline residues in vivo has
not been conclusively demonstrated, this chaperone has been shown
to bind nascent polypeptides (Hesterkamp et al., 1996; Valent et al.,
1995), and more recent studies suggest that TF cooperates with another
nascent-chain binding chaperone, DnaK, to stabilize nascent chains and
assist de novo protein folding (Deuerling et al., 1999; Teter et al., 1999)
(Fig. 1). Although TF and DnaK are each dispensible for E. coli viabil-
ity, disruption of both leads to synthetic lethality. This suggests that
their functions, which are both directed toward assisting the folding of
nascent chains, are complementary and partially overlapping (reviewed
by Pfanner, 1999). Archaea lack a TF homologue, although all encode
at least one FKBP-type PPI (Section XII.A), and it has been noted that
those archaea which do possess an Hsp70 chaperone system also have
a PPI of the cyclophilin family (see Maruyama and Furutani, 2000, and
references within). The recently sequenced T. acidophilum genome represents an exception to this apparent codistribution of archaeal PPI and
DnaK, however (Ruepp et al., 2000). Regardless, it is unclear whether
archaeal PPI(s) are associated with the ribosome like TF, and it is possible that a ribosome-associated eukaryotic-like factor (NAC) may play
an analogous function in archaea.

B. NASCENT POLYPEPTIDE-ASSOCIATED COMPLEX

Eukaryotes, but not bacteria, encode a heterodimeric $(\alpha\beta)$ protein
complex that is known as the nascent polypeptide-associated complex,
or NAC (Wiedmann et al., 1994). Its reported functions are to prevent signal recognition particle (SRP) binding to inappropriate nascent chains
(i.e., those lacking a signal peptide) and prevent ribosomes translat-
ing nonsecretory polypeptides from interacting with the ER membrane
(Section XI.B) (reviewed by Rassow and Pfanner, 1996). Because it is

the first protein complex to interact with nascent chains after extrusion from the ribosome, it may also have a chaperone function comparable to TF in assisting *de novo* protein folding (Wang *et al.*, 1995).

The existence of an archaeal protein complex homologous to NAC has not been generally recognized, although one subunit was reported to exist in close proximity to a *T. acidophilum* zinc-containing ferrodoxin (Cosper *et al.*, 1999). Sequence homology searches using various eukaryotic α- and β-NAC protein sequences reveal that the eukaryotic α subunit of NAC has a highly conserved protein homologue in archaea (e.g., the hypothetical protein MTH177 from *M. thermoautotrophicum* is an α-NAC subunit). The β-NAC subunit either is not encoded in archaeal genomes or has diverged sufficiently to escape detection. An excellent study by the Wiedmann laboratory (Beatrix *et al.*, 2000) describing the properties of eukaryotic NAC revealed that although the separate subunits contribute to nascent-chain binding, and the α subunit alone can act to prevent ribosome binding to the ER membrane, both subunits are required for preventing SRP interactions with the nascent chain. Nevertheless, α-NAC can interact with nucleic acids on its own, including ribosomal RNA (Beatrix *et al.*, 2000). It is thus conceivable that archaeal α-NAC may function either alone or in association with a yet unrecognized partner in protein targeting and possibly in stabilizing nascent chains. Since the α-NAC subunit contains a domain with homology to the Hsp40/DnaJ chaperone and interacts with Hsp70 in eukaryotes (Beatrix *et al.*, 2000), eukaryotic and archaeal NAC may cooperate with the Hsp70 system (if present) during *de novo* protein folding in a manner similar to the synergistic action of TF with DnaK.

C. Other Possible Ribosome-Associated Chaperones

The small heat-shock protein HrpA of a *Mycobacterium*, a chaperone that also has archaeal homologues (Section VIII), apparently interacts with ribosomes (Tabira *et al.*, 1998), but this represents the only known example of such an association between this family of chaperones and ribosomes. Interestingly, *S. cerevisiae* possess one class of Hsp70 chaperone, Ssb, that is tightly associated with ribosomes (Pfund *et al.*, 1998). The physiological and functional significance of this is clear, given that ribosome binding is mediated by zuotin, a known Hsp40/DnaJ cofactor of Hsp70 (Yan *et al.*, 1998). Nonetheless, there is no evidence that Hsp70 systems form an integral part of the protein synthesis machinery in other eukaryotes, bacteria, or archaea that possess Hsp70. Since Hsp70 interacts with nascent polypeptide chains (see below), a close proximity to the polypeptide exit tunnel may not be obligatory.

V. The Hsp70 Chaperone System

The Hsp70 molecular chaperone family has received a tremendous amount of attention, due to its important and diverse functions under both normal and cellular stress conditions (Morimoto *et al.*, 1997; Schröder *et al.*, 1993). Its role in folding newly synthesized proteins in the eukaryotic and bacterial cytosol, as well as in eukaryotic organelles and endoplasmic reticulum, is largely believed to be essential (Bukau and Horwich, 1998; Craig *et al.*, 1995; Hartl, 1996; Rothman, 1989). Yet the recent sequencing of several archaeal genomes is challenging our present view of the importance of an Hsp70 system in *de novo* protein folding pathways.

A. Hsp70: A Nearly Ubiquitous Molecular Chaperone System

Searching for DnaK/Hsp70 homologues in fully sequenced archaeal genomes thus far reveals that four of seven archaea lack this molecular chaperone: *Archaeoglobulus fulgidus, M. jannaschii, Pyrococcus horikoshii,* and *A. pernix* are devoid of DnaK, whereas *Halobacterium* sp., *M. thermoautotrophicum,* and *T. acidophilum* possess one. In contrast, all eukaryotes contain multiple copies of Hsp70, and all sequenced bacterial genomes sport at least one DnaK homologue and its two known obligatory cofactors, DnaJ and GrpE. As expected, archaea that have DnaK invariably have DnaJ and GrpE homologues. It is striking that many of the above-mentioned archaea live in similar environments and that all are from the Kingdom Euryarchaeota (except for *A. pernix*, which is a crenarchaeote)—thus, the lack or presence of DnaK is not indicative of phylogenetic or environmental differences in the archaeal domain. The proposal that DnaK may have been acquired by some archaea via lateral gene transfers with bacteria therefore seems highly probable (Gribaldo *et al.*, 1999). Indeed, archaeal Hsp70 homologues are more closely related to the bacterial family of chaperones than they are to eukaryotic cytosolic homologues (Gupta and Golding, 1993). Given the importance of Hsp70 in *de novo* folding pathways, its absence in many archaea has correctly been described as paradoxical (Macario and Conway de Macario, 1999). The question is, can this paradox be resolved?

B. Role in *de Novo* Protein Folding and Other Cellular Processes

Hsp70 was first shown in 1990 by the Welch group to bind a substantial fraction of nascent polypeptide chains. Based on the previous finding that Hsp70 was implicated in posttranslational protein assembly and protein translocation across membranes, the authors proposed that Hsp70 facilitates the proper folding of newly made proteins (Beckmann *et al.*, 1990). Numerous additional reports of eukaryotic Hsp70 binding to proteins cotranslationally were subsequently published (Eggers *et al.*, 1997; Frydman *et al.*, 1994; Nelson *et al.*, 1992; Pfund *et al.*, 1998; Thulasiraman *et al.*, 1999), and after doubts that bacterial DnaK functioned in this capacity (e.g., see Fenton and Horwich, 1997), two papers describing the interaction of DnaK with nascent chains were published (Deuerling *et al.*, 1999; Teter *et al.*, 1999). In addition, both eukaryotic and bacterial Hsp70 proteins play important roles in maintaining the structural integrity of proteins under stress conditions by preventing aggregation and refolding (Arsene *et al.*, 2000; Hartl, 1996). Other assigned functions for Hsp70 include roles in protein translocation, degradation of unstable proteins and regulation of signaling molecule functions (Clarke, 1996; Hayes and Dice, 1996; Helmbrecht *et al.*, 2000; Mayer and Bukau, 1998; Morimoto *et al.*, 1997).

Hsp70 proteins consist of two major domains, namely, an ATPase domain that is structurally similar to actin (Bork *et al.*, 1992; Flaherty *et al.*, 1990; Harrison *et al.*, 1997; Sriram *et al.*, 1997) and a peptide binding domain (Zhu *et al.*, 1996). Hsp70/DnaK interacts with substrate proteins by binding short extended segments displaying a core of three to five hydrophobic amino acids (Blond-Elguindi *et al.*, 1993; Rudiger *et al.*, 1997; Zhu *et al.*, 1996). The structures of the peptide binding domain with and without bound peptide have been solved (Morshauser *et al.*, 1999; Zhu *et al.*, 1996). The hydrophobic peptide is bound in an extended conformation within a β-sandwich subdomain that is closed off by side loops and an α-helical "lid" (Fig. 2A). The affinity of substrates for DnaK is modulated by conformational changes in the ATPase domain (reviewed by Bukau and Horwich, 1998; Mayer and Bukau, 1998). When charged with ATP, DnaK has a relatively high on and off rate of substrate binding and release, which corresponds to an opening of the α-helical segment (Palleros *et al.*, 1993; Schmid *et al.*, 1994). DnaJ, a cofactor of DnaK, is itself a molecular chaperone that can transfer substrates to DnaK (Langer *et al.*, 1992) (Fig. 2B). The crystal structure of Hsc20 (a DnaJ homologue) has now been determined (Cupp-Vickery and Vickery, 2000). The protein contains two separate domains, a J domain representative of DnaJ/Hsp40 proteins and a protein-binding domain that may act as a scaffold to facilitate delivery of proteins to Hsc66, a DnaK homologue. DnaJ/Hsp40 has the additional role of accelerating the ATP hydrolysis rate of DnaK, locking the bound substrate in the cleft (reviewed by Kelley, 1999; Mayer and Bukau, 1998). Substrate release then depends on GrpE, the DnaK nucleotide exchange factor (Harrison *et al.*, 1997).

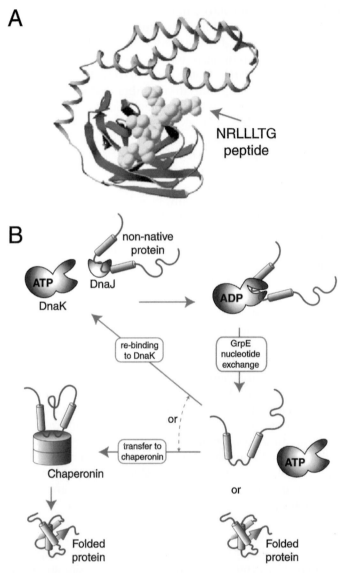

FIG. 2. Structure of a DnaK–peptide complex and functional cycle of DnaK/Hsp70 systems. (A) Hsp70 consists largely of peptide binding and ATPase domains separated by a short linker region. The DnaK–peptide structure shows the peptide containing a hydrophobic core (NH$_2$–NRLLLTG–COOH) tightly bound in a β-sandwich subdomain bordered by loops and an α-helical "lid" segment. (B) DnaJ, itself a molecular chaperone, transfers nonnative proteins to the ATP-bound, substrate-accepting form of DnaK. DnaJ promotes DnaK ATP hydrolysis, which transduces information to the peptide binding domain and increases its affinity for the substrate. GrpE, a nucleotide exchange factor, allows the substrate to leave DnaK, either in a folded conformation or for subsequent folding by Hsp70 again or other chaperones (e.g., a chaperonin).

On the whole, the functional cycle and substrate binding ability of Hsp70 is consistent with its role in protein folding, where it transiently interacts with and stabilizes short hydrophobic stretches in nascent or denatured proteins—allowing the substrate to fold to the native state or transferring it for further folding by a chaperonin (Fig. 2B).

C. THE ARCHAEAL HSP70 SYSTEM

As with its eukaryal and bacterial counterparts, the archaeal Hsp70 system from *Methanosarcina mazei* is induced under stress (heat-shock) conditions, possibly indicating a role in preventing the aggregation and refolding of denatured proteins (Lange *et al.*, 1997; Macario *et al.*, 1991). The strict cooccurrence of archaeal DnaK with DnaJ and GrpE confirms that the chaperone system is functional (Conway de Macario *et al.*, 1994). Other than these observations, little is known about the cellular function(s) of archaeal Hsp70; in fact, its absence in many archaeal clades suggests that it plays a nonessential, albeit probably beneficial function in this domain (Gribaldo *et al.*, 1999). A further implication is that either its function is not required for stabilizing nascent chains or another cellular protein(s) has(have) a similar function(s) in archaea (Macario and Conway de Macario, 1999). While the first possibility cannot be discounted, the recent analysis of a novel protein complex termed prefoldin/GimC provides evidence that a different, conserved archaeal chaperone may assist *de novo* protein folding in a manner that is consistent with an Hsp70-like function.

VI. Prefoldin

A. TWO PATHS OF DISCOVERY

Despite the availability of complete genome sequences from various organisms, numerous open reading frames that may encode chaperones await characterization, either after selection based on sequence similarity (where, to a large degree, function can be anticipated) or after unforeseen discovery in genetic, expression, or biochemical screens. A novel molecular chaperone family was recently uncovered by such gene-hunting screens.

Using *S. cerevisiae* harboring a mutation in γ-tubulin, the Schiebel laboratory searched for genes whose disruption lead to synthetic lethality (Geissler *et al.*, 1998). As could be expected, two genes encoding spindle-pole body proteins known to interact with γ-tubulin were isolated. Five additional genes encoding a family of coiled-coil proteins were also found. Biochemical and genetic analyses of the GIM proteins, an acronym for Genes Involved in Microtubule biogenesis, revealed that they assemble into a common heterooligomeric complex and are involved in the biogenesis of tubulins and probably also actins. A genetic interaction with CCT was also observed, suggesting potential chaperone-like functions in cytoskeletal protein folding/assembly (Geissler *et al.* 1998).

Shortly thereafter, the Cowan laboratory reported on their biochemical isolation of a bovine testis protein chaperone that could form a stable binary complex with unfolded actin and tubulin and release them to the cytosolic chaperonin CCT for folding (Vainberg *et al.*, 1998). Remarkably, the heterooligomeric complex contained orthologues of the five yeast GIM proteins plus one additional related protein that was later observed in the purified yeast GIM complex (Siegers *et al.*, 1999). The apparent necessity for the chaperone-bound actin to be transferred to CCT for folding suggested a function and a name for the complex: prefoldin (PFD), a play-on-word for pre(before)-folding (Vainberg *et al.*, 1998).

B. CLASSIFICATION AND DISTRIBUTION OF PREFOLDIN PROTEINS

1. Eukaryotes

All eukaryotes examined to date possess six PFD/GIM proteins, named PFD1 to -6 or GIM1 to -6 (Geissler *et al.*, 1998; Leroux *et al.*, 1999; Siegers *et al.*, 1999; Vainberg *et al.*, 1998). Orthologues of all six genes are found in *S. cerevisiae, C. elegans, Drosophila melanogaster, Arabidopsis thaliana,* and mammals. For consistency, a unified PFD nomenclature is used here.

2. Archaea

Inspection of archaeal genome sequences reveals the presence of two types of PFD subunits: one is most closely related to eukaryotic PFD5, and the other is most similar to PFD6. Sequence comparisons and phylogenetic analyses show that the six eukaryal PFD proteins can be grouped into two separate classes, α and β. PFD3 and -5 belong to the α class, whereas PFD1, -2, -4, and -6 all fall into the β class (Fig. 3A) (Leroux *et al.*, 1999). Based on this grouping, archaeal PFD5 was named PFDα, and PFD6 is referred to as PFDβ. Interestingly, *M. jannaschii* contains two divergent PFDα subunits, similar to the situation in eukaryotes (Leroux *et al.*, 1999). Crystallographic, biochemical, and genetic studies on archaeal PFD confirmed the classification as well as the structural and functional relatedness of archaeal and eukaryal PFD complexes (see below).

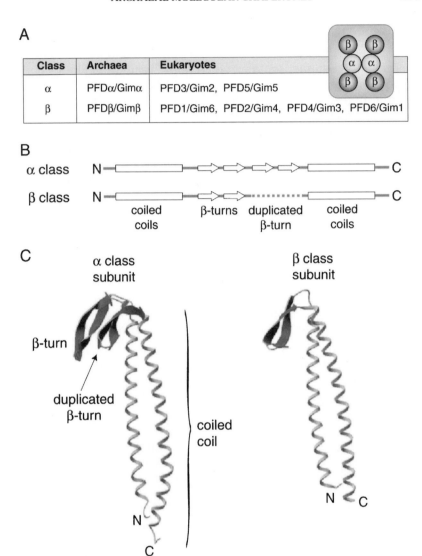

FIG. 3. Classification, domain structures, and tertiary structures of prefoldin subunits. (A) Primary sequence and phylogenetic analyses reveal that archaeal and eukaryal prefoldin subunits can be grouped into two separate classes (α and β). Archaea generally possess one member from each class (PFDα and PFDβ). Eukaryotic prefoldin has two members of the α class and four members of the β class. A schematic representation shows the organization of archaeal and eukaryal prefoldin subunits into a hexameric complex. (B) Secondary structure predictions demonstrate that all prefoldin subunits possess a central core of β-turns that is flanked by long, uninterrupted coiled coils. α-class subunits contain an additional β-turn involved in dimerization. (C) The crystal structures of archaeal prefoldin subunits agree well with the secondary structure predictions and show that the coiled coil regions form intramolecularly.

3. Bacteria

An α-class PFD gene (PFD5) is present in *Aquifex aeolicus,* but *Thermotoga maritima,* another thermophilic bacterium, lacks a coding region homologous to PFD. Given that PFD is not present in any other bacteria, it seems likely that the *Aquifex* sequence was acquired via lateral gene transfer (Leroux *et al.,* 1999). It is somewhat perplexing that its coding region not only has been retained, but has not degenerated relative to archaeal and eukaryal PFD5; it appears to encode a functional subunit, but without a β-class subunit, it would not be expected to be functional as a chaperone (see Section VI.D.3). Contrary to the previous report that *A. pernix* lacked an α-class PFD subunit (Leroux *et al.,* 1999), this organism does appear to encode a divergent PFDα protein, lending further proof that a functional archaeal complex requires at least one α-class, and one β-class subunit. The possibility that an individual PFD subunit support a chaperone function in *A. aeolicus* must be investigated before any firm conclusions are drawn.

4. Cooccurrence of Prefoldin and Group II Chaperonins and the Emergence of Eukaryotes

As alluded to previously, CCT and PFD share the task of assisting the biogenesis of actins and tubulins and appear to do so collaboratively. The significance of this complementarity in function is further highlighted by the fact that PFD and Group II chaperonins (CCT in the eukaryotic cytosol and thermosome in archaea) coexist strictly within the same cellular compartments (Leroux *et al.,* 1999; Vainberg *et al.,* 1998). Furthermore, the expansion in the number of eukaryotic PFD subunits parallels that of CCT. There are six PFD subunits and eight CCT subunits in eukaryotes, compared to typically two PFD subunits and one or two thermosome subunits in archaea. The increase in subunit complexity of eukaryotic CCT is likely to be required for actin and tubulin folding, a function not supported by the homooligomeric bacterial GroEL/GroES chaperonin system (Leroux and Hartl, 2000b; Tian *et al.,* 1995). Similarly, archaeal PFD cannot replace eukaryotic prefoldin *in vivo* (Leroux *et al.,* 1999).

Because of the essential role of CCT as well as PFD in cytoskeletal protein biogenesis, it is tempting to speculate that the coevolution and subunit diversification of CCT and PFD may have been instrumental to the emergence of an ur-eukaryote from its progenitor. Since a cytoskeleton is one of the most important and defining attribute of eukaryotes compared with prokaryotes, it can be argued that without archaeal-like chaperones (PFD and the CCT-related thermosome), eukaryotes may never have come into existence (Leroux and Hartl, 2000b; see also Willison, 1999). On the other hand, it is widely accepted that eukaryotes were the joint product of a symbiotic event between an archaeum and a bacterium (Doolittle, 1999). It is thus noteworthy that most bacteria contain both FtsA and FtsZ, cell-division proteins that are structurally related to actin and tubulin, respectively (Lowe and Amos, 1998; van Den Ent and Lowe, 2000). Archaea, on the other hand, apparently do not make use of FtsA, although nearly all have an FtsZ homologue [*A. pernix* is an exception (Faguy and Doolittle, 1999)]. Thus, the transfer of bacterial FtsA to an ur-eukaryotic genome that already possessed an FtsZ homologue (perhaps from archaea) may have been a critical piece of the puzzle needed to evolve the first 'modern' eukaryotic cell (Faguy and Doolittle, 1998).

C. Function of Prefoldin in Eukaryotes and Archaea

Most of the *in vivo* functional data on PFD have been obtained from studies of the *S. cerevisiae* chaperone complex, with additional information derived from cell-free translation studies. Conversely, essentially all of the biochemical and structural studies have made use of archaeal PFD, which represents a simplified model system amenable to structure–function studies.

1. Eukaryotic Prefoldin: A Specialized Molecular Chaperone?

Disruption of individual PFD genes most noticeably cause cytoskeletal protein phenotypes, including hypersensitivity to the microtubule depolymerizing drug benomyl and to the actin inhibitor latrunculin A, as well as osmotic and cold sensitivity (Geissler *et al.*, 1998; Vainberg *et al.*, 1998). Similar phenotypes are also observed with many temperature-sensitive mutants of CCT subunits at the restrictive temperature (reviewed by Stoldt *et al.*, 1996). Unlike the lethality observed with yeast CCT gene disruptions (Ursic and Culbertson, 1991; Chen *et al.*, 1994), however, yeast survive null mutations in one or more PFD subunits (Geissler *et al.*, 1998; Vainberg *et al.*, 1998; Siegers *et al.*, 1999).

In an effort to understand the *in vivo* function of PFD, the folding pathway of actin was characterized with respect to folding kinetics and interaction with chaperones in CCT or PFD mutant backgrounds. The folding of actin can be monitored by performing "pull-downs" with beads cross-linked to deoxyribonuclease I (DNase I), which tightly binds folded actin (Lazarides and Lindberg, 1974). Pulse-chase experiments showed that newly synthesized actin from yeast and mammalian cells folds with a half-time of approximately 1 min (Siegers *et al.*, 1999). Because CCT binds and releases newly made actin with the same half-time, the actin must be in a native or near-native conformation when it is released from the chaperonin. Heterologous expression of a GroEL mutant "trap" that can bind nonnative actin (and other proteins) but not release it has no effect on the folding of newly made actin (Siegers *et al.*, 1999; Thulasiraman *et al.*, 1999). In contrast, newly synthesized actin made under nonpermissive conditions in a temperature-sensitive CCT mutant, or in the presence of an amino acid analogue (which causes protein misfolding), can be seen to interact with the GroEL chaperonin trap (Siegers *et al.*, 1999; Thulasiraman *et al.*, 1999). These findings suggest that in normal circumstances, the chaperonin can specifically sequester nonnative forms of the actin from the bulk cytosol, preventing off-pathway folding reactions that would make folding inefficient.

What happens to actin biogenesis when PFD function is abrogated? The result is that as with mutant CCT, the chaperonin trap can bind to newly produced actin (Siegers *et al.*, 1999). DNase I binding studies further revealed that the rate of actin folding was slowed about five-fold, directly demonstrating that PFD enhances the efficiency of actin folding *in vivo*. Importantly, the rate of actin release from CCT was diminished to the same extent (Siegers *et al.*, 1999). Because both yeast and mammalian PFD are able to interact physically with CCT (Siegers *et al.*, 1999; Vainberg *et al.*, 1998), the above results imply that PFD behaves as a cofactor of the eukaryotic chaperonin in ways that are similar to how GroES acts as a cofactor of GroEL: PFD sequesters nonnative proteins from the cytosol in conjunction with the chaperonin, and it accelerates the folding efficiency of the ATP-dependent, cylindrical chaperone. Whether PFD caps CCT during folding, in effect redirecting nonfolded forms of the chaperonin substrates back to the chaperonin, or accelerates the chaperonin ATPase activity, is unknown and must be further investigated.

Although PFD appears to function in close cooperation with the chaperonin, it appears likely that it can also interact directly with stalled, truncated actin and tubulin nascent chains produced in a rabbit reticulocyte translation system somewhat before CCT binds to these chains (Hansen *et al.*, 1999). Thus, it is likely that PFD interacts with its substrates in a cotranslational manner, as does Hsp70/DnaK, and then transfers the nonnative proteins to a chaperonin for folding. The same study also provided evidence that PFD is a specialized chaperone that interacts strictly with actins and tubulins, although it may assist the biogenesis of other target proteins. One likely candidate is the tumor suppressor von Hippel–Lindau (VHL), which interacts with PFD and requires CCT for its biogenesis (folding/assembly) (Tsuchiya *et al.*, 1996).

2. Archaeal Prefoldin: A General Chaperone Like Hsp70?

If eukaryotic PFD facilitates the biogenesis of actins and tubulins in cooperation with CCT, what is PFD doing in archaea? This is a fair question: prokaryotic genomes do not encode these cytoskeletal proteins, and thus archaeal PFD must interact with a different set of cellular proteins. One possible target would be the tubulin homologue, FtsZ, but newly made *M. thermoautotrophicum* FtsZ does not appear to interact with PFD from the same organism in an *in vitro* translation system (unpublished observation). Prefoldin does bind a wide variety of denatured proteins *in vitro*, however (Leroux *et al.*, 1999). Denatured lysozyme, mitochondrial rhodanese, actin, *T. acidophilum* glucose dehydrogenase, and firefly luciferase can be prevented from aggregating, or can form complexes, with archaeal PFD. This suggests that the archaeal chaperone may have the ability to recognize a broad range of nonnative proteins in the cell (Leroux *et al.*, 1999). It will be important to analyze what the physiological substrates of PFD (and the chaperonin) are in archaea.

There is an intriguing parallel between the function of Hsp70 and that of PFD during *de novo* protein folding. Both chaperones interact with nascent chains and can release incompletely folded polypeptides to a chaperonin for folding to the native state (Hansen *et al.*, 1999; Leroux *et al.*, 1999; Vainberg *et al.*, 1998). The similarity does not extend to their mechanism of action, as Hsp70 binds and releases hydrophobic peptide fragments in an ATP- and cofactor-dependent reaction cycle. There is no evidence that PFD uses nucleotides for binding or releasing its substrates. Moreover, it is possible that the PFD hexamer does not bind to short extended hydrophobic segments of nonnative proteins as Hsp70 does but, rather, interacts with either whole proteins or subdomains of larger proteins (Hansen *et al.*, 1999; Siegert *et al.*, 2000; see also below). Regardless of their precise functional mechanism, the overall contribution of Hsp70 and PFD to a chaperonin-dependent protein folding pathway may be similar. Indeed, this hypothesis may help to explain the apparent paradox that the otherwise ubiquitous Hsp70 chaperone system is not present in all archaea, as PFD may replace it. Where both archaeal DnaK and PFD coexist, as is the case in eukaryotes, their functions may overlap, perhaps in a manner similar to that of DnaK and the ribosome-bound chaperone trigger factor. Exploring the relative contribution of DnaK and PFD to *de novo* protein folding in archaea will necessitate disrupting either gene, or both, simultaneously. Such genetic manipulations should eventually become possible (Tumbula and Whitman, 1999).

D. STRUCTURE AND FUNCTIONAL MECHANISM OF PREFOLDIN

1. Structural Model for Archaeal and Eukaryal Prefoldin

Producing a basic model for the quaternary structure of PFD was greatly facilitated by the fact that the two archaeal subunits (α and β) could be produced in *E. coli* and assembled into a functional complex *in vitro* by simply mixing the two proteins together (Leroux *et al.,* 1999). The complex formed has, by size exclusion chromatography, the same apparent molecular weight (~200 kDa) as do native archaeal PFD and eukaryotic PFD. The archaeal α and β subunits in the complex were found to be at a 1:2 ratio, and the molecular mass of the complex was determined to be 84 kDa by sedimentation velocity analysis. Thus, archaeal PFD is an $\alpha_2\beta_4$ hexamer (Leroux *et al.,* 1999), and the presence of any significant amount of mixed oligomeric complexes (e.g., $\alpha_3\beta_3$) was ruled out by mass spectrometric analysis of the native complex (Fändrich *et al.,* 2000). Cross-linking analyses revealed that the archaeal hexamer is apparently assembled from PFDβ monomers that interact with a PFDα dimer "core" (Leroux *et al.,* 1999). These data are consistent with a model in which eukaryotic and archaeal PFD complexes contain a dimer core of α-class subunits (PFD3 and -5, or PFDα) flanked by four β-class subunits (PFD1, -2, -4 and -6, or PFDβ) (Fig. 3A).

2. Structure of Archaeal Prefoldin

Sequence analysis and secondary structure predictions strongly suggested that all PFD subunits are built of a central β-strand containing region flanked by two coiled coils of approximately the same length (Fig. 3B) (Leroux *et al.,* 1999). Elucidating the crystal structure of *M. thermoautotrophicum* PFD bore out these predictions (Siegert *et al.,* 2000). The α-class subunit has two β-hairpins, one more than the β-class subunit. This extra β-hairpin, believed to have been duplicated during evolution from the neighboring strands, couples another subunit to form the PFDα dimer core. From these β-hairpins extend two unbroken α-helices that form coiled coils 60–70 Å in length (Fig. 3C). Coiled coils are found in many other proteins, such as tropomyosin and DNA polymerase I (see review by Lupas, 1996) and possess a degenerate heptad repeat (*a*–*g*) with usually hydrophobic residues in the *a* and *d* positions.

The PFD hexamer has a singular quaternary structure not previously observed in other molecular chaperones or protein complexes in general (Fig. 4). The overall shape is best described as that of a jellyfish or octopus with six "tentacles." The body, or core, of the complex consists of a unique double β-barrel structure formed by the β-hairpins of all six subunits. From this base protrude six tentacle-like coiled coils formed by the joining of the N- and C-terminal α-helical regions. The coiled coils make contacts with residues from the β-barrel near the base but are otherwise structurally independent from each other. Indeed, the packing of PFD in the crystal provides evidence that the coiled coils are likely to have significant flexibility in solution (Siegert *et al.,* 2000).

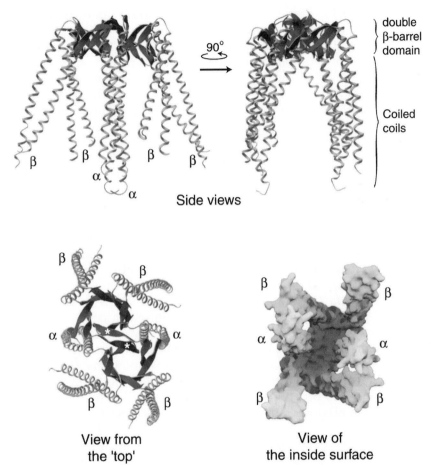

FIG. 4. Quaternary structure of prefoldin from the archaeum *M. thermoautotrophicum.* The overall organization of the hexameric complex is unique and is best described as that of a jellyfish or octopus with six "tentacles." The body, or core, of the complex is an oligomerization domain of eight β-turns forming a double β-barrel structure. The two β-strands labeled with white stars allow α-subunit dimerization; other β-strands form the rest of the β-barrel (as shown in the "view from the 'top'" structure). From this core protrude six α-helical coiled coils that are well solvated with hydrophilic residues and are structurally independent of each other. A molecular surface representation of the cavity formed by the coiled coils (lower-right structure) is shown.

3. Molecular Mechanism of Prefoldin

Many chaperones expose hydrophobic residues that are involved in binding to apolar side chains in nonnative substrate proteins (Fenton *et al.,* 1994; Xu *et al.,* 2000; Zhu *et al.,* 1996). Since archaeal PFD binds a wide range of unfolded proteins, any apolar patches on PFD might reflect nonspecific binding sites. Molecular surface representations of the PFD structure show that most of the solvent-accessible surface of PFD is hydrophilic in character (Siegert *et al.,* 2000). There are, however, two areas where hydrophobic patches are exposed: one is at the bottom of the "cavity" formed by the six coiled coils (Fig. 4; "view of the inside surface"), and the others are near the distal ends of the coiled coils, where partial unwinding of the helices exposes part of the hydrophobic interface.

Truncating the distal regions (approximately four helical turns) of the four PFDβ coiled coils completely abrogates the ability of the chaperone complex to bind and stabilize an unfolded protein (Siegert *et al.,* 2000). Similarly, truncation of the two PFDα coiled coils (leaving the PFDα subunits intact) results in a severe reduction in chaperone function. These observations suggest that PFD makes use of multisubunit (multivalent) interactions of the coiled-coil end regions with unfolded proteins. Although the functional mechanism of PFD is unique, chaperonins are also believed to stabilize nonnative proteins on their apical domains through interactions with multiple subunits (Farr *et al.,* 2000; Llorca *et al.,* 1999a, 2000). As with CCT, which has eight types of apical domain surfaces on which actins, tubulins, and other proteins can interact specifically, eukaryotic PFD has six subunits that could expose actin- and tubulin-specific binding sites. The expected flexibility of the tentacles in solution may allow PFD to adapt to protein substrates of varying sizes or conformations. Indeed, archaeal PFD can bind and stabilize a small denatured protein (lysozyme, 14 kDa), as well as unfolded firefly luciferase, a large, 62-kDa protein.

In conclusion, PFD appears to interact with and stabilize newly synthesized proteins before transfering them to a chaperonin for subsequent folding to the native state. Although eukaryotic PFD appears to interact with and cooperate with the chaperonin during the folding reaction, there is, so far, no evidence that archaeal PFD interacts with the archaeal chaperonin (Leroux *et al.,* 1999).

VII. Chaperonins

Chaperonins belong to the Hsp60 family of molecular chaperones, a group of ~60 kDa proteins that typically assemble into double-ring toroidal structures and are involved in assisting protein folding and assembly (Bukau and Horwich, 1998; Gutsche *et al.,* 1999; Hartl, 1996; Sigler *et al.,* 1998). A protein complex known as Rubisco subunit Binding Protein, or RuBP, was the first chaperonin to be characterized (reviewed by Ellis and Hemmingsen, 1991). It was only in 1988, however, that Hemmingsen *et al.* (1988) uncovered a striking similarity between RuBP and the bacterial protein, GroEL, and coined the term "chaperonin." Although much of what is known today about chaperonins is based on studies of the chaperonin system from *E. coli,* there is now strong evidence that the Group II chaperonins from archaea and eukarya, which are closely related, probably function in a markedly different manner.

A. Two Phylogenetically Distinct Chaperonin Families

1. Group I Chaperonins

Together with the mitochondrial homologue Hsp60, GroEL and RuBP are classified under the unifying category of Group I chaperonins (Willison and Kubota, 1994). All have a common bacterial origin, and an oligomeric structure consisting of heptameric rings, and function in cooperation with a saucer-shaped cofactor, GroES or Hsp10, that covers the chaperonin cylindrical cavity during the ATP-driven protein folding reaction (Fig. 5, top) (reviewed by Braig, 1998; Fenton and Horwich, 1997; Hartl, 1996; Sigler *et al.,* 1998). Of these, the *E. coli* GroEL and GroES chaperonin system has emerged as the archetype for understanding chaperonin-mediated protein folding.

2. Group II Chaperonins

A few years after the recognition of chaperonins as a novel chaperone family, the archaeal chaperonin counterpart was discovered by accident; upon lysing the thermophilic archaeum *Pyrodictium occultum* and examining its contents by electron microscopy, double-ring toroids that were shown to possess an ATPase activity and were speculated to be chaperonin-like archaeal protein complexes, were observed (Phipps *et al.,* 1991). The sequence of this protein was not known, but that of another related toroidal ATPase from *Sulfolobus shibatae* was determined and found to possess the ability to bind unfolded proteins, as expected for a molecular chaperone (Trent *et al.,* 1991). Surprisingly, it showed limited sequence homology to Group I chaperonins but was more closely related to the eukaryotic protein *t*-complex polypeptide-1 (TCP-1).

The function of the *tcp-1* gene, found in lower eukaryotes to mammals, was not clear at the time, although it was shown to be involved in microtubule-mediated processes in *S. cerevisiae* and speculated that it may participate in the assembly of microtubules (Ursic and Culbertson, 1991). Three separate publications in 1992 then reported that the *tcp-1* gene product was part of a heterooligomeric chaperonin complex

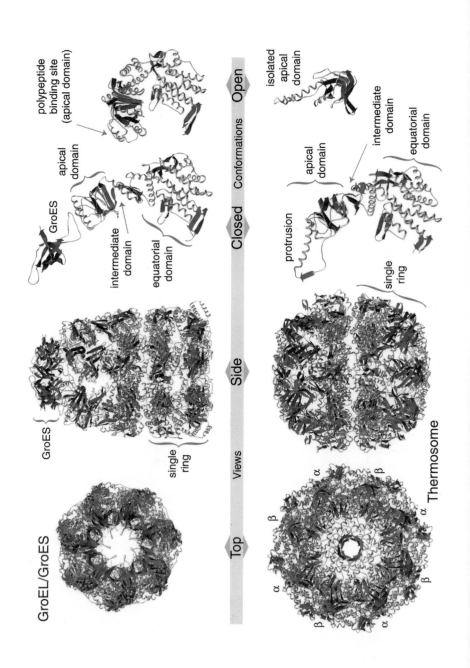

GroEL/GroES

GroES

single ring

Top   Side   Views

Thermosome

single ring

α   β

β   β

α   α

β   α

Open   Closed   Conformations

polypeptide binding site (apical domain)

apical domain

GroES

intermediate domain

equatorial domain

isolated apical domain

intermediate domain

equatorial domain

apical domain

protrusion

capable of assisting the folding of actin and tubulin (Frydman *et al.*, 1992; Gao *et al.*, 1992; Lewis *et al.*, 1992). The archaeal and eukaryotic cytosolic chaperonins are now commonly called thermosome and CCT (chaperonin-containing TCP-1; also termed TRiC, TCP-1 ring complex, and c-cpn), respectively. Some surprises came with the discoveries of thermosome and CCT: the rings contained eight or nine subunits, rather than seven for Group I chaperonins, and CCT was built from a total of eight related protein subunits (GroEL is homo-oligomeric) (reviewed by Gutsche *et al.*, 1999). Archaeal thermosome contains one or two subunits, although a third subunit may be present in *Sulfolobus* chaperonins (Archibald *et al.*, 1999). Thermosome subunits are approximately 40% identical to members of the CCT family (Kubota *et al.*, 1994). These observations indicate that there is another type of chaperonin (termed Group II) in the archaeal and eukaryal cytosols (Willison and Kubota, 1994). So far, only the crystal structure of the thermosome from *T. acidophilum* has been solved (Ditzel *et al.*, 1998) (Fig. 5, bottom).

A captivating story will likely emerge from studies on a novel, thermosome-related chaperonin in humans. Recently, mutations in a previously uncharacterized chaperonin gene were found to be linked to the McKusick–Kaufman (MKKS) and Bardet–Biedl (BBS) syndromes, which together predispose individuals to several ailments, including retinal dystrophy, polydactyly (additional fingers and toes), diabetes, and cardiomyopathy (Katsanis *et al.*, 2000; Slavotinek *et al.*, 2000; Stone *et al.*, 2000). The oligomeric structure, cellular function, and role of the Group II chaperonin in developmental pathways affecting MKKS and BBS patients will need to be clarified.

B. CELLULAR FUNCTIONS OF CHAPERONINS

1. Structure and Function of the GroEL/GroES (Group I) Chaperonin System

The basic functional mechanism of Group I chaperonins has been resolved largely as a result of observing the structural dynamics of

---

FIG. 5. Structural similarities and differences among a Group I chaperonin system, GroEL/GroES from *E. coli* (top), and an archaeal Group II chaperonin ($\alpha\beta$-thermosome) from *T. acidophilum* (bottom). The first two structures on the left represent top and side views of the two types of chaperonins, respectively. The two structures on the right are single chaperonin subunits in the closed and open conformations, respectively. In the case of GroEL, the closed structure has GroES bound to it, whereas the open structure is apo-GroEL. In the thermosome, the closed structure has so-called "protrusions" covering the central cavity, whereas in the open form, the protrusions face up and leave the cavity open (only the apical domain structure is available).

GroEL during the different stages of its ATP- and GroES-driven folding cycle, as well as by site-directed mutagenesis, biochemical, and biophysical studies (reviewed by Fenton and Horwich, 1997; Hartl, 1996; Sigler et al., 1998). Crystal structures of apo-GroEL, of GroEL bound to a non-hydrolyzable ATP analogue, and of the active GroEL/GroES protein folding complex (Boisvert et al., 1996; Braig et al., 1994; Xu et al., 1997) define two basic states of the chaperonin: an acceptor state, which can bind nonnative substrates, and a protein folding complex that forms a closed folding compartment (Fig. 5). In contrast to DnaK, which binds short hydrophobic segments in nascent polypeptides, GroEL interacts with polypeptides that are fully synthesized and are best described as molten globules (i.e., incompletely folded, compact proteins which display a significant amount of secondary and tertiary structure) (Hayer-Hartl et al., 1994; Martin et al., 1991). The chaperonin binding site has been mapped to the apical domain, which exposes hydrophobic residues that line the opening of the cylinder (Fenton et al., 1994) (Fig. 5). The acceptor state can be either apo-GroEL or GroEL having ADP and GroES bound on the trans side of the chaperonin (the GroEL ring opposite the substrate binding cis ring) (Sigler et al., 1998) (Fig. 6). After polypeptide binding, seven ATP molecules occupy the cis equatorial (ATPase) domains of the chaperonin, and this is followed by the coaxial docking of the GroES heptameric ring to the cis ring. Large conformational changes in the cis ring accompany ATP and GroES binding, resulting in the displacement of the bound substrate to the interior of a cavity (Fig. 6).

The large en bloc conformational changes in the chaperonin, an upward movement and rotation of the apical domain within the bilobal protein, produce an interior surface that is considerably more hydrophilic compared to its original state (Xu et al., 1997). The polar microenvironment provided by the chaperonin, in effect comparable to creating a situation of infinite protein dilution where aggregation is prevented, is believed to be conducive for protein folding. While the sequestration of aggregation-prone folding intermediates may be the primary driving force behind chaperonin-assisted protein folding, there is now also evidence that a partial unfolding of the polypeptide or rearrangement of protein domains may contribute to the folding process by "unscrambling" kinetically-trapped folding intermediates (Coyle et al., 1999; Shtilerman et al., 1999).

Much of what is known about the function of Group I chaperonins is derived from studies using various model substrates which are not typically found in bacteria. Recently, an effort to catalog some of the endogenous (natural) substrates of GroEL has been made. Large-scale

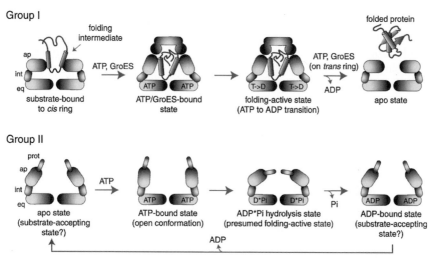

FIG. 6. Functional cycles of Group I and Group II chaperonins. For simplicity, only the cis rings of the two chaperonins are shown; in the case of Group II chaperonins, the conformational changes occurring in the opposite (trans) ring are unclear. Group I chaperonins (e.g., *E. coli* GroEL) bind polypeptides via mostly hydrophobic residues lining the cavity, either in the apoprotein (apo) state or with ADP and GroES bound in the trans ring. ATP and GroES binding causes large conformational changes and results in the displacement of the bound substrate into a hydrophilic cavity conducive for protein folding. Following ATP hydrolysis and binding of ATP and GroES on the trans ring, the protein is released in a folded or near-native form. A tentative ATPase cycle of Group II chaperonins, as deduced from studies on the thermosome, is shown at the bottom. The apo and ADP-bound states of the chaperonin are likely to be substrate-accepting states (substrate not shown). Upon binding ATP, the thermosome appears to enlarge in size, probably as a result of an "opening" of the apical domains and protrusions. Closure of the chaperonin cavity by the protrusions is likely to occur during the hydrolysis of ATP (represented as ADP*Pi), after which P$_i$ is released and the chaperonin is ready for another round of folding.

immunoprecipitations of GroEL–substrate complexes led to the identification of ~50 proteins involved in varied cellular processes and provided further evidence that about 10% of all *E. coli* proteins exploit the chaperonin for their biogenesis (Houry *et al.*, 1999; see also Ewalt *et al.*, 1997). Experimentally identifying some of the natural substrates for all known chaperones, as was done for GroEL and DnaK as well (Mogk *et al.*, 1999), may help us understand not only the specificity of chaperones but also what structural determinants make certain proteins more likely to require chaperone assistance. For example, a motif consisting of two or more $\alpha\beta$-folds was found in many of the GroEL substrates, suggesting that such motif-bearing proteins may be more prone to misfolding or aggregating when produced *de novo* (Houry *et al.*, 1999).

2. The Group II Chaperonin System

Most Group II chaperonins accommodate eight subunits in each ring and, thus, display eightfold symmetry in the crystal structure or under electron microscopic examination (Ditzel *et al.*, 1998; Llorca *et al.*, 1999b). *Sulfolobus* thermosomes possess nine-membered rings (Marco *et al.*, 1994), consistent with the recent finding that the organism has three thermosome proteins, likely assembling in a $(\alpha\beta\gamma)_3$ arrangement (Archibald *et al.*, 1999). The most surprising difference between Group I and Group II chaperonins is that the latter proteins possess an insertion in their apical domain (Bosch *et al.*, 2000; Ditzel *et al.*, 1998; Klumpp *et al.*, 1997). The crystal structure of the isolated apical domain from the *T. acidophilum* thermosome revealed a helical protrusion from the apical domain that is clearly absent in Group I chaperonins (Fig. 5, compare top and bottom structures) (Klumpp *et al.*, 1997). The suggestion that the protrusion could serve as a gate mimicking the presence of a GroES-like cofactor was correct, as the crystal structure of the whole $\alpha\beta$-thermosome from the same organism revealed the same protrusion closing the chaperonin chamber in an iris-like fashion (Ditzel *et al.*, 1998) (Fig. 5). Interestingly, the different crystal structures of the thermosome α and β apical domains, in addition to the closed thermosome structure, show that the protrusions exhibit some conformational variability (Bosch *et al.*, 2000). Overall, the built-in lid of Group II chaperonins provides further evidence that chaperonin function is closely linked to their ability to sequester nonnative substrates completely from the bulk cytosol. As with GroEL, the inside of the thermosome cavity is hydrophilic and is thus conducive to protein folding (Ditzel *et al.*, 1998).

a. Conformational Cycle of Group II Chaperonins. The eukaryotic cytosolic chaperonin CCT is predicted to have the same overall fold as the thermosome except for the fact that it is built from eight related subunits. All CCT subunits appear to possess the protrusions seen in the thermosome (Klumpp *et al.*, 1997). Clearly, the archaeal Group II chaperonin is an excellent model system for understanding CCT function. Apart from a number of structural studies performed with CCT (see Carrascosa *et al.*, 2001; Liou and Willison, 1997; Llorca *et al.*, 1998, 1999b; reviewed by Gutsche *et al.*, 1999), much of what is known about the conformational changes that occur during its ATPase cycle has been obtained from the analysis of the *T. acidophilum* thermosome (Gutsche *et al.*, 2000a–c). An outline of the ATP-driven thermosome cycle is shown in Fig. 6 (bottom). Little information regarding substrate–thermosome interactions is available. It is likely that the apo- or ADP-bound chaperonin can interact with substrates, based on studies of the eukaryotic cytosolic chaperonin (Melki and Cowan, 1994; Melki et al., 1997). Upon ATP binding, a marked enlargement of the chaperonin cylinder appears to take place (Gutsche et al., 2000a). The closed conformation of the chaperonin, which is likely to be the active-folding state, has been suggested to occur immediately after the hydrolysis of the γ phosphate in ATP and before release of the phosphate (Gutsche et al., 2000a) (Fig. 6). The large change in conformation resulting from the cleavage of the bond has no parallel in Group I chaperonins, although the latter's active-folding state coincides with ATP hydrolysis (Sigler et al., 1998) (Fig. 6).

b. A Different Polypeptide Binding Site?   While the apical domain extensions present in Group II chaperonins are highly suggestive of a GroES-like function in cytosolic compartments lacking GroES, their proposed role in substrate binding, based on their hydrophobic character (Ditzel et al., 1998; Klumpp et al., 1997), remains to be demonstrated experimentally. This proposition stems from the observation that the hydrophobic apical domain residues of GroEL, which mediate substrate binding, are not conserved and are not all hydrophobic in both thermosome and CCT subunits (Ditzel et al., 1998; Klumpp et al., 1997). In the case of the eukaryotic chaperonin, an electron microscopic reconstruction of CCT bound to nonnative actin shows that the binding site is likely to be below the apical domain protrusions (Llorca et al., 1999a) (i.e., below the protrusion as seen in the open conformation of the isolated thermosome apical domain) (Fig. 5). Furthermore, CCT–substrate interactions would not be expected to involve only hydrophobic residues given the apparent specificity of the interactions with actins and tubulins (Leroux and Hartl, 2000b). It has also been shown that polar, as well as charged residues, play a role in substrate binding to GroEL (e.g., see Chen and Sigler, 1999).

   There are at present few substrate binding studies published for the thermosome that provide valuable information regarding its mode of interaction with nonnative proteins or refolding mechanism. In one study, Guagliardi et al. (1995) found that the Sulfolobus solfataricus archaeal chaperonin possesses two surfaces (I and II) that are spatially separated and are differentially affected by ATP binding—in this case, surface I is apparently hidden, whereas surface II is open for substrate binding. Whether the protrusions form one of these surfaces remains to be seen. Similarly, there is at present no information on thermosome substrates, although the same thermosome was shown to prevent the aggregation of various proteins in vitro, as well as promote the folding of these proteins (Guagliardi et al., 1994). Thermosomes from other organisms have also been shown to possess the ability to prevent the aggregation of various proteins (reviewed by Gutsche *et al.,* 1999). The relative abundance of the archaeal chaperone suggests that it may be important in folding a large number of cellular proteins. CCT, the eukaryotic homologue, was once believed to fold only actins and tubulins, but there is growing evidence that its substrate repertoire is not limited to these cytoskeletal proteins and that it may interact with the same proportion of proteins as does GroEL in bacteria (reviewed by Leroux and Hartl, 2000b).

VIII. Small Heat-Shock Proteins

Small Hsps have at times been referred to as "junior chaperones" (Jakob and Buchner, 1994), perhaps not only because of their lower molecular weights, but because their mechanism of action is less well studied and understood relative to that of their "senior" counterparts, Hsp70 and chaperonins. This designation is not justified, as they are highly ubiquitous chaperones, forming a diverse family of 12- to 43-kDa proteins that are present in the eukaryotic cytosol and organelles (mitochondria and chloroplasts), in most bacteria, and in all archaea examined to date (Table 1).

A. Function of Small Hsps

1. Eukaryotic and Bacterial Small Hsps

The common *in vitro* functional attribute of small Hsps from the three domains is their ability to bind and stabilize efficiently proteins that are unfolded and would otherwise embark on an aggregation pathway (Chang *et al.,* 1996; Horwitz, 1992; Jakob *et al.,* 1993; Kim *et al.,* 1998b; Leroux *et al.,* 1997b). In the cell, their function is important in protecting organisms under stress conditions and contributing to thermotolerance (adaptation to more severe stresses) (Landry *et al.,* 1989; Rollet *et al.,* 1992; Yeh *et al.,* 1994). Not surprisingly, many small Hsps (although not all) are highly induced under stress conditions (e.g., see Stringham *et al.,* 1992). Eukaryotic small Hsps such as Hsp25, Hsp27, and α-crystallins appear to perform additional, specific functions. For example, α-crystallins play a critical role in maintaining the structural and functional integrity of the eye lens (Horwitz, 2000), and mutations in the gene not only lead to cataract formation but also to desmin-related myopathy as a result of the accumulation of aggregated desmin (an intermediate filament) in muscles (reviewed by Clark and Muchowski, 2000).

While the ability of small Hsps to protect nonnative proteins from aggregating is well established, their involvement in *refolding* proteins is controversial (Chang *et al.*, 1996; Leroux *et al.*, 1997b). Complicating matters further, there is evidence that a small Hsp (αB-crystallin) may use ATP to enhance this process (Muchowski and Clark, 1998; Muchowski *et al.*, 1999a), but such an intriguing possibility will require substantiation by examining this property in a wide range of family members. Paradoxically, ATP inhibits the ability of a plant small Hsp to refold citrate synthase (Smykal *et al.*, 2000)—the very same substrate that was shown to be more efficiently folded in the presence of ATP by human αB-crystallin (Muchowski and Clark, 1998). The genome sequence of the archaeum *T. acidophilum* has been reported to encode three small Hsps, including one (Ta0437) containing an ATPase domain homologous to *E. coli,* ArsA, the catalytic component of a pump that extrudes arsenite and antimony from the cytosol (Ruepp *et al.*, 2000). Inspection of the protein sequence, however, reveals no homology to the α-crystallin domain, signifying that it is not a *bona fide* small Hsp.

A plausible mechanism for the function of small Hsps has recently been suggested. Three independent research groups found that the refolding of substrates bound to small Hsps necessitated additional components, in the form of an Hsp70 chaperone and possibly also a chaperonin. Thus, denatured firefly luciferase bound to plant Hsp18 is efficiently refolded only in the presence of rabbit reticulocyte lysate (a rich source of chaperones) and ATP (Lee *et al.*, 1997), mammalian Hsp25-bound substrate is reactivated in the presence of Hsp70 and ATP (Ehrnsperger *et al.*, 1997), and a bacterial small Hsp (IbpB) releases bound substrate to the DnaK/DnaJ/GrpE chaperone system, after which folding to the native state requires GroEL/GroES (Veinger *et al.*, 1998). The cellular function of small Hsps could therefore be to capture and stabilize efficiently a large pool of denatured proteins that can be subsequently reactivated following the cellular stress (Leroux and Hartl, 2000a). In cases where unfolding proteins escape such a protective mechanism, Nature has also evolved a means to recover the function of the aggregated proteins (see Section IX.A).

2. Archaeal Small Hsp Function

There is only one publication that addresses the function of an archaeal small Hsp, a 16.5-kDa protein from *M. jannaschii* (Kim *et al.*, 1998b). Consistent with the known function of small Hsps, Hsp16.5 can suppress the aggregation of a wide range of proteins under heat-shock conditions: two purified proteins (single-chain monellin and citrate synthase) and the majority of proteins from an *E. coli* extract were protected from precipitating at high temperatures in the presence of the chaperone but not in its absence. Assuming that archaeal small Hsps also need to cooperate with an Hsp70 chaperone system to refold bound proteins, what happens in archaea lacking this ATP-dependent chaperone system? One possibility is that archaeal proteins are transferred to thermosome for folding or that another type of chaperone can be used. To address this question, it would be interesting to purify a refolding factor from archaea lacking Hsp70 using an assay that consists of a refoldable substrate bound to a small Hsp.

B. STRUCTURE OF SMALL HSPS

The majority of small Hsps examined to date form large protein complexes that have highly polydisperse oligomeric states. The complexes range between about 200 and 800 kDa and form roughly globular structures as judged by electron microscopic examination (Behlke *et al.*, 1991; Haley *et al.*, 2000). The overall organization of all small Hsps is identical and can be divided into three domains (Caspers *et al.*, 1995) (Fig. 7A). The N-terminal domain is poorly conserved between family members and is highly variable in size. The central region of small Hsps is invariably the α-crystallin domain, an ~85-amino acid region that displays strong sequence conservation (Ingolia and Craig, 1982). The C-terminal region of small Hsps is often termed an "extension" and, as with the N-terminal domain, varies in size and is highly divergent in sequence.

The large sizes and dynamic structures of small Hsps have made them extremely difficult targets for structural analysis by X-ray crystallography or NMR. For example, attempts to solve the structure of an apparently monomeric small Hsp failed (Leroux *et al.*, 1997a, unpublished observation), and the structure of a seemingly well-defined nonameric small Hsp from a *Mycobacterium* has not been reported (Chang *et al.*, 1996). However, not long after the sequence of the first archaeal small Hsp was revealed by sequencing the genome of *M. jannaschii*, the first crystal structure of a small Hsp was reported (Kim *et al.*, 1998a) (Figs. 7B and C). What likely made this feat possible is that Hsp16.5 is a homogeneous multimer whose structure is "locked" into a crystallizable form below the optimal growth temperature of the thermophilic host (60°C). Similarly, the crystallization of *M. thermoautotrophicum* PFD (Siegert *et al.*, 2000) was likely possible because of the same reason.

Hsp16.5 assembles into a spherical structure containing 24 subunit monomers. The protein complex displays two-, three-, and fourfold symmetry and displays six square and eight triangular windows (two views of the complex are shown in Figs. 7B and C). The inside of the sphere is occupied with the N-terminal region immediately preceeding the conserved α-crystallin domain (Fig. 7D), although it is likely

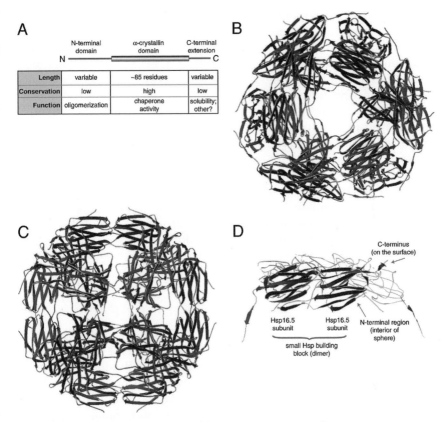

FIG. 7. Small heat-shock protein domain organization and quaternary structure. (A) All members of the family possess a conserved central region about 85 residues in length termed the α-crystallin domain. This domain is flanked by an N-terminal domain of variable size required for multimerization and by a nonconserved C-terminal extension. (B) View of the small Hsp spherical cage built of 24 subunits, shown along the threefold axis of symmetry. (C) Small Hsp oligomer shown along the fourfold axis of symmetry. (D) View of the small Hsp dimeric building block in a surface cutaway view of the oligomer. The N- and C-terminal regions face the interior and exterior of the spherical complex, respectively.

to be disordered and was not observed in the crystal structure. This N-terminal region is involved in multimerization, as its removal from *C. elegans* Hsp16-2 and α-crystallin results in smaller complexes of significantly smaller sizes (dimers–tetramers) (Leroux *et al.,* 1997b; Merck *et al.,* 1992). The C-terminal extensions of individual archaeal small Hsp monomers are exposed on the surface of the oligomer, where they make contacts with neighboring subunits (Fig. 7D). The role of this extension in stabilizing the protein complex appears fairly certain, although there are mixed reports as to whether it plays a crucial function in binding unfolded proteins (Fernando and Heikkila, 2000; Leroux *et al.*, 1997b). The highly conserved α-crystallin domain is clearly important for substrate binding, however (Bova *et al.*, 1999; Dai *et al.*, 2000; Muchowski *et al.*, 1999b). Studies aimed at identifying substrate binding sites, such as locating where the binding of the hydrophobic probe bis-ANS occurs, suggest that a number of regions in this domain may be involved in binding to hydrophobic regions in nonnative proteins (reviewed by Leroux and Hartl, 2000a). The ratio of small Hsp monomer to substrate protein is 1:1 for the archaeal chaperone (Kim *et al.*, 1998b), as well as for α-crystallin, bacterial, plant, and nematode small Hsps (Chang *et al.*, 1996; Das *et al.*, 1996; Lee *et al.*, 1997, and unpublished observations). Together, the above findings are consistent with small Hsps binding proteins on the denaturation pathway mainly through surface interactions with the α-crystallin domain (Haslbeck *et al.*, 1999; Lee *et al.*, 1997), stabilizing proteins for subsequent refolding by ATP-dependent chaperones.

C. CHAPERONES WITH THE IMMUNOGLOBULIN FOLD

It should be noted that the structure of Hsp16.5 monomers resembles the immunoglobulin fold and is also related to that of the bacterial periplasmic chaperones FimC/PapD (Holmgren and Branden, 1989; Kim *et al.*, 1998b). The two chaperones are clearly not homologous, however. The periplasmic chaperones do not form large oligomeric structures, and unlike the general chaperone function of small Hsps, they have evolved the highly specialized task of chaperoning pilus subunits to their site of assembly on the outer membrane (reviewed by Sauer *et al.*, 1999). The functional mechanism of FimC/PapD chaperones is now well understood and is very elegant: their immunoglobulin-like structure lacks a peripheral β-strand that accepts a donor strand from a pilus subunit and releases it to the pilus structure for assembly with another β-strand (Choudhury *et al.*, 1999; Sauer *et al.*, 1999). Interestingly, the small Hsp dimeric building block (Fig. 7D) also includes an extension which contacts a neighboring subunit via a β-strand interaction (Saibil, 2000). Remarkably, the secreted mammalian chaperone termed clusterin appears to interact with a large number of extracellular protein substrates and contains a region with limited sequence homology to the α-crystallin domain (Wilson and Easterbrook-Smith, 2000). Although the structure of clusterin, a 75- to 80-kDa heterodimeric protein containing a coiled-coil region, is strikingly different from that of small Hsps, its mechanism of action may be similar.

IX. Hsp100/Clp/AAA ATPase Chaperones: Disassembly, Folding, Unfolding, and Role in Degradation

The AAA, or closely related AAA$^+$, family of ATPases plays important roles in a multitude of cellular processes (for recent reviews, see Langer, 2000; Patel and Latterich, 1998; Vale, 2000). The AAA ATPases (ATPases Associated with different cellular Activities) form a diverse family of usually hexameric ring complexes found in all three domains of life. They share one or two evolutionary conserved ~250-amino acid sequences that contain Walker A and B motifs essential for ATP binding and hydrolysis. Numerous members have now been shown to possess molecular chaperone activity, an intriguing feature given that many (although not all) are also associated with protease complexes.

A. The Hsp100/Clp Family

S. cerevisiae encodes a hexameric molecular chaperone of the Hsp100 family (Hsp104) that apparently has few close homologues except for *Arabidopsis thaliana* Hsp101 (Lee *et al.,* 1994), although the sequence- and structure-related HslU and Clp members (e.g., ClpA, ClpB, and ClpX) are found in most bacteria and are also present in eukaryotes (Porankiewicz *et al.,* 1999). Susan Lindquist's laboratory first reported the importance of Hsp104 in thermotolerance or, more specifically, in protecting yeast from thermal and other types of cellular stresses (Sanchez and Lindquist, 1990; Sanchez *et al.,* 1992). Her group later showed that the *in vivo* function of Hsp104 function is not to prevent protein aggregation during stresses (as do small Hsps, Hsp70, and chaperonins) but, rather, is to mediate the disassembly/resolubilization of aggregated proteins (Glover and Lindquist, 1998; Parsell *et al.,* 1994). The Hsp70 chaperone system is required in such protein "revivals" (Glover and Lindquist, 1998). Importantly, this cooperation between chaperones is also observed with bacterial ClpB, which can resolubilize aggregated proteins in cooperation with the DnaK chaperone system (Goloubinoff *et al.,* 1999; Mogk *et al.,* 1999; Motohashi *et al.,* 1999; Zolkiewski, 1999). Furthermore, two additional Clp members, ClpA and ClpX, are known to facilitate the disassembly of various protein complexes into active monomers in a manner analogous to how Hsp104 and ClpB disassemble aggregated proteins and are, thus, classified as molecular chaperones [see Schirmer *et al.* (1996) and Porankiewicz *et al.* (1999) and references within].

ClpA and ClpX also associate in a symmetry-mismatched fashion with a heptameric protease termed ClpP (Grimaud *et al.,* 1998). Together, the chaperone–protease complexes are involved in the degradation of various substrates (Horwich *et al.,* 1999; Wickner *et al.,* 1999). The emerging view as to how this occurs is that the chaperone binds to the protein to be degraded and, in an ATP-dependent process, promotes its unfolding and threading into the cylindrical protease compartment for degradation (Singh *et al.,* 1999, 2000; Weber-Ban *et al.,* 1999). As discussed below, such a function is similar to that of the proteasomes of archaea and eukaryotes. Crystal structures of a bacterial Hsp100 family chaperone, HslU, have recently been solved in a complex with HslV, its proteasome-like partner (see Fig. 8 for the structure) (Bochtler *et al.,* 2000; Sousa *et al.,* 2000).

Oddly, archaea appear to lack Hsp100/Clp family members (Hsp100, Clp, or HslU). This is somewhat surprising, given that all sequenced archaeal genomes contain SppA, a ClpP-like protease (Bolhuis *et al.,* 1999). Similarly, archaeal genomes contain the HslV protease but not HslU, its chaperone partner. It is thus likely that other AAA ATPases present in archaea fulfill the chaperone function in such chaperone–protease systems.

B. ARCHAEAL AAA ATPASES

1. Proteasome Activating Nucleotidase (PAN)

The major protease of eukaryotic cells, termed proteasome, specifically degrades proteins that are covalently bound to ubiquitin, a protein tag that is absent from prokaryotes. The 26S proteasome consists of a 20S protease particle assembled from four stacked heptameric rings formed from seven different α and β subunits ($\alpha_7\beta_7\beta_7\alpha_7$). The first proteasome structure to be solved (Lowe *et al.,* 1995), from *T. acidophilum,* is closely related to its eukaryotic counterpart (Groll *et al.,* 1997) except for its simpler subunit composition (it has single α and β subunits) (Fig. 8). The eukaryotic 19S regulatory particle, associated with both ends of the 20S core, is the site where ubiquitylated proteins bind and are presumed to be unfolded and threaded into the degradation chamber (Ferrell *et al.,* 2000). It includes six AAA ATPases (Rpt 1–6) that are likely to form a ring complex. These proteins, believed to have unfolding activity, were recently shown also to exhibit chaperone-like activity in suppressing the aggregation and refolding of denatured citrate synthase (Braun *et al.,* 1999).

A homologue with approximately 45% sequence identity to the six eukaryotic Rpt subunits also exists in archaea (Zwickl *et al.,* 1999). The initial characterization of Rpt from *M. jannaschii* revealed that it associates with the archaeal proteasome and greatly stimulates its proteolytic

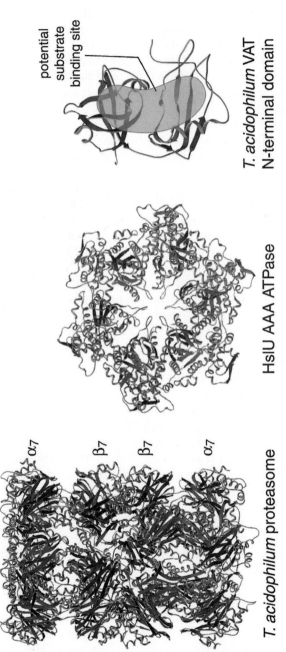

potential substrate binding site

T. acidophilum proteasome

α_7
β_7
β_7
α_7

HslU AAA ATPase

T. acidophilum VAT N-terminal domain

Fig. 8. Archaeal proteasome and AAA ATPase molecular chaperones. The proteasome from *T. acidophilum* consists of four stacked heptameric rings with the core β subunits involved in proteolysis. Bacterial HslU is the first Hsp100/AAA ATPase structure to be solved. This protein assembles into a hexamer, as with most or all AAA ATPases, and associates with a protease termed HslV in a manner similar to that in which the archaeal PAN chaperone associates with the proteasome. The NMR structure of VAT, an AAA ATPase chaperone from *T. acidophilum*, reveals a groove on the surface of the molecule (highlighted) that is positively charged and has been proposed by Coles *et al.* (1999) to represent the substrate binding site. *T. acidophilum* lacks PAN, and thus the VAT hexamer or its related homologue, VAT-2, may replace it.

activity (Wilson *et al.*, 2000; Zwickl *et al.*, 1999). It was termed PAN, for Proteasome-Activating Nucleotidase (Zwickl *et al.*, 1999). A recent study from the Goldberg laboratory demonstrated that as with the 19S regulatory particle, PAN has the ability to prevent the aggregation of unfolded proteins as well as to promote their refolding (Benaroudj and Goldberg, 2000). The authors suggest that this chaperone function may occur independently of the proteasome. In association with the proteasome, PAN can also unfold proteins in an ATP-dependent manner, as with the ClpAP and ClpXP chaperone–protease complexes, and this process is linked to substrate degradation by the proteasome (Benaroudj and Goldberg, 2000).

2. Valosine-Containing Protein (VCP)-like ATPase (VAT)

Interestingly, *T. acidophilum* does not contain PAN, raising the possibility that another chaperone may play a similar, or redundant, function (Ruepp *et al.*, 2000). One of the candidates is VAT or its close homologue, VAT-2 (Golbik *et al.*, 1999). Archaeal VAT is related to yeast Cdc48 and mammalian p97, AAA ATPases involved in membrane fusion processes and ubiquitin-dependent degradation, and to an unwinding factor that functions in DNA replication (reviewed by Patel and Latterich, 1998; Vale, 2000). VAT has both foldase and unfoldase activities, depending on the activity of the ATPase, which can be modulated by altering the Mg^{2+} concentration (Golbik *et al.*, 1999).

The crystal structure of the hexameric p97 AAA ATPase ring complex has been solved (Zhang *et al.*, 2000), and cryoelectron microscopy has revealed that upon ATP binding, substantial conformational changes to the complex occur (Rouiller *et al.*, 2000). The NMR structure of the N-terminal domain of VAT, lacking the two AAA ATPase domains, has also been solved (Fig. 8) (Coles *et al.*, 1999). The kidney-shaped N-terminal domain structure is nearly identical to that of the homologous domains in p97, NSF, and Sec 18 [the latter two proteins are eukaryotic AAA ATPases (Babor and Fass, 1999; May *et al.*, 1999)]. The authors propose that a positively charged cleft formed along the surface of the N-terminal domain, which has chaperone activity on its own, may be the substrate binding site (highlighted in the VAT structure in Fig. 8).

3. Lon Protease

Two proteases containing AAA ATPase domains, Lon and FtsH, are found in bacteria and eukaryotic organelles (Suzuki *et al.*, 1997). The single polypeptide chains are likely to possess both molecular chaperone activity (mediating the disassembly, folding, and oligomerization of proteins) as well as their confirmed proteolytic activity (Suzuki *et al.*,

1997). All archaeal genomes sequenced to date, except that of *A. pernix,* encode Lon homologues. There is no evidence of FtsH homologues, which are associated with membranes and mediate the destruction of membrane (and possibly other) proteins (Langer, 2000).

In conclusion, protease-associated AAA ATPases are molecular chaperones that likely serve two opposed cellular functions using one common mechanism to accomplish their goals. ATP binding/hydrolysis is used to drive conformational rearrangements in the hexamer, and this leads to either the partial or the total unfolding of the tethered protein substrate. Such an action can cause the disassembly of protein aggregates and their folding, either with or without the participation of other molecular chaperones. Conversely, unfolding can facilitate the transfer of the bound protein to the linked protease for degradation. The molecular basis for sorting out the fate of the substrates, an apparent polypeptide quality control mechanism which is very poorly understood, should prove to be a fascinating area for future research. The use of simpler archaeal model systems for these studies could prove to be highly valuable.

X. Hsp90

Hsp90 is an abundant chaperone present in bacteria and in the cytosol and endoplasmic reticulum of eukaryotes. It has been studied extensively in eukaryotes, where its diverse functions include modulating the conformation of signaling proteins, assisting protein folding under stress conditions, and facilitating protein degradation (see the excellent review by Csermely *et al.,* 1998). In *E. coli,* the Hsp90 homologue known as HtpG appears to play an important role in stress tolerance (possibly in collaboration with other chaperones, such as DnaK and ClpB) (Tanaka and Nakamoto, 1999; Thomas and Baneyx, 1998; Thomas and Baneyx, 2000). The complete absence of an Hsp90-like chaperone in archaea is thus somewhat surprising. On the other hand, small Hsps, thermosome, prefoldin, and AAA ATPase chaperones may perform analogous functions in protecting archaea from cellular stress. An absolutely essential role for HtpG in extreme environments is unlikely, as the hyperthermophilic bacteria *A. aeolicus* and *T. maritima* do not encode this chaperone (Deckert *et al.,* 1998; Nelson *et al.,* 1999).

XI. Protein Translocation

A large fraction of the proteins synthesized in the cytosol must traverse lipid bilayers to reach their final destination. These extracytoplasmic proteins are normally translocated across a proteinaceous core complex consisting of evolutionarily conserved components. Further-more, a cellular machinery must be present to ensure that the proteins are maintained in a translocation-competent (extended) conformation for translocation (reviewed by Schatz and Dobberstein, 1996). There are two major translocation machineries found in the three domains, a Sec-dependent system and a signal recognition particle (SRP)-dependent system.

A. The Sec-Dependent Translocation System

In bacteria, SecY, SecE, and SecG form the core, integral mem-brane components of the translocation apparatus. In the archaeal do-main, these three components are more closely related to their eukary-otic homologues than they are to the bacterial proteins (Eichler, 2000; Pohlschröder et al., 1997). Indeed, cross-domain comparisons proved to be helpful in identifying two of the proteins: no apparent homologues of bacterial SecY and SecG could be found in archaea, although proteins with sequence similarity to the eukaryotic $Sec61\alpha$ and $Sec61\beta$ homo-logues were uncovered (Pohlschröder et al., 1997).

In the Sec-dependent pathway of bacteria, a molecular chaperone known as SecB targets newly synthesized preproteins to SecA, an ATPase which associates with SecYEG (Driessen et al., 1998). The re-cently solved crystal structure of SecB revealed a homotetramer with a hydrophobic groove along its surface and an acidic region at the ex-treme end of the molecule that is likely to be the SecA binding site (Xu et al., 2000). Since many archaea do not possess an Hsp70 sys-tem that may function analogously to SecB, it is surprising that archaea also lack the SecB chaperone (only M. jannaschii has one, apparently obtained via lateral gene transfer). Furthermore, bacterial SecA func-tions as a translocation ATPase that provides the energy during protein translocation (Lill et al., 1989). In eukaryotes, various Hsp70 homo-logues accomplish this energy-dependent function (Schatz and Dob-berstein, 1996). It is thus highly intriguing that archaea also lack SecA (Eichler, 2000; Pohlschröder et al., 1997). The search for an archaeal pro-tein with a SecA or Hsp70-like function in translocation could reveal a novel protein(s) that may be conserved in bacteria and/or eukaryotes and be essential in archaea.

B. The SRP-Dependent Translocation System

A second protein translocation apparatus operates in all organ-isms (Eichler, 2000; Pohlschröder et al., 1997). In eukaryotes, nascent polypeptides bearing an endoplasmic reticulum (ER) signal sequence are targeted to the ER membrane via the signal recognition particle (SRP). Eukaryotic SRP is a ribonucleic acid–protein complex that binds the nascent polypeptide, causing the ribosome to arrest translation, and interacts with its membrane-bound receptor (Bui and Strub, 1999). The bacterial equivalents of this system are simpler and comprise Ffh, a homologue of the 54-kDa SRP subunit and FtsY, a homologue of the SRP receptor α subunit (Dalbey and Kuhn, 2000).

Archaea possess at least two components homologous to eukaryal SRP, namely, SRP19 and SRP54 (Althoff et al., 1994; Moll et al., 1999). In addition, all sequenced archaea possess SRP receptors (α subunit). Together, the archaeal SRP-dependent translocation system is more similar to its eukaryotic counterpart than it is to the bacterial system (Pohlschröder et al., 1997). This should not be surprising, given that the discrimination between bacterial SRP and Sec-dependent pathways invoves the selective recognition of nascent chains by SRP and trigger factor, the ribosome-bound molecular chaperone (Beck et al., 2000). TF is not present in archaea (see Section IV.A), but a subunit of NAC, which is similarly involved with SRP in discriminating nascent chains that lack or contain a targeting signals in eukaryotes, is also present in archaea (Section IV.B). Analyzing the minimal protein translocation machinery of archaea may thus be extremely useful for our understanding of the eukaryotic translocation system.

XII. Protein Folding Catalysts

A. PEPTIDYL-PROLYL ISOMERASE

Three classes of peptidyl-prolyl isomerases (PPIases), cyclophilin, FK506 binding protein (FKBP), and parvulin, have been characterized in eukaryotes and bacteria (Maruyama and Furutani, 2000). Trigger factor is a chaperone and PPI with limited sequence homology to FKBP (Callebaut and Mornon, 1995). Similarly, the periplasmic parvulin of E. coli, SurA, also has the ability to interact with nonnative proteins (Behrens et al., 2001). Thus, some PPIases have both the ability to catalyze prolyl residue cis–trans isomerization and to function as a "classical" molecular chaperone.

In archaea, FKBP-type PPIases are ubiquitous, parvulin-type PPIases are absent, and only two members of the cyclophilins are known [one from Halobacterium cutirubrum has been characterized and M. thermoautotrophicum encodes a homologue (reviewed by Maruyama and Furutani, 2000)]. Although the role of archaeal PPIases remains obscure, it is noteworthy that an archaeal FKBP has been found to exhibit chaperone-like activity in refolding denatured mitochondrial rhodanese, a commonly used substrate for molecular chaperone assays (prevention of aggregation and renaturation) (Furutani *et al.,* 2000). Interestingly, the chaperone activity of the 17-kDa protein from *M. thermoautotrophicum* was found to be independent of its PPI activity (Furutani *et al.,* 2000), mirroring what has been observed with *E. coli* TF (Scholz *et al.,* 1997). Whether such archaeal chaperones interact with nascent chains, or function mostly under physiological conditions or cellular stresses, remains to be seen.

B. Protein Disulfide Isomerase

Protein disulfide bond formation is a rate-limiting step in protein folding that is catalyzed by protein disulfide isomerases (PDIases). These protein folding catalysts form a diverse family of proteins found in both prokaryotes and eukaryotes (Ferrari and Soling, 1999). The crystal structure of a PDI member from the hyperthermophilic archaeon *Pyrococcus furiosus* revealed a structural similarity to eukaryotic PDI, having two domains containing the thioredoxin fold (Ren *et al.,* 1998). Archaea possess a variety of proteins containing thioredoxin domains (e.g., TrxA, which is one of the few chaperones conserved in *all* sequenced genomes), but few biochemical studies on archaeal PDI proteins have been reported. In one important study, it was shown that a PDI from *M. jannaschii* showed greatly enhanced thermotolerance and somewhat higher specific activity relative to its *E. coli* homologue (Lee *et al.,* 2000). Interestingly, an *E. coli* PDI, DsbG, has recently been shown to prevent the aggregation of misfolded proteins (Shao *et al.,* 2000). At least two archaea (*A. fulgidus* and *Halobacterium sp.* NRC-1) also have DsbG homologues, apparently acquired via lateral gene transfer events. As with archaeal PPIases, the *in vivo* function of PDIases has not been demonstrated and it is not known whether they are essential cellular components. Clearly, establishing genetic systems for performing gene knockouts in archaea remains an achievable goal that will provide essential insights into the importance of cellular proteins involved in protein folding.

XIII. Protein Folding in Extremophiles and the Protective Function of Molecular Chaperones

A. Adaptation to Extreme Environments

Hyperthermophilic, halophilic (salt-loving), and psychrophilic (cold-loving) proteins are naturally adapted to the cellular milieu inside their host. Surprisingly, only relatively minor changes in the amino acid compositions and structures of proteins are required to adapt proteins to such extreme conditions (reviewed by Scandurra *et al.,* 2000). For example, proteins from thermophilic archaea may possess any one (or more) of the following characteristics relative to their mesophilic counterparts: a somewhat more compact and hydrophobic core, shorter solvent-exposed loops, and slightly larger polar surface areas. On the other hand, the stability of native structures in extreme environments can be tremendously affected by single amino acid substitutions. With this in mind, what happens when a particular protein is not in a native, stably folded state in an inhospitable cellular environment?

B. General Molecular Chaperone Requirements

The question as to whether proteins found in extremophiles require additional safeguarding from molecular chaperones seems sensible (Ladenstein and Antranikian, 1998). Proteins may be more prone to misfold and aggregate under exceptionally inclement conditions. Molecular chaperones have been shown to be key players in an organism's defense against environmental insults (Morimoto *et al.,* 1997). The types of agents which induce a stress response are highly varied (e.g., temperatures significantly higher or lower than the optimal growth temperature of the organism, heavy metals, free oxygen radicals, and other reactive substances) but have one universal characteristic: they are proteotoxic and can cause protein denaturation and aggregation (Ananthan *et al.,* 1986). Chaperones have the unique ability to stabilize and refold denatured proteins, as well as to rescue protein aggregates (Leroux and Hartl, 2000a).

Interestingly, however, the overall molecular chaperone complement of extremophilic archaea does not appear to be specifically geared for handling nonnative proteins compared with mesophilic organisms. Archaea do possess a stress response, however (reviewed by Conway de Macario and Macario, 1994, 2000; Macario *et al.,* 1999). In one striking case, a chaperone has been found to be highly induced under heat-shock conditions. Remarkably, the thermosome from *P. occultum* was found to consitute up to almost 75% of the total cellular proteins at the heat-shock temperature of 108°C (Phipps *et al.,* 1991). The expression of the chaperonin is not induced to such high levels in other archaea, however, arguing against the possibility that protection from this family of chaperones is an absolute requirement in extremophilic archaea. Moreover, archaea often lack Hsp70, which plays an important role under stress conditions in bacteria and eukaryotes, and archaeal PFD does not appear to be heat inducible (unpublished observations). In these circumstances, small Hsps and AAA ATPases may play an important role in protecting archaea from stress conditions, especially given that they are absolutely conserved.

C. OXIDATIVE STRESS

The exposure of cells to reactive oxygen species, such as hydrogen peroxide (H_2O_2) and oxygen radicals, can lead to protein damage (e.g., inappropriate disulfide bond formation, side chain oxidation) and thus instability and aggregation in the cell (McDuffee et al., 1997). Heat stress itself results in oxidative stress, perhaps as a consequence of the thermal denaturation of antioxidant proteins (Davidson et al., 1996). In bacteria, a 33-kDa stress-induced protein, Hsp33, was recently demonstrated to be a molecular chaperone activated under oxidizing cellular conditions (Jakob et al., 1999). Homology searches reveal that this protein is relatively well conserved in bacteria, with homologues also present in the T. maritima and A. aeolicus thermophiles, but is absent from eukaryotes and archaea. It is conceivable that other archaeal molecular chaperones may protect archaeal proteins from terminal inactivation due to oxidative damage. As a first line of defense during oxidative stress, archaea also possess enzymes that can eliminate reactive oxygen species [i.e., superoxide dismutase, catalase and peroxidase (Cannio et al., 2000)]. Interestingly, a catalase–peroxidase bifunctional enzyme is shared between bacteria and archaea, presumably as a result of lateral gene transfer (Faguy and Doolittle, 2000).

XIV. Future Outlook

Archaea possess many eukaryotic-like proteins or protein complexes whose structures have been solved. This has been the case for a few molecular chaperones, such as the Group II chaperonin, PFD, and the small Hsp. Some archaeal protein complexes have no direct counterparts in bacteria and are more readily amenable to structure–function studies than their eukaryotic homologues because of their simpler subunit structures. Thus, as more eukaryotic proteins involved in protein biogenesis and degradation are uncovered, it will be fruitful to look toward the abundance of sequenced archaeal genomes in search of related proteins, for both structural and functional characterization. We can also look forward to novel developments in archaeal genetic methodologies to determine the functions of the proteins within the organism itself, as opposed to simply in vitro. On the whole, the combined analysis and comparison of archaeal, eukaryotic, and bacterial homologues not only helps to shed light on specific proteins themselves, but also provides us with a broader perspective on how organisms have evolved similar (or different) biological systems to support fundamental cellular processes.

REFERENCES

Althoff, S., Selinger, D., and Wise, J. A. (1994). *Nucleic Acids Res.* **22**, 1933–1947.
Ananthan, J., Goldberg, A. L., and Voellmy, R. (1986). *Science* **232**, 522–524.
Andersson, S. G., Zomorodipour, A., Andersson, J. O., Sicheritz-Ponten, T., Alsmark, U. C., Podowski, R. M., Naslund, A. K., Eriksson, A. S., Winkler, H. H., and Kurland, C. G. (1998). *Nature* **396**, 133–140.
Anfinsen, C. B. (1973). *Science* **181**, 223–230.
Archibald, J. M., Logsdon, J. M., and Doolittle, W. F. (1999). *Curr. Biol.* **9**, 1053–1056.
Arsene, F., Tomoyasu, T., and Bukau, B. (2000). *Int. J. Food Microbiol.* **55**, 3–9.
Babor, S. M., and Fass, D. (1999). *Proc. Natl. Acad. Sci. USA* **96**, 14759–14764.
Beatrix, B., Sakai, H., and Wiedmann, M. (2000). *J. Biol. Chem.* **275**, 37838–37845.
Beck, K., Wu, L. F., Brunner, J., and Muller, M. (2000). *EMBO J.* **19**, 134–143.
Beckmann, R. P., Mizzen, L. E., and Welch, W. J. (1990). *Science* **248**, 850–854.
Behlke, J., Lutsch, G., Gaestel, M., and Bielka, H. (1991). *FEBS Lett.* **288**, 119–122.
Behrens, S., Maier, R., de Cock, H., Schmid, F. X., and Gross, C. A. (2001). *EMBO J.* **20**, 285–294.
Beissinger, M., and Buchner, J. (1998). *Biol. Chem.* **379**, 245–259.
Benaroudj, N., and Goldberg, A. L. (2000). *Nature Cell Biol.* **2**, 833–839.
Blond-Elguindi, S., Cwirla, S. E., Dower, W. J., Lipshutz, R. J., Sprang, S. R., Sambrook, J. F., and Gething, M. J. (1993). *Cell* **75**, 717–728.
Bochtler, M., Hartmann, C., Song, H. K., Bourenkov, G. P., Bartunik, H. D., and Huber, R. (2000). *Nature* **403**, 800–805.
Boisvert, D. C., Wang, J., Otwinowski, Z., Horwich, A. L., and Sigler, P. B. (1996). *Nature Struct. Biol.* **3**, 170–177.
Bolhuis, A., Matzen, A., Hyyrylainen, H. L., Kontinen, V. P., Meima, R., Chapuis, J., Venema, G., Bron, S., Freudl, R., and van Dijl, J. M. (1999). *J. Biol. Chem.* **274**, 24585–24592.
Bork, P., Sander, C., and Valencia, A. (1992). *Proc. Natl. Acad. Sci. USA* **89**, 7290–7294.
Bosch, G., Baumeister, W., and Essen, L. O. (2000). *J. Mol. Biol.* **301**, 19–25.
Bova, M. P., Yaron, O., Huang, Q., Ding, L., Haley, D. A., Stewart, P. L., and Horwitz, J. (1999). *Proc. Natl. Acad. Sci. USA* **96**, 6137–6142.
Braig, K. (1998). *Curr. Opin. Struct. Biol.* **8**, 159–165.
Braig, K., Otwinowski, Z., Hegde, R., Boisvert, D. C., Joachimiak, A., Horwich, A. L., and Sigler, P. B. (1994). *Nature* **371**, 578–586.
Braun, B. C., Glickman, M., Kraft, R., Dahlmann, B., Kloetzel, P. M., Finley, D., and Schmidt, M. (1999). *Nature Cell Biol.* **1**, 221–226.
Brodsky, J. L., and McCracken, A. A. (1999). *Semin. Cell. Dev. Biol.* **10**, 507–513.
Bui, N., and Strub, K. (1999). *Biol. Chem.* **380**, 135–145.
Bukau, B., Hesterkamp, T., and Luirink, J. (1996). *Trends Cell Biol.* **6**, 480–486.
Bukau, B., and Horwich, A. L. (1998). *Cell* **92**, 351–366.
Callebaut, I., and Mornon, J. P. (1995). *FEBS Lett.* **374**, 211–215.
Cannio, R., Fiorentino, G., Morana, A., Rossi, M., and Bartolucci, S. (2000). *Front. Biosci.* **5**, D768–D779.
Carrascosa, J. L., Llorca, O., and Valpuesta, J. M. (2001). *Micron* **32**, 43–50.

Caspers, G. J., Leunissen, J. A., and de Jong, W. W. (1995). *J. Mol. Evol.* **40**, 238–248.
Cavicchioli, R., Thomas, T., and Curmi, P. M. (2000). *Extremophiles* **4**, 321–331.
Chang, Z., Primm, T. P., Jakana, J., Lee, I. H., Serysheva, I., Chiu, W., Gilbert, H. F., and Quiocho, F. A. (1996). *J. Biol. Chem.* **271**, 7218–7223.
Chen, L. L., and Sigler, P. B. (1999). *Cell* **99**, 757–768.
Chen, X., Sullivan, D. S., and Huffaker, T. C. (1994). *Proc. Natl. Acad. Sci. USA* **91**, 9111–9115.
Choudhury, D., Thompson, A., Stojanoff, V., Langermann, S., Pinkner, J., Hultgren, S. J., and Knight, S. D. (1999). *Science* **285**, 1061–1066.
Chuang, S. E., and Blattner, F. R. (1993). *J. Bacteriol.* **175**, 5242–5252.
Clark, J. I., and Muchowski, P. J. (2000). *Curr. Opin. Struct. Biol.* **10**, 52–59.
Clarke, A. R. (1996). *Curr. Opin. Struct. Biol.* **6**, 43–50.
Coles, M., Diercks, T., Liermann, J., Groger, A., Rockel, B., Baumeister, W., Koretke, K. K., Lupas, A., Peters, J., and Kessler, H. (1999). *Curr. Biol.* **9**, 1158–1168.
Conway de Macario, E., Dugan, C. B., and Macario, A. J. (1994). *J. Mol. Biol.* **240**, 95–101.
Conway de Macario, E., and Macario, A. J. (1994). *Trends Biotechnol.* **12**, 512–518.
Conway de Macario, E., and Macario, A. J. (2000). *Front. Biosci.* **5**, D780–D786.
Cosper, N. J., Stalhandske, C. M., Iwasaki, H., Oshima, T., Scott, R. A., and Iwasaki, T. (1999). *J. Biol. Chem.* **274**, 23160–23168.
Coyle, J. E., Texter, F. L., Ashcroft, A. E., Masselos, D., Robinson, C. V., and Radford, S. E. (1999). *Nature Struct. Biol.* **6**, 683–690.
Craig, E., Ziegelhoffer, T., Nelson, J., Laloraya, S., and Halladay, J. (1995). *Cold Spring Harbor Symp. Quant. Biol.* **60**, 441–449.
Csermely, P., Schnaider, T., Soti, C., Prohaszka, Z., and Nardai, G. (1998). *Pharmacol. Ther.* **79**, 129–168.
Cupp-Vickery, J. R., and Vickery, L. E. (2000). *J. Mol. Biol.* **304**, 835–845.
Cyr, D. M., Langer, T., and Douglas, M. G. (1994). *Trends Biochem. Sci.* **19**, 176–181.
Dai, H., Mao, Q., Yang, H., Huang, S., and Chang, Z. (2000). *J. Protein Chem.* **19**, 319–326.
Dalbey, R. E., and Kuhn, A. (2000). *Annu. Rev. Cell Dev. Biol.* **16**, 51–87.
Das, K. P., Petrash, J. M., and Surewicz, W. K. (1996). *J. Biol. Chem.* **271**, 10449–10452.
Davidson, J. F., Whyte, B., Bissinger, P. H., and Schiestl, R. H. (1996). *Proc. Natl. Acad. Sci. USA* **93**, 5116–5121.
Deckert, G., Warren, P. V., Gaasterland, T., Young, W. G., Lenox, A. L., Graham, D. E., Overbeek, R., Snead, M. A., Keller, M., Aujay, M., Huber, R., Feldman, R. A., Short, J. M., Olsen, G. J., and Swanson, R. V. (1998). *Nature* **392**, 353–358.
Deuerling, E., Schulze-Specking, A., Tomoyasu, T., Mogk, A., and Bukau, B. (1999). *Nature* **400**, 693–696.
Ding, L., and Candido, E. P. M. (2000). *Biochem. J.* **351**, 13–17.
Dinner, A. R., Sali, A., Smith, L. J., Dobson, C. M., and Karplus, M. (2000). *Trends Biochem. Sci.* **25**, 331–339.
Ditzel, L., Lowe, J., Stock, D., Stetter, K. O., Huber, H., Huber, R., and Steinbacher, S. (1998). *Cell* **93**, 125–138.
Doolittle, W. F. (1999). *Science* **284**, 2124–2129.
Driessen, A. J., Fekkes, P., and van der Wolk, J. P. (1998). *Curr. Opin. Microbiol.* **1**, 216–222.
Eggers, D. K., Welch, W. J., and Hansen, W. J. (1997). *J. Biol. Chem.* **272**, 19645–19648.
Ehrnsperger, M., Graber, S., Gaestel, M., and Buchner, J. (1997). *EMBO J.* **16**, 221–229.
Eichler, J. (2000). *Eur. J. Biochem.* **267**, 3402–3412.
Ellgaard, L., Molinari, M., and Helenius, A. (1999). *Science* **286**, 1882–1888.
Ellis, R. J. (1987). *Nature* **328**, 378–379.

Ellis, R. J., and Hartl, F. U. (1996). *FASEB J.* **10**, 20–26.

Ellis, R. J., and Hartl, F. U. (1999). *Curr. Opin. Struct. Biol.* **9**, 102–110.

Ellis, R. J., and Hemmingsen, S. M. (1989). *Trends Biochem. Sci.* **14**, 339–342.

Ellis, R. J., and van der Vies, S. M. (1991). *Annu. Rev. Biochem.* **60**, 321–347.

Ewalt, K. L., Hendrick, J. P., Houry, W. A., and Hartl, F. U. (1997). *Cell* **90**, 491–500.

Faguy, D. M., and Doolittle, W. F. (1998). *Curr. Biol.* **8**, R338–R341.

Faguy, D. M., and Doolittle, W. F. (1999). *Curr. Biol.* **9**, R883–R886.

Faguy, D. M., and Doolittle, W. F. (2000). *Trends Genet.* **16**, 196–197.

Fändrich, M., Tito, M. A., Leroux, M. R., Rostom, A. A., Hartl, F. U., Dobson, C. M., and Robinson, C. V. (2000). *Proc. Natl. Acad. Sci. USA* **97**, 14151–14155.

Farr, G. W., Furtak, K., Rowland, M. B., Ranson, N. A., Saibil, H. R., Kirchhausen, T., and Horwich, A. L. (2000). *Cell* **100**, 561–573.

Feldman, D. E., and Frydman, J. (2000). *Curr. Opin. Struct. Biol.* **10**, 26–33.

Fenton, W. A., and Horwich, A. L. (1997). *Protein Sci.* **6**, 743–760.

Fenton, W. A., Kashi, Y., Furtak, K., and Horwich, A. L. (1994). *Nature* **371**, 614–619.

Fernando, P., and Heikkila, J. J. (2000). *Cell Stress Chaperones* **5**, 148–159.

Ferrari, D. M., and Soling, H. D. (1999). *Biochem J.* **339**, 1–10.

Ferrell, K., Wilkinson, C. R., Dubiel, W., and Gordon, C. (2000). *Trends Biochem. Sci.* **25**, 83–88.

Flaherty, K. M., DeLuca-Flaherty, C., and McKay, D. B. (1990). *Nature* **346**, 623–628.

Fraser, C. M., Gocayne, J. D., White, O., *et al.* (1995). *Science* **270**, 397–403.

Frydman, J., Nimmesgern, E., Erdjument-Bromage, H., Wall, J. S., Tempst, P., and Hartl, F. U. (1992). *EMBO J.* **11**, 4767–4778.

Frydman, J., Nimmesgern, E., Ohtsuka, K., and Hartl, F. U. (1994). *Nature* **370**, 111–117.

Furutani, M., Ideno, A., Iida, T., and Maruyama, T. (2000). *Biochemistry* **39**, 2822.

Gao, Y., Thomas, J. O., Chow, R. L., Lee, G. H., and Cowan, N. J. (1992). *Cell* **69**, 1043–1050.

Geissler, S., Siegers, K., and Schiebel, E. (1998). *EMBO J.* **17**, 952–966.

Gething, M. J., and Sambrook, J. (1992). *Nature* **355**, 33–45.

Glass, J. I., Lefkowitz, E. J., Glass, J. S., Heiner, C. R., Chen, E. Y., and Cassell, G. H. (2000). *Nature* **407**, 757–762.

Glover, J. R., and Lindquist, S. (1998). *Cell* **94**, 73–82.

Golbik, R., Lupas, A. N., Koretke, K. K., Baumeister, W., and Peters, J. (1999). *Biol. Chem.* **380**, 1049–1062.

Goloubinoff, P., Mogk, A., Zvi, A. P., Tomoyasu, T., and Bukau, B. (1999). *Proc. Natl. Acad. Sci. USA* **96**, 13732–13737.

Gothel, S. F., and Marahiel, M. A. (1999). *Cell. Mol. Life Sci.* **55**, 423–436.

Gottesman, S., Wickner, S., and Maurizi, M. R. (1997). *Genes Dev.* **11**, 815–823.

Gribaldo, S., Lumia, V., Creti, R., Conway de Macario, E., Sanangelantoni, A., and Cammarano, P. (1999). *J. Bacteriol.* **181**, 434–443.

Grimaud, R., Kessel, M., Beuron, F., Steven, A. C., and Maurizi, M. R. (1998). *J. Biol. Chem.* **273**, 12476–12481.

Groll, M., Ditzel, L., Lowe, J., Stock, D., Bochtler, M., Bartunik, H. D., and Huber, R. (1997). *Nature* **386**, 463–471.

Guagliardi, A., Cerchia, L., Bartolucci, S., and Rossi, M. (1994). *Protein Sci.* **3**, 1436–1443.

Guagliardi, A., Cerchia, L., and Rossi, M. (1995). *J. Biol. Chem.* **270**, 28126–28132.

Gupta, R. S., and Golding, G. B. (1993). *J. Mol. Evol.* **37**, 573–582.

Gutsche, I., Essen, L. O., and Baumeister, W. (1999). *J. Mol. Biol.* **293**, 295–312.

Gutsche, I., Holzinger, J., Rossle, M., Heumann, H., Baumeister, W., and May, R. P. (2000a). *Curr. Biol.* **10**, 405–408.

Gutsche, I., Mihalache, O., and Baumeister, W. (2000b). *J. Mol. Biol.* **300**, 187–196.

Gutsche, I., Mihalache, O., Hegerl, R., Typke, D., and Baumeister, W. (2000c). *FEBS Lett.* **477,** 278–282.

Haley, D. A., Bova, M. P., Huang, Q. L., McHaourab, H. S., and Stewart, P. L. (2000). *J. Mol. Biol.* **298,** 261–272.

Hansen, W. J., Cowan, N. J., and Welch, W. J. (1999). *J. Cell Biol.* **145,** 265–277.

Harrison, C. J., Hayer-Hartl, M., Di Liberto, M., Hartl, F., and Kuriyan, J. (1997). *Science* **276,** 431–435.

Hartl, F. U. (1996). *Nature* **381,** 571–579.

Haslbeck, M., Walke, S., Stromer, T., Ehrnsperger, M., White, H. E., Chen, S., Saibil, H. R., and Buchner, J. (1999). *EMBO J.* **18,** 6744–6751.

Hayer-Hartl, M. K., Ewbank, J. J., Creighton, T. E., and Hartl, F. U. (1994). *EMBO J.* **13,** 3192–3202.

Hayes, S. A., and Dice, J. F. (1996). *J. Cell Biol.* **132,** 255–258.

Helmbrecht, K., Zeise, E., and Rensing, L. (2000). *Cell Prolif.* **33,** 341–365.

Hemmingsen, S. M., Woolford, C., van der Vies, S. M., Tilly, K., Dennis, D. T., Georgopoulos, C. P., Hendrix, R. W., and Ellis, R. J. (1988). *Nature* **333,** 330–334.

Hesterkamp, T., and Bukau, B. (1996a). *FEBS Lett.* **389,** 32–34.

Hesterkamp, T., and Bukau, B. (1996b). *FEBS Lett.* **385,** 67–71.

Hesterkamp, T., Hauser, S., Lutcke, H., and Bukau, B. (1996). *Proc. Natl. Acad. Sci. USA* **93,** 4437–4441.

Hightower, L. E. (1980). *J. Cell. Physiol.* **102,** 407–427.

Holmgren, A., and Branden, C. I. (1989). *Nature* **342,** 248–251.

Horwich, A. L., Weber-Ban, E. U., and Finley, D. (1999). *Proc. Natl. Acad. Sci. USA* **96,** 11033–11040.

Horwich, A. L., and Willison, K. R. (1993). *Philos. Trans. R. Soc. Lond. B. Biol. Sci.* **339,** 313–325; discussion 325–326.

Horwitz, J. (1992). *Proc. Natl. Acad. Sci. USA* **89,** 10449–10453.

Horwitz, J. (2000). *Semin. Cell. Dev. Biol.* **11,** 53–60.

Houry, W. A., Frishman, D., Eckerskorn, C., Lottspeich, F., and Hartl, F. U. (1999). *Nature* **402,** 147–154.

Ingolia, T. D., and Craig, E. A. (1982). *Proc. Natl. Acad. Sci. USA* **79,** 2360–2364.

Jakob, U., and Buchner, J. (1994). *Trends Biochem. Sci.* **19,** 205–211.

Jakob, U., Gaestel, M., Engel, K., and Buchner, J. (1993). *J. Biol. Chem.* **268,** 1517–1520.

Jakob, U., Muse, W., Eser, M., and Bardwell, J. C. (1999). *Cell* **96,** 341–352.

Karplus, M. (1997). *Fold Des* **2,** S69–S75.

Katsanis, N., Beales, P. L., Woods, M. O., Lewis, R. A., Green, J. S., Parfrey, P. S., Ansley, S. J., Davidson, W. S., and Lupski, J. R. (2000). *Nature Genet.* **26,** 67–70.

Kawarabayasi, Y., Hino, Y., Horikawa, H. *et al.* (1999). *DNA Res.* **6,** 83–101, 145–152.

Kelley, W. L. (1998). *Trends Biochem. Sci.* **23,** 222–227.

Kelley, W. L. (1999). *Curr. Biol.* **9,** R305–R308.

Kim, K. K., Kim, R., and Kim, S. H. (1998a). *Nature* **394,** 595–599.

Kim, R., Kim, K. K., Yokota, H., and Kim, S. H. (1998b). *Proc. Natl. Acad. Sci. USA* **95,** 9129–9133.

Klumpp, M., Baumeister, W., and Essen, L. O. (1997). *Cell* **91,** 263–270.

Kolb, V. A., Makeyev, E. V., Kommer, A., and Spirin, A. S. (1995). *Biochem. Cell. Biol.* **73,** 1217–1220.

Kubota, H., Hynes, G., Carne, A., Ashworth, A., and Willison, K. (1994). *Curr. Biol.* **4,** 89–99.

Kuwajima, K., and Arai, M. (2000). *In* "Frontiers in Molecular Biology. Mechanisms of Protein Folding" (R. Pain, Ed.), 2nd ed., pp. 138–174. Oxford University Press, Oxford.

Ladenstein, R., and Antranikian, G. (1998). *Adv. Biochem. Eng. Biotechnol.* **61,** 37–85.
Landry, J., Chretien, P., Lambert, H., Hickey, E., and Weber, L. A. (1989). *J. Cell Biol.* **109,** 7–15.
Lange, M., Macario, A. J., Ahring, B. K., and Conway de Macario, E. (1997). *Curr. Microbiol.* **35,** 116–121.
Langer, T. (2000). *Trends Biochem. Sci.* **25,** 247–251.
Langer, T., Lu, C., Echols, H., Flanagan, J., Hayer, M. K., and Hartl, F. U. (1992). *Nature* **356,** 683–689.
Laskey, R. A., Honda, B. M., Mills, A. D., and Finch, J. T. (1978). *Nature* **275,** 416–420.
Lazarides, E., and Lindberg, U. (1974). *Proc. Natl. Acad. Sci. USA* **71,** 4742–4746.
Lee, Y. R., Nagao, R. T., and Key, J. L. (1994). *Plant Cell* **6,** 1889–1897.
Lee, G. J., Roseman, A. M., Saibil, H. R., and Vierling, E. (1997). *EMBO J.* **16,** 659–671.
Lee, D. Y., Ahn, B. Y., and Kim, K. S. (2000). *Biochemistry* **39,** 6652–6659.
Leroux, M. R., and Hartl, F. U. (2000a). *In* "Frontiers in Molecular Biology. Mechanisms of Protein Folding" (R. Pain, Ed.), 2nd ed., pp. 364–405. Oxford University Press, Oxford.
Leroux, M. R., and Hartl, F. U. (2000b). *Curr. Biol.* **10,** R260–R264.
Leroux, M. R., Ma, B. J., Batelier, G., Melki, R., and Candido, E. P. (1997a). *J. Biol. Chem.* **272,** 12847–12853.
Leroux, M. R., Melki, R., Gordon, B., Batelier, G., and Candido, E. P. (1997b). *J. Biol. Chem.* **272,** 24646–24656.
Leroux, M. R., Fändrich, M., Klunker, D., Siegers, K., Lupas, A. N., Brown, J. R., Schiebel, E., Dobson, C. M., and Hartl, F. U. (1999). *EMBO J.* **18,** 6730–6743.
Levinthal, C. (1968). *J. Chim. Phys.* **65,** 44–45.
Lewis, V. A., Hynes, G. M., Zheng, D., Saibil, H., and Willison, K. (1992). *Nature* **358,** 249–252.
Lill, R., Cunningham, K., Brundage, L. A., Ito, K., Oliver, D., and Wickner, W. (1989). *EMBO J.* **8,** 961–966.
Linder, B., Jin, Z., Freedman, J. H., and Rubin, C. S. (1996). *J. Biol. Chem.* **271,** 30158–30166.
Lindquist, S., and Craig, E. A. (1988). *Annu. Rev. Genet.* **22,** 631–677.
Liou, A. K., and Willison, K. R. (1997). *EMBO J.* **16,** 4311–4316.
Llorca, O., Smyth, M. G., Marco, S., Carrascosa, J. L., Willison, K. R., and Valpuesta, J. M. (1998). *J. Biol. Chem.* **273,** 10091–10094.
Llorca, O., McCormack, E. A., Hynes, G., Grantham, J., Cordell, J., Carrascosa, J. L., Willison, K. R., Fernandez, J. J., and Valpuesta, J. M. (1999a). *Nature* **402,** 693–696.
Llorca, O., Smyth, M. G., Carrascosa, J. L., Willison, K. R., Radermacher, M., Steinbacher, S., and Valpuesta, J. M. (1999b). *Nature Struct. Biol.* **6,** 639–642.
Llorca, O., Martin-Benito, J., Ritco-Vonsovici, M., Grantham, J., Hynes, G. M., Willison, K. R., Carrascosa, J. L., and Valpuesta, J. M. (2000). *EMBO J.* **19,** 5971–5979.
Lowe, J., and Amos, L. A. (1998). *Nature* **391,** 203–206.
Lowe, J., Stock, D., Jap, B., Zwickl, P., Baumeister, W., and Huber, R. (1995). *Science* **268,** 533–539.
Lupas, A. (1996). *Trends Biochem. Sci.* **21,** 375–382.
Macario, A. J., and Conway de Macario, E. (1999). *Genetics* **152,** 1277–1283.
Macario, A. J., and Conway de Macario, E. (2001). *Front. Biosci.* **6,** 262–283.
Macario, A. J., Dugan, C. B., and Conway de Macario, E. (1991). *Gene* **108,** 133–137.
Macario, A. J., Lange, M., Ahring, B. K., and Conway de Macario, E. (1999). *Microbiol. Mol. Biol. Rev.* **63,** 923–967.
Marco, S., Urena, D., Carrascosa, J. L., Waldmann, T., Peters, J., Hegerl, R., Pfeifer, G., Sack-Kongehl, H., and Baumeister, W. (1994). *FEBS Lett.* **341,** 152–155.

Martin, J., Langer, T., Boteva, R., Schramel, A., Horwich, A. L., and Hartl, F. U. (1991). *Nature* **352**, 36–42.

Maruyama, T., and Furutani, M. (2000). *Front. Biosci.* **5**, D821–D836.

May, A. P., Misura, K. M., Whiteheart, S. W., and Weis, W. I. (1999). *Nat. Cell. Biol.* **1**, 175–182.

Mayer, M. P., and Bukau, B. (1998). *Biol. Chem.* **379**, 261–268.

McDuffee, A. T., Senisterra, G., Huntley, S., Lepock, J. R., Sekhar, K. R., Meredith, M. J., Borrelli, M. J., Morrow, J. D., and Freeman, M. L. (1997). *J. Cell Physiol.* **171**, 143–151.

Melki, R., and Cowan, N. J. (1994). *Mol. Cell Biol.* **14**, 2895–2904.

Melki, R., Batelier, G., Soulie, S., and Williams, R. C., Jr. (1997). *Biochemistry* **36**, 5817–5826.

Merck, K. B., De Haard-Hoekman, W. A., Oude Essink, B. B., Bloemendal, H., and De Jong, W. W. (1992). *Biochim. Biophys. Acta* **1130**, 267–276.

Minton, A. P. (2000). *Curr. Opin. Struct. Biol.* **10**, 34–39.

Mogk, A., Tomoyasu, T., Goloubinoff, P., Rudiger, S., Roder, D., Langen, H., and Bukau, B. (1999). *EMBO J.* **18**, 6934–6949.

Moll, R., Schmidtke, S., and Schafer, G. (1999). *Eur. J. Biochem.* **259**, 441–448.

Morimoto, R. I., Kline, M. P., Bimston, D. N., and Cotto, J. J. (1997). *Essays Biochem.* **32**, 17–29.

Morshauser, R. C., Hu, W., Wang, H., Pang, Y., Flynn, G. C., and Zuiderweg, E. R. (1999). *J. Mol. Biol.* **289**, 1387–1403.

Motohashi, K., Watanabe, Y., Yohda, M., and Yoshida, M. (1999). *Proc. Natl. Acad. Sci. USA* **96**, 7184–7189.

Muchowski, P. J., and Clark, J. I. (1998). *Proc. Natl. Acad. Sci. USA* **95**, 1004–1009.

Muchowski, P. J., Hays, L. G., Yates, J. R., 3rd, and Clark, J. I. (1999a). *J. Biol. Chem.* **274**, 30190–30195.

Muchowski, P. J., Wu, G. J., Liang, J. J., Adman, E. T., and Clark, J. I. (1999b). *J. Mol. Biol.* **289**, 397–411.

Nelson, K. E., Clayton, R. A., Gill, S. R., *et al.* (1999). *Nature* **399**, 323–329.

Nelson, R. J., Ziegelhoffer, T., Nicolet, C., Werner-Washburne, M., and Craig, E. A. (1992). *Cell* **71**, 97–105.

Ng, W. V., Kennedy, S. P., Mahairas, G. G., *et al.* (2000). *Proc. Natl. Acad. Sci. USA* **97**, 12176–12181.

Pahl, A., Brune, K., and Bang, H. (1997). *Cell Stress Chaperones* **2**, 78–86.

Palleros, D. R., Reid, K. L., Shi, L., Welch, W. J., and Fink, A. L. (1993). *Nature* **365**, 664–666.

Parsell, D. A., Kowal, A. S., Singer, M. A., and Lindquist, S. (1994). *Nature* **372**, 475–478.

Patel, S., and Latterich, M. (1998). *Trends Cell Biol.* **8**, 65–71.

Pfanner, N. (1999). *Curr. Biol.* **9**, R720–R724.

Pfund, C., Lopez-Hoyo, N., Ziegelhoffer, T., Schilke, B. A., Lopez-Buesa, P., Walter, W. A., Wiedmann, M., and Craig, E. A. (1998). *EMBO J.* **17**, 3981–3989.

Phipps, B. M., Hoffmann, A., Stetter, K. O., and Baumeister, W. (1991). *EMBO J.* **10**, 1711–1722.

Pohlschröder, M., Prinz, W. A., Hartmann, E., and Beckwith, J. (1997). *Cell* **91**, 563–566.

Porankiewicz, J., Wang, J., and Clarke, A. K. (1999). *Mol. Microbiol.* **32**, 449–458.

Radford, S. E. (2000). *Trends Biochem. Sci.* **25**, 611–618.

Rassow, J., and Pfanner, N. (1996). *Curr. Biol.* **6**, 115–118.

Rassow, J., Voos, W., and Pfanner, N. (1995). *Trends Cell Biol.* **5**, 207–212.

Ren, B., Tibbelin, G., de Pascale, D., Rossi, M., Bartolucci, S., and Ladenstein, R. (1998). *Nature Struct. Biol.* **5**, 602–611.

Rollet, E., Lavoie, J. N., Landry, J., and Tanguay, R. M. (1992). *Biochem. Biophys. Res. Commun.* **185,** 116–120.

Rothman, J. E. (1989). *Cell* **59,** 591–601.

Rouiller, I., Butel, V. M., Latterich, M., Milligan, R. A., and Wilson-Kubalek, E. M. (2000). *Mol. Cell* **6,** 1485–1490.

Rudiger, S., Germeroth, L., Schneider-Mergener, J., and Bukau, B. (1997). *EMBO J.* **16,** 1501–1507.

Ruepp, A., Graml, W., Santos-Martinez, M. L., Koretke, K. K., Volker, C., Mewes, H. W., Frishman, D., Stocker, S., Lupas, A. N., and Baumeister, W. (2000). *Nature* **407,** 508–513.

Rumbley, J., Hoang, L., Mayne, L., and Englander, S. W. (2001). *Proc. Natl. Acad. Sci. USA* **98,** 105–112.

Saibil, H. (2000). *Curr. Opin. Struct. Biol.* **10,** 251–258.

Sanchez, Y., and Lindquist, S. L. (1990). *Science* **248,** 1112–1115.

Sanchez, Y., Taulien, J., Borkovich, K. A., and Lindquist, S. (1992). *EMBO J.* **11,** 2357–2364.

Sauer, F. G., Futterer, K., Pinkner, J. S., Dodson, K. W., Hultgren, S. J., and Waksman, G. (1999). *Science* **285,** 1058–1061.

Sauer, F. G., Knight, S. D., Waksman, G. J., and Hultgren, S. J. (2000). *Semin. Cell. Dev. Biol.* **11,** 27–34.

Scandurra, R., Consalvi, V., Chiaraluce, R., Politi, L., and Engel, P. C. (2000). *Front. Biosci.* **5,** D787–D795.

Schatz, G., and Dobberstein, B. (1996). *Science* **271,** 1519–1526.

Schiene, C., and Fischer, G. (2000). *Curr. Opin. Struct. Biol.* **10,** 40–45.

Schirmer, E. C., Glover, J. R., Singer, M. A., and Lindquist, S. (1996). *Trends Biochem. Sci.* **21,** 289–296.

Schmid, D., Baici, A., Gehring, H., and Christen, P. (1994). *Science* **263,** 971–973.

Scholz, C., Stoller, G., Zarnt, T., Fischer, G., and Schmid, F. X. (1997). *EMBO J.* **16,** 54–58.

Schröder, H., Langer, T., Hartl, F. U., and Bukau, B. (1993). *EMBO J.* **12,** 4137–4144.

Shao, F., Bader, M. W., Jakob, U., and Bardwell, J. C. (2000). *J. Biol. Chem.* **275,** 13349–13352.

Shtilerman, M., Lorimer, G. H., and Englander, S. W. (1999). *Science* **284,** 822–825.

Siegers, K., Waldmann, T., Leroux, M. R., Grein, K., Shevchenko, A., Schiebel, E., and Hartl, F. U. (1999). *EMBO J.* **18,** 75–84.

Siegert, R., Leroux, M. R., Scheufler, C., Hartl, F. U., and Moarefi, I. (2000). *Cell* **103,** 621–632.

Sigler, P. B., Xu, Z., Rye, H. S., Burston, S. G., Fenton, W. A., and Horwich, A. L. (1998). *Annu. Rev. Biochem.* **67,** 581–608.

Singh, S. K., Guo, F., and Maurizi, M. R. (1999). *Biochemistry* **38,** 14906–14915.

Singh, S. K., Grimaud, R., Hoskins, J. R., Wickner, S., and Maurizi, M. R. (2000). *Proc. Natl. Acad. Sci. USA* **97,** 8898–8903.

Slavotinek, A. M., Stone, E. M., Mykytyn, K., Heckenlively, J. R., Green, J. S., Heon, E., Musarella, M. A., Parfrey, P. S., Sheffield, V. C., and Biesecker, L. G. (2000). *Nature Genet.* **26,** 15–16.

Smykal, P., Masin, J., Hrdy, I., Konopasek, I., and Zarsky, V. (2000). *Plant J.* **23,** 703–713.

Sousa, M. C., Trame, C. B., Tsuruta, H., Wilbanks, S. M., Reddy, V. S., and McKay, D. B. (2000). *Cell* **103,** 633–643.

Sriram, M., Osipiuk, J., Freeman, B., Morimoto, R., and Joachimiak, A. (1997). *Structure* **5,** 403–414.

Stephens, R. S., Kalman, S., Lammel, C., Fan, J., Marathe, R., Aravind, L., Mitchell, W., Olinger, L., Tatusov, R. L., Zhao, Q., Koonin, E. V., and Davis, R. W. (1998). *Science* **282**, 754–759.

Stoldt, V., Rademacher, F., Kehren, V., Ernst, J. F., Pearce, D. A., and Sherman, F. (1996). *Yeast* **12**, 523–529.

Stoller, G., Rucknagel, K. P., Nierhaus, K. H., Schmid, F. X., Fischer, G., and Rahfeld, J. U. (1995). *EMBO J.* **14**, 4939–4948.

Stoller, G., Tradler, T., Rucknagel, K. P., Rahfeld, J. U., and Fischer, G. (1996). *FEBS Lett.* **384**, 117–122.

Stone, D. L., Slavotinek, A., Bouffard, G. G., Banerjee-Basu, S., Baxevanis, A. D., Barr, M., and Biesecker, L. G. (2000). *Nature Genet.* **25**, 79–82.

Stringham, E. G., Dixon, D. K., Jones, D., and Candido, E. P. (1992). *Mol. Biol. Cell* **3**, 221–233.

Suzuki, C. K., Rep, M., Maarten van Dijl, J., Suda, K., Grivell, L. A., and Schatz, G. (1997). *Trends Biochem. Sci.* **22**, 118–123.

Tabira, Y., Ohara, N., Kitaura, H., Matsumoto, S., Naito, M., and Yamada, T. (1998). *Res. Microbiol.* **149**, 255–264.

Tanaka, N., and Nakamoto, H. (1999). *FEBS Lett.* **458**, 117–123.

Teter, S. A., Houry, W. A., Ang, D., Tradler, T., Rockabrand, D., Fischer, G., Blum, P., Georgopoulos, C., and Hartl, F. U. (1999). *Cell* **97**, 755–765.

Thomas, J. G., and Baneyx, F. (1998). *J. Bacteriol.* **180**, 5165–5172.

Thomas, J. G., and Baneyx, F. (2000). *Mol. Microbiol.* **36**, 1360–1370.

Thulasiraman, V., Yang, C. F., and Frydman, J. (1999). *EMBO J.* **18**, 85–95.

Tian, G., Vainberg, I. E., Tap, W. D., Lewis, S. A., and Cowan, N. J. (1995). *Nature* **375**, 250–253.

Trent, J. D., Nimmesgern, E., Wall, J. S., Hartl, F. U., and Horwich, A. L. (1991). *Nature* **354**, 490–493.

Tsuchiya, H., Iseda, T., and Hino, O. (1996). *Cancer Res.* **56**, 2881–2885.

Tumbula, D. L., and Whitman, W. B. (1999). *Mol. Microbiol.* **33**, 1–7.

Ursic, D., and Culbertson, M. R. (1991). *Mol. Cell Biol.* **11**, 2629–2640.

Vainberg, I. E., Lewis, S. A., Rommelaere, H., Ampe, C., Vandekerckhove, J., Klein, H. L., and Cowan, N. J. (1998). *Cell* **93**, 863–873.

Vale, R. D. (2000). *J. Cell Biol.* **150**, F13–F19.

Valent, Q. A., Kendall, D. A., High, S., Kusters, R., Oudega, B., and Luirink, J. (1995). *EMBO J.* **14**, 5494–5505.

van Den Ent, F., and Lowe, J. (2000). *EMBO J.* **19**, 5300–5307.

Veinger, L., Diamant, S., Buchner, J., and Goloubinoff, P. (1998). *J. Biol. Chem.* **273**, 11032–11037.

Vivares, C. P., and Metenier, G. (2000). *Curr. Opin. Microbiol.* **3**, 463–467.

Wang, C. C., and Tsou, C. L. (1998). *FEBS Lett.* **425**, 382–384.

Wang, S., Sakai, H., and Wiedmann, M. (1995). *J. Cell Biol.* **130**, 519–528.

Weber-Ban, E. U., Reid, B. G., Miranker, A. D., and Horwich, A. L. (1999). *Nature* **401**, 90–93.

Wickner, S., Maurizi, M. R., and Gottesman, S. (1999). *Science* **286**, 1888–1893.

Wiedmann, B., Sakai, H., Davis, T. A., and Wiedmann, M. (1994). *Nature* **370**, 434–440.

Willison, K. R. (1999). *In* "Molecular Chaperones and Protein Folding Catalysts" (B. Bukau, Ed.), Vol. 1, pp. 555–571. Hardwood Academic, London.

Willison, K. R., and Kubota, H. (1994). *In* "The Biology of Heat Shock Proteins and Molecular Chaperones" (R. I. Morimoto, A. Tissieres, and C. Georgopoulos, Eds.), pp. 299–312. Cold Spring Harbor Laboratory Press, Cold Spring Harbor, NY.

Wilson, H. L., Ou, M. S., Aldrich, H. C., and Maupin-Furlow, J. (2000). *J. Bacteriol.* **182,** 1680–1692.

Wilson, M. R., and Easterbrook-Smith, S. B. (2000). *Trends Biochem. Sci.* **25,** 95–98.

Xu, Z., Horwich, A. L., and Sigler, P. B. (1997). *Nature* **388,** 792–798.

Xu, Z., Knafels, J. D., and Yoshino, K. (2000). *Nature Struct. Biol.* **7,** 1172–1177.

Yan, W., Schilke, B., Pfund, C., Walter, W., Kim, S., and Craig, E. A. (1998). *EMBO J.* **17,** 4809–4817.

Yeh, K. W., Jinn, T. L., Yeh, C. H., Chen, Y. M., and Lin, C. Y. (1994). *Biotechnol. Appl. Biochem.* **19,** 41–49.

Zhang, X., Shaw, A., Bates, P. A., Newman, R. H., Gowen, B., Orlova, E., Gorman, M. A., Kondo, H., Dokurno, P., Lally, J., Leonard, G., Meyer, H., van Heel, M., and Freemont, P. S. (2000). *Mol. Cell* **6,** 1473–1484.

Zhu, X., Zhao, X., Burkholder, W. F., Gragerov, A., Ogata, C. M., Gottesman, M. E., and Hendrickson, W. A. (1996). *Science* **272,** 1606–1614.

Zolkiewski, M. (1999). *J. Biol. Chem.* **274,** 28083–28086.

Zwickl, P., Ng, D., Woo, K. M., Klenk, H. P., and Goldberg, A. L. (1999). *J. Biol. Chem.* **274,** 26008–26014.

Archaeal Proteasomes: Proteolytic Nanocompartments of the Cell

Julie A. Maupin-Furlow, Steven J. Kaczowka, Mark S. Ou,
and Heather L. Wilson

Department of Microbiology and Cell Science
University of Florida
Gainesville, Florida 32611-0700

I. Introduction

Energy-dependent proteolysis is not only vital to the elimination of defective proteins but also central to the specific regulation of essential functions in both procaryotic and eukaryotic cells. In eukaryotes, processes regulated by energy-dependent proteases include cell cycle control, transcription, metabolism, differentiation, DNA repair, responses to stress, apoptosis, and the generation of peptides presented on major histocompatibility complex class I molecules to the immune system (Peters, 1994; Hilt and Wolf, 1996; Coux *et al.*, 1996; Ciechanover *et al.*, 2000). Similarly in eubacteria these proteases have central roles in systems including stress responses, sporulation, cell division, restriction modification, mRNA stability, swarming, and biofilm formation (Gottesman, 1999). The roles of energy-dependent proteases in the archaea are also presumed to be integral to cell survival but remain to be established.

ADVANCES IN APPLIED MICROBIOLOGY, VOLUME 50

There is only a small group of energy-dependent proteases known, and they classify into four general families based on the primary sequence of their proteolytic active sites. Proteases classifying to the first family appear to have a single threonine residue active site at the amino terminus. They include the 20S proteasomes found in eukarya, archaea, and Gram-positive actinomycetes, all of which are presumed to associate with energy-dependent regulatory components. The simpler eubacterial HslUV protease, composed of the HslU ATPase and HslV protease, is also included in the proteasome family. The Clp family includes the ClpAP and ClpXP proteases, in which ClpP, a serine protease, associates with either ClpA or ClpX ATPases of differing substrate specificities. The Lon family members have ATPase and serine protease domains on the same polypeptide and have little primary sequence similarity to the serine active site of the Clp proteases. The FtsH family consists of membrane-associated Zn^{2+}-dependent metalloproteases with ATPase domains on the same polypeptide. Homologues from all four of these families are fairly widespread among the eubacteria and play critical roles in the cytosol and organelles of eukaryotes. Based on genome sequences, however, FtsH and Clp proteases are not found in the archaea. Instead, the proteasome may be responsible for energy-dependent degradation of the majority of proteins in the archaeal cytosol. An archaeal Lon-related serine protease is apparently anchored in the cell membrane and may function in the energy-dependent hydrolysis of hydrophobic and/or membrane-associated proteins (Maupin-Furlow et al., 2000; Ruepp et al., 2000).

Although these four families of energy-dependent proteases share limited primary sequence identity, they appear to have converged to a self-compartmentalized quaternary structure or nanocompartment (Lupas et al., 1997a; Aravind and Ponting, 1998). Detailed structural analysis reveals that the proteolytic active sites of Clp and proteasome family members are isolated in a cylindrical chamber and are separated from the cytosol or membrane by narrow, ring-shaped openings no wider than the width of an α-helix (Löwe et al., 1995; Wang et al., 1997; Bochtler et al., 1997; Groll et al., 1997). Likewise, ring-shaped structures have been observed for FtsH (Shotland et al., 1997). Thus, the degradation of biological substrates is presumed to require additional energy-dependent proteins or protein domains to mediate substrate recognition, unfolding, and translocation into the protease compartment. Many of these energy-dependent components that participate in proteolysis may also have independent roles as chaperones. This is believed to provide a proofreading step following initial binding of protein substrates to ensure the proper selection of proteins destined for processive hydrolysis (Wickner et al., 1999).

This review focuses on the structure and function of proteasomes, with special emphasis on those found in the archaea. The relationships of proteasomes to other energy-dependent proteases including the Clp, FtsH, and Lon protease families are integrated into the discussion. In addition, the architecture, distribution, catalytic mechanism, and assembly of the core 20S proteasome and associated regulatory proteins are included. Recent findings on how cells target proteins for degradation and what roles energy-dependent proteases play in cell physiology are also communicated.

II. 20S Proteasomes

A. ARCHITECTURE

Proteasomes are composed of a proteolytic core, the 20S proteasome, which associates with ATPase regulatory components to form an energy-dependent protease complex. In eukaryotes, the 19S cap regulators are ATPase complexes that associate with 20S proteasomes to form the 26S proteasome. Similarly the archaeal PAN or proteasome-activating nucleotidase, which is related to the ATPase subunits of the 19S cap, appears to associate with and stimulate protein hydrolysis by 20S proteasomes in the archaea (Zwickl et al., 1999; Wilson et al., 2000). The overall architecture of the 20S proteasome core is highly conserved from bacteria to human (Dahlmann et al., 1989). It is cylindrical, with a length of 15 nm and a diameter of 11 to 12 nm, and consists of four stacked rings with a central channel (Fig. 1). 20S proteasomes are composed of 28 subunits that are related in primary sequence and classify into two groups: α- and β-type (Coux et al., 1994). Immunoelectron microscopy and average projection image analysis reveal that 20S proteasomes have a sevenfold symmetry with α-type subunits located in the two outer rings and β-type subunits forming the two inner rings with each ring composed of seven subunits (Grizwa et al., 1991; Lupas et al., 1994). The number of different subunits that form 20S proteasomes varies among organisms. The majority of archaeal and eubacterial 20S proteasomes have fairly simple compositions of single α and β subunits. In contrast, eukaryal 20S proteasomes are composed of seven α-type and seven β-type subunits. Each eukaryal subunit is located at a defined position in the heptameric rings to form 20S complexes with twofold symmetry (Kopp et al., 1993; Kopp et al., 1997). Immunoelectron microscopy and protein cross-linking studies coupled with recent biochemical analysis using N-ethylmaleimide suggest that the arrangement of the different subunits of the human 20S proteasome is similar to that of yeast (Kopp et al., 1997; Hochstrasser et al., 1999). X-ray crystal structures of 20S proteasomes

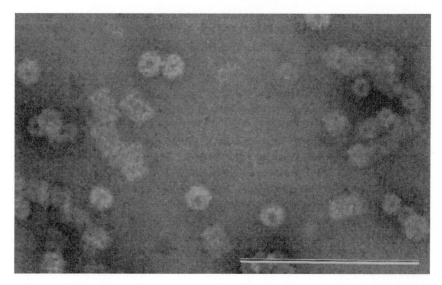

FIG. 1. Transmission electron micrograph of 20S proteasomes from the archaeon *Haloferax volcanii*. Typical four-stacked rings are visible in the side-on views. The central channel is visible in the end-on views. The particles, negatively stained with uranyl acetate, have a length of 15 nm and a diameter of 12 nm. Bar, 100 nm.

from the archaeon *Thermoplasma acidophilum* and eukaryote *Saccharomyces cerevisiae* have been determined to 3.4- and 2.4-Å resolution, respectively (Löwe *et al.*, 1995; Groll *et al.*, 1997). This detailed structural information provides tremendous insight into the mechanism of protein degradation mediated by these elaborate complexes and enables predictions to be made on the atomic structure of related 20S proteasomes such as those from the archaeon *Methanosarcina thermophila* (Fig. 2).

Ironically, the *T. acidophilum* 20S proteasome is similar in overall structure to the molecular chaperone GroEL that promotes protein folding (Zwickl *et al.*, 1990; Weissman *et al.*, 1995). Both have sevenfold rotational symmetry around a large central cavity that is accessible only through an entry port at each end of the particle. There are fundamental differences, however, between these two complexes. For GroEL, the hydrophobic substrate-binding sites are immediately interior to 4.5-nm channel openings at each end of the cylinder (Braig *et al.*, 1994; Boisvert *et al.*, 1996; Xu *et al.*, 1997). These dimensions may protect GroEL from self-association while enabling nonnative polypeptides to enter the chamber and bind sites used to refold proteins. The archaeal 20S proteasome, in contrast, has only 1.3-nm openings at each end that appear to protect most folded proteins from entering the central cavity that harbors the proteolytic active sites (Löwe *et al.*, 1995).

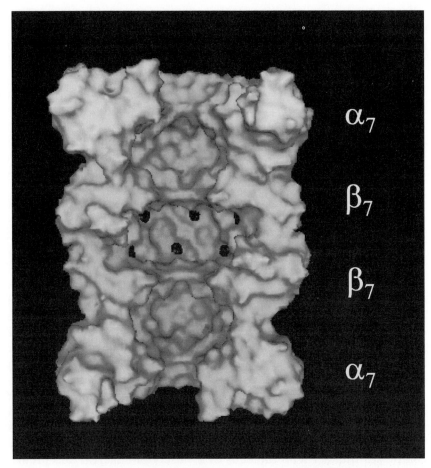

α_7

β_7

β_7

α_7

FIG. 2. Predicted structure of the *Methanosarcina thermophila* 20S proteasome. The image is an atomic model based on the crystal structure coordinates of the 20S proteasomes from *Thermoplasma acidophilum* and yeast. The active sites involved in peptide bond hydrolysis are indicated in black. Computational results were obtained using software programs from Molecular Simulations Inc. (http://www/msi.com), and from the Swiss Institute of Bioinformatics (http://www.expasy.ch).

In general, the yeast 20S proteasome is similar in overall structure to that of the archaeal 20S proteasome. Both contain three inner cavities including two antechambers of 59 nm^3 and a central chamber of 84 nm^3 (Hegerl *et al.*, 1991; Löwe *et al.*, 1995; Groll *et al.*, 1997). The antechambers are each formed by single α- and β-rings and have 2.2-nm openings to the central chamber formed by the two β-rings. The central chamber harbors the 6 to 14 active sites responsible for hydrolysis of peptide bonds. *In vitro* 20S proteasomes are able to hydrolyze unfolded proteins in an energy-independent mechanism. It is unclear how these unfolded substrates are translocated in the absence of ATP from the α-ring openings to the active sites formed by the β subunits that are 8 to 10 nm away. The role of the antechambers also remains to be determined since other self-compartmentalized proteases such as ClpP and HslV do not have antechambers yet still function in the cell (Wang et al., 1997; Bochtler et al., 1997). It is possible that clusters of hydrophobic residues lining the antechamber may preserve the substrate in an unfolded state and direct the polypeptide chain to the active site clefts of the central chamber (Baumeister et al., 1998). The antechambers may also play a role in restricting access to the central chamber of the 20S proteasome (Bochtler et al., 1999).

Although there are numerous similarities between the archaeal and the yeast 20S proteasomes, there are also detectable differences in structure (Löwe et al., 1995; Groll et al., 1997; Baumeister et al., 1998). In contrast to the simpler archaeal enzyme, the yeast 20S proteasome forms a larger number of intersubunit contacts that may facilitate the rather complicated assembly of 14 different subunits. There are also side windows of 1.3 nm in the yeast structure that may be used for rapid discharge of degradation products; these windows are not apparent in the archaeal enzyme. Most notably the amino termini of the yeast α subunits appear to form tightly closed constrictions or axial gates that appear to block the substrate entry ports at the ends of the 20S proteasome cylinder. In contrast, the α amino termini of the archaeal 20S proteasome are disordered and do not block the axial openings. Thus it was proposed that activators, known to associate with yeast 20S proteasomes, promote an open gate conformation for protein substrate entry. Consistent with this mechanism, 19S cap (PA700) and 11S (PA28, PA26, or REG) regulatory complexes interact with the ends of a variety of eukaryal 20S proteasomes and stimulate the hydrolysis of peptides and proteins (Ma et al., 1992; Dubiel et al., 1992; Adams et al., 1998a; Strickland et al., 2000). A recent crystal structure analysis of 11S regulators in complex with a 20S proteasome suggests that activation loops of the regulator induce conformational changes to open the axial gates formed by the α subunits of 20S proteasomes (Whitby et al., 2000). Furthermore, eukaryal 20S proteasomes purified in the presence of glycerol are often in a latent state and require heat, chaotrophic agents, or hydrostatic pressure for activation; presumably these agents mediate a conformational change that opens the axial gates (McGuire et al., 1989; Weitman and Etlinger, 1992; Arribas et al., 1994; Gardrat et al., 1999).

Analysis of yeast 20S proteasomes using atomic force microscopy (AFM), however, suggests that there is no need for the binding of additional proteins to open the axial gates of yeast 20S proteasomes

(Osmulski and Gaczynska, 2000). Although the majority (75%) of the gates exist in a closed state, the remaining complexes are in an open gate conformation. The addition of a chymotrypsin-like peptide substrate [i.e., N-succinyl-leu-leu-val-Amc (Amc), 7-amino-4-methyl-coumarin] shifts the equilibrium of uninhibited 20S proteasomes towards the open conformation. Thus, it is possible that the detection of only the closed form in the crystal structure is due to the thermodynamic preference and/or abundance of the "idle" state. It appears that the axial gates of the yeast 20S proteasome are continuously undergoing conformational changes. Both the presence of unblocked catalytic sites and cleavable substrate shift the equilibrium of the axial gates to the open state. The regulatory complexes may lock the gates into the open position. Interestingly, ligand-induced conformational changes are also evident in the side view of the yeast 20S proteasomes as observed by AFM, although the rationale for these changes remains to be determined (Osmulski and Gaczynska, 2000). Plasticity in structure is also evident in the rabbit 20S proteasome based on the stimulatory effects of hydrostatic pressure on the caseinolytic activity of the enzyme (Gardrat et al., 1999).

The narrow openings for substrate access to the active sites of 20S proteasomes reveal that additional enzymes or cofactors are probably needed to unfold substrate proteins prior to hydrolysis. In fact, most native proteins are resistant to hydrolysis by 20S proteasomes in vitro (Wenzel and Baumeister, 1993; Dick et al., 1994; Wenzel and Baumeister, 1995), whereas polypeptides unfolded by heat (Dahlmann et al., 1992; Maupin-Furlow et al., 1998; Wilson et al., 2000), oxidative damage (Fagen et al., 1986; Pacifici et al., 1989; Sacchetta et al., 1990; Wenzel and Baumeister, 1993; Pacifici et al., 1993), or disulfide bond reduction (Mykles and Haire, 1991; Cardozo et al., 1992; Dick et al., 1994; Wenzel and Baumesister, 1995) are usually degraded. Covalent attachment of 2-nm-diameter Nanogold particles to unfolded insulin prevents its degradation by 20S proteasomes, and electron micrographs reveal that this bulky label blocks insulin from entering the proteasome channel (Wenzel and Baumeister, 1995). It appears that the dimensions of the cylinder openings protect the cell from the random and nonspecific hydrolysis of proteins. These dimensions may also restrict the size of products that are released from the complex to short polypeptides of 3 to 30 amino acids which have little secondary structure (Akopain et al., 1997; Kisselev et al., 1998).

B. Distribution and Types of 20S Proteasome Subunits

20S proteasomes appear to be universally distributed among the archaea and eukarya. In contrast, genuine eubacterial proteasomes have been identified only in the high-G+C Gram-positive actinomycetes, which may have acquired the gene(s) by horizontal transfer. All 20S proteasome subunits are related and classify into either α- or β-type superfamilies based on their relationship to the single α and β subunits of the *T. acidophilum* 20S proteasome, the first to be completely sequenced (Zwickl *et al.*, 1992a; Coux *et al.*, 1994). This simple two-subunit ($\alpha\beta$) 20S proteasome composition appears to be widespread among other archaea including *Archaeoglobus fulgidus, Methanobacterium thermoautotrophicum* ΔH, *Methanococcus jannaschii, Methanosarcina thermophila,* and *Halobacterium* sp. NRC-1 (Maupin-Furlow *et al.*, 2000). Single $\alpha\beta$–20S proteasomes are also common among the actinomycetes including *Frankia* strain ACN14a/+s-R, *Mycobacterium smegmatis, Mycobacterium tuberculosis, Streptomyces coelicolor,* and most *Rhodococcus* species (De Mot *et al.*, 1999).

There are exceptional prokaryotes, however, that synthesize up to four 20S proteasome subunits related to either the α- or the β-type subfamilies. The eubacterium *Rhodococcus erythropolis* strain N186/21 synthesizes a single 20S proteasome composed of two α-type and two β-type subunits in which the paralogs are over 80% identical (Tamura *et al.*, 1995; Zühl *et al.*, 1997a). Several archaea including *Haloferax volcanii, Aeropyrum pernix,* and two *Pyrococcus* species encode three 20S proteasome subunits related to either the α- or the β-type proteins (Maupin-Furlow *et al.*, 2000). The archaeal paralogs are only 42 to 57% identical, suggesting that they probably have separate roles and do not serve redundant functions in these organisms. These significant differences in primary sequence also suggest that gene duplication events that may have generated these paralogs occurred earlier in the archaea than in *R. erythropolis.*

Analysis of 20S proteasomes isolated from *H. volcanii* reveals that a single β subunit and two α-type subunit paralogs are synthesized in this halophilic archaeon. Purification of native 20S proteasomes in combination with homologous production of epitope-tagged derivatives of the individual subunits in *H. volcanii* reveals that at least two different 20S proteasomes are synthesized. These include complexes of α1, α2, and β subunits as well as complexes of α1 and β subunits (Wilson *et al.*, 1999; Kaczowka and Maupin-Furlow, unpublished). Based on genome sequence, several *Pyrococcus* species including *P. abysii, P. horikoshii,* and *P. furiosus* appear to synthesize a single α-type and two β-type proteasomal proteins (Kawarabayasi *et al.*, 1998; Heilig, 2000) (*P. furiosus* unfinished genome, http://www.ncbi.nlm.nih. gov/Microb_blast/credits/2261.html). Both β-type proteins are probably cleaved to generate mature proteins with a conserved amino-terminal threonine residue that forms the proteolytic active site. Interestingly,

20S proteasomes purified from *P. furiosus* are composed of a single type of α and β subunit ($\alpha\beta1$) based on protein sequence analysis of SDS-PAGE-separated proteins (Bauer *et al.*, 1997). Whether or not the second β subunit ($\beta2$) is synthesized and incorporated into active 20S proteasomes as $\alpha\beta1\beta2$ or $\alpha\beta2$ complexes remains to be determined. The archaeon *A. pernix* also appears to produce a single α- and two β-type proteasomal proteins based on genome sequence (Kawarabayasi *et al.*, 1999). Interestingly, only one of two β proteins is likely to be processed to expose the active-site amino-terminal threonine residue involved in peptide bond hydrolysis. It is still unclear why some prokaryotes produce multiple 20S proteasomes of different subunit composition or 20S proteasomes of greater subunit complexity than others.

The eukaryal 20S proteasomes are highly complex, with seven α-type and seven β-type proteins assembling into an ordered complex of twofold symmetry (Kopp *et al.*, 1993, 1997). Nomenclature for these subunits has recently been adopted (e.g., $\alpha1$ to $\alpha7$ and $\beta1$ to $\beta7$). The complete yeast genome sequence and biochemical studies of isolated proteasomes suggest that lower eukaryotes produce a single 20S proteasome of 14 different subunits (Heinemeyer *et al.*, 1994). Sequence analysis of the *Arabidopsis* genome reveals a set of 23 genes encoding 20S proteasome subunits that classify into 14 groups corresponding to the $\alpha1$ to $\alpha7$ and $\beta1$ to $\beta7$ subunits from yeast. In contrast to yeast, where each of the 14 subunits is encoded by a single gene, many of the *Arabidopsis* subunits (i.e., $\alpha1$ to $\alpha6$, $\beta2$ to $\beta4$) have a corresponding paralogue with greater than 90% identity. It is speculated that these highly identical isoforms serve redundant functions, which may reflect the absolute necessity of these gene products for development and cellular responses to environmental changes (Fu *et al.*, 1999).

Isoforms of β-type subunits have also been identified in higher eukaryotes including mammals and zebrafish; however, they are only 70% identical and evidence suggests that they serve different roles in the cell (Tanaka and Kasahara, 1998; Murray *et al.*, 1999). 20S proteasomes of different subunit composition have been purified from higher eukaryotes including a general housekeeping proteasome that is constitutively produced in most mammalian cell lines and an immunoproteasome that is produced at high levels in cells involved in the immune system as well as other cell types after IFN-γ induction (Tanaka and Kasahara, 1998; Murray *et al.*, 1999). Differences between these two 20S proteasomes include replacement of the constitutive $\beta1$, $\beta2$, and $\beta5$ subunits by the IFN-γ-inducible $\beta1i$, $\beta2i$, and $\beta5i$ to generate a proteasome with altered peptidase and protease activities. The constitutive $\beta1$, $\beta2$, and $\beta5$ subunits are likely required for general proteolysis, whereas the inducible isoforms may stimulate the digestion of foreign proteins for presentation by the MHC class I pathway to the immune system (Driscoll *et al.*, 1993; Gaczynska *et al.*, 1993; Brown *et al.*, 1993; Früh *et al.*, 1994). Two distinct forms of β1i have been identified in isolated 20S proteasomes; one corresponds to the "inactive" primary translation product of the gene, and the other to the active form that is processed at the amino terminus (Martinez and Monaco, 1993).

Most eubacteria do not synthesize 20S proteasomes and instead have an HslV protease that is a self-compartmentalized Ntn-threonine hydrolase and a divergent member of the β-type proteasome subfamily (Rohrwild *et al.*, 1996; Yoo *et al.*, 1997; Lupas *et al.*, 1997b). HslV is a cylindrical homooligomer of two stacked hexameric rings with putative substrate openings of about 0.7 nm at each end of the cylinder axis (Bochtler *et al.*, 1997). Proteasomal α-type proteins have not been identified in most eubacteria including those whose genomes have been completely sequenced (De Mot *et al.*, 1999). Instead, a six- to seven-membered-ring HslU ATPase interacts directly with HslV and is required for the hydrolysis of peptide and protein substrates in an ATP-dependent reaction (Rohrwild *et al.*, 1996). The occurrence of HslUV and 20S proteasomes among all organisms appears to be mutually exclusive, and some organisms such as the cyanobacterium *Synechocystis* sp. PCC6803 as well as the eubacterial pathogens *Mycoplasma genitalium*, *Mycoplasma pneumoniae*, *Chlamydia trachomatis*, and *Treponema pallidum* do not encode either protease (De Mot *et al.*, 1999).

C. MECHANISM OF PEPTIDE BOND HYDROLYSIS

Small synthetic peptide substrates are often used to characterize the peptidase activities of purified 20S proteasomes. Eukaryal 20S proteasomes have three major peptidase activities including hydrolysis of peptide bonds carboxyl to hydrophobic, basic, and acidic amino acid residues [chymotrypsin-like, trypsin-like, and postglutamyl peptide hydrolyzing (PGPH) or caspase-like activities, respectively] (Rivett, 1993; Orlowski *et al.*, 1993; Cardozo, 1993). Mutagenesis and cocrystallization of the yeast 20S proteasome with specific inhibitors suggest that β1, β2, and β5 subunits are responsible for the caspase-, trypsin-, and chymotrypsin-like activities, respectively (Heinemeyer *et al.*, 1993; Arendt and Hochstrasser, 1997; Groll *et al.*, 1997; Dick *et al.*, 1998). The catalytic role of the remaining four β subunits, if any, is unclear. The 20S proteasomes isolated from many eukaryotes have two additional peptidase activities including cleavage between small neutral amino acids and after branched chain residues (SNAAP and BrAAP activities) (Orlowski *et al.*, 1993; Mykles, 1996). Mutations in the β1 subunit of yeast (Dick *et al.*, 1998) and inhibition of 20S proteasomes isolated from bovine pituitary with 3,4-dichloroisocoumarin (Cardozo *et al.*, 1999) strongly suggest the involvement of the PGPH activity in the cleavage of BrAAP substrates. Mammalian 20S proteasomes also undergo a significant change in subunit composition after IFN-γ induction, which results in replacement of the constitutive "active" β subunits with the three β-subunit paralogs, β1i (LMP2), β2i (LMP10/MECL-1), and β5i (LMP7), which are also catalytically active (Glynne *et al.*, 1991; Martinez and Monaco, 1991; Ortiz-Navarrete *et al.*, 1991; Nandi *et al.*, 1996; Hisamatsu *et al.*, 1996). This alters the cleavage specificity of 20S proteasomes (Driscoll *et al.*, 1993; Gaczynska *et al.*, 1993, 1996; Eleuteri *et al.*, 1997; Orlowski *et al.*, 1997; Cardozo and Kohanski, 1998) and may facilitate the generation of antigenic peptides (Fehling *et al.*, 1994; Van Kaer *et al.*, 1994; Cerundolo *et al.*, 1995; Sibille *et al.*, 1995). Peptidase activities of 20S proteasomes are further enhanced and modified by association with the IFN-γ-inducible 11S regulatory complex (Ustrell *et al.*, 1995; Groettrup *et al.*, 1996; Mykles, 1996).

Prokaryotic 20S proteasomes are relatively simple in the types of peptides hydrolyzed compared to those of eukaryotes. Most have only chymotrypsin-like peptidase activity including those of the eubacteria *Rhodococcus erythropolis* (Tamura *et al.*, 1995), *Streptomyces coelicolor* (Nagy *et al.*, 1998), *Frankia* strain ACN14a/ts-r (Pouch *et al.*, 2000) as well as the archaea *T. acidophilum* (Dahlmann *et al.*, 1992) and *H. volcanii* (Wilson *et al.*, 1999). The 20S proteasomes of the methanoarchaea *Methanosarcina thermophila* (Maupin-Furlow *et al.*, 1998) and *Methanococcus jannaschii* (Wilson *et al.*, 2000) are somewhat unusual, with high levels of both chymotrypsin- and caspase-like peptidase activities. Only low levels of trypsin-like peptidase activity have been observed for archaeal proteasomes (Akopian *et al.*, 1997; Maupin-Furlow *et al.*, 1998; Wilson *et al.*, 2000). A few of the archaeal enzymes are stimulated by SDS (Akopian *et al.*, 1997; Maupin-Furlow *et al.*, 1998) but none appear to purify in a latent state, suggesting that the openings of the α-rings do not significantly restrict access of peptides and unfolded proteins to the β-subunit active sites.

It should be noted that analysis of the peptide products generated during the degradation of proteins by 20S proteasomes reveals redundant cleavage patterns which do not appear to correlate with preferences for these synthetic peptide substrates (Wenzel *et al.*, 1994; Dick *et al.*, 1994; Leibovitz *et al.*, 1994, 1995; Ehring *et al.*, 1996). Typically the peptide substrates have a scissile bond that is an amide bond between an amino acid and an aromatic amine (e.g., Amc, 7-amino-4-methyl-coumarin), which differs structurally from true peptide bonds. Thus, peptidase activities are useful in providing insight into how small peptides bind and are hydrolyzed by the active sites formed by the β-type subunits but do not fully reflect the mechanisms of protein hydrolysis.

Several β-type subunits of 20S proteasomes belong to the amino-terminal (Ntn) hydrolase family (Brannigan et al., 1995) which also includes HslV (Bochtler et al., 1997), penicillin acylase (Duggleby et al., 1995), glutamine-phosphoribosyl-pyrophosphate amidotrans-ferase (Smith et al., 1994), lysosomal aspartyl glucosaminidase (Oinonen et al., 1995), asparagine synthase (Larsen et al., 1999), and glucosamine-6-phosphate synthase (Isupov et al., 1996). The common feature of this family is the generation of an amino-terminal residue that is used as an active-site nucleophile in hydrolysis and that is exposed after autocatalytic removal of a propeptide (Seemüller et al., 1996; Maupin-Furlow et al., 1998). Among the β-type proteasomal subunits undergoing such maturation, an amino-terminal Thr is exposed which has a γ-oxygen that acts as the nucleophile and an α-amino group that is the likely proton acceptor in the hydrolysis of peptide bonds (Löwe et al., 1995; Fenteany et al., 1995; Seemüller et al., 1995a; Groll et al., 1997; Maupin-Furlow et al., 1998). In addition, a salt bridge formed by conserved Lys33 and Asp/Glu17 residues of the mature β subunit may also be involved in accepting the Thr1 side-chain proton through a charge-relay system.

The 20S proteasomes are classified as Ntn hydrolases as determined by mutagenesis, inhibitor, and crystallography studies. Initially, all Ser, Cys, and His residues of the T. acidophilum 20S proteasome were systematically modified to Ala with no change in protease activity (Seemüller et al., 1995b). This revealed that the 20S proteasome was not an unusual type of serine protease as suggested previously by inhibitor studies (Mason, 1990; Dahlmann et al., 1992; Djaballah et al., 1992; Orlowski et al., 1993). Further evidence for the mechanism of proteolysis came from two simultaneous studies on the T. acidophilum 20S proteasome. The amino-terminal Thr of the β subunit was modified to an Ala, which resulted in abolishment of the chymotrypsin-like peptidase activity (Seemüller et al., 1995a), and the 20S proteasome was cocrystallized with a peptide aldehyde, acetyl-leu-leu-norleucinal, which bound to the active site and inhibited peptidase activity (Löwe et al., 1995). The results of these studies were in agreement with the finding that a hydrolyzed form of the antibiotic lactacystin, clasto-lactacystin β-lactone, covalently bound to the amino-terminal Thr residues of a subset of β-type subunits to inhibit several peptidase activities of a mammalian 20S proteasome (Fenteany et al., 1995; Dick et al., 1996). Amino acid sequence alignment of eukaryal β-type subunits predicts that only three of the seven constitutive β-type subunits are "active" and harbor the residues involved in peptide bond hydrolysis (Seemüller et al., 1995a). Cocrystallization of the yeast 20S proteasome with peptidase inhibitors suggests that the caspase-, trypsin-, and chymotrypsin-like activities are mediated by the $\beta 1$, $\beta 2$, and $\beta 5$ subunits, respectively (Groll *et al.*, 1997). These results are consistent with the previous *in vivo* mutagenesis studies that suggested that pairs of active and "inactive" β-type subunits cooperate to catalyze the three peptidase activities (Heinemeyer *et al.*, 1991, 1993; Hilt *et al.*, 1993; Enenkel *et al.*, 1994; Chen and Hochstrasser, 1996; Gueckel *et al.*, 1998; Dick *et al.*, 1998). In contrast, the methanoarchaeal 20S proteasomes have 14 identical β subunits catalyzing both chymotrypsin- and caspase-like peptide activities (Maupin-Furlow *et al.*, 1998; Wilson *et al.*, 2000).

There is a tendency to assume that identical active β subunits of proteasomes have the same kinetic constants and activities; however, there are several studies that suggest otherwise. For example, the two active sites of the eukaryal enzyme that cleave PGPH substrates have different biochemical properties with sigmoidal kinetics consistent with positive cooperativity. One site has a much higher affinity for substrate than the other site, and these two catalytic components respond differently to metal ions and peptide aldehydes (Arribas and Castaño, 1990; Orlowski *et al.*, 1991; Djaballah and Rivett, 1992). Likewise, the two chymotrypsin-like peptidase activities of rat liver 20S proteasomes can be distinguished based on their different susceptibilities to inhibition by peptidyl chloromethyl ketones (Djaballah *et al.*, 1992; Reidlinger *et al.*, 1997). In general, subunit interactions appear to modify the kinetics of substrate binding as well as the specificity of the proteolytic active sites within the proteasome complex.

As discussed above, it is suspected that the amino group of Thr1 is the primary proton acceptor of 20S proteasomes in the hydrolysis of peptide bonds, but this remains to be directly determined. The Lys33 residue is near this catalytic nucleophile but is predicted to be protonated at physiological pH. Additionally, Lys33 is not needed for the trypsin-like peptidase activity of the *Methanosarcina* 20S proteasome, suggesting that it is not essential for peptide bond hydrolysis (Maupin-Furlow *et al.*, 1998). In contrast, amino-α-acetylation of propeptide-deleted β subunits inhibits peptidase activity in yeast, providing the most definitive evidence that the Thr1 α-amino group is involved in the catalytic mechanism of the proteasome (Arendt and Hochstrasser, 1999). Water molecules may also play a role in proteolysis as a proton shuttle between the amino-terminal Thr γ-oxygen and the α-amino group based on their location in the yeast 20S proteasome crystal structure (Bochtler *et al.*, 1999).

Based on these studies, a mechanism similar to serine proteases is envisioned. It is likely that the hydroxyl group of the active-site threonine residue initiates hydrolysis by attacking the carbonyl of the peptide bond. This results in the formation of a tetrahedral intermediate that collapses into an acyl-enzyme intermediate and releases the peptide product generated downstream of the cleavage site. Then nucleophilic attack of the acyl-enzyme intermediate by water yields free proteasome enzyme, and the second peptide product upstream of the cleavage site is released. Why threonine functions as the active-site nucleophile in proteasomes was recently investigated by comparing the kinetics of Thr1Ser, Thr1Cys, and Thr1Ala mutants to the unmodified proteasome. Based on this and earlier studies it appears that the Thr1Cys and Thr1Ala mutants are completely inactive in peptide and protein degradation. The Thr1Ser mutant, in contrast, is relatively unaffected in cleavage of peptides synthesized with aromatic amine reporter groups (e.g., Amc) but diminished severalfold in hydrolysis of proteins, peptide esters, and short polypeptides (Seemüller *et al.*, 1995a; Maupin-Furlow *et al.*, 1998; Kisselev *et al.*, 2000). Thus, the rate-limiting step in a proteasome with a serine nucleophile appears to be the collapse of the tetrahedral intermediate into the acyl-enzyme since the group amino to the cleavage site in the substrate is so influential in hydrolytic activity of the Thr1Ser mutant proteasome. The additional methyl group of the threonine-vs-serine side chain is somehow necessary to support the rapid rates of protein breakdown catalyzed by the proteasome. Furthermore, in contrast to the Thr1Ser mutant, the rate-limiting step in proteolysis by the wild-type proteasome appears to be the entry of the protein into the proteasome chamber or translocation of substrate between active sites, since the rate of peptide bond cleavage decreases with increasing chain length of unfolded polypeptide (Kisselev *et al.*, 2000).

Proteasomes appear to hydrolyze unfolded polypeptides via a processive mechanism in the cell, i.e., they completely degrade a substrate protein to oligopeptides before attacking another protein molecule (Akopian *et al.*, 1997; Kisselev *et al.*, 1999a). The majority of proteins are likely hydrolyzed by processive mechanisms in the cell to avoid the accumulation of partially digested proteins and peptide fragments that would otherwise be harmful. Processive degradation appears to be an intrinsic feature of 20S proteasomes and may be accounted for by a trapping of substrate inside the cylinder. *In vitro*, 20S proteasomes alone hydrolyze relatively small, unfolded substrate proteins by a processive mechanism in the absence of ATP or cofactors. This has been demonstrated using a fluorescent derivative of bovine β-casein, a 209-amino acid globular protein containing five or six residues of fluorescein isothiocyanate (FITC) per molecule, as a substrate protein (Akopian *et al.*, 1997; Kisselev *et al.*, 1999a). The *T. acidophilum* 20S proteasome as well as rabbit muscle 26S and 20S proteasomes were incubated with a molar excess of FITC-labeled casein and the fluorescent peptide products of the reaction were analyzed by reversed-phase HPLC. These enzymes each generated a specific pattern of products independent of time, which suggests a processive mechanism of degradation. Although association of the eukaryal 20S proteasome with the 19S cap in the presence of ATP modified the pattern of peptides generated, it did not influence processive hydrolysis of FITC-labeled casein (Kisselev et al., 1999a). The kinetics of hydrolysis of the 30-amino acid insulin B-chain by 20S proteasomes isolated from human red blood cells (Dick et al., 1991) and enolase 1 by yeast 20S proteasomes (Mathew and Morimoto, 1998) support this model. The related Lon and ClpAP proteases of E. coli also degrade proteins processively; however, unlike 20S proteasomes, the mechanism appears to be linked to ATP hydrolysis (Goldberg, 1992; Maurizi, 1992; Thompson et al., 1994). Interestingly, some proteins are not degraded by 20S proteasomes in a processive mechanism and instead are cleaved into large fragments. This may occur in vivo during the cotranslational processing of NF-κB p105 and related NF-κB p100 to the smaller p50 and p52 proteins (Palombella et al., 1994; Sears et al., 1998; Lin et al., 1998; Coux and Goldberg, 1998; Orian et al., 1999; Heusch et al., 1999). Furthermore, in vitro degradation of unfolded S-carboxamidomethylated lysozyme of 129 amino acids by bovine 20S proteasomes is nonprocessive, with the release of intermediate fragments of masses greater than 7 kDa (about 60 amino acids) that are later hydrolyzed to smaller peptides of 6 to 20 amino acids (Wang et al., 1999a). Thus, processive hydrolysis of some protein substrates may be more efficient when 20S proteasomes are associated with energy-requiring complexes such as the eukaryal 19S cap, archaeal PAN, or eubacterial ARC (see discussion of AAA$^+$ proteins). Consistent with this possibility, the estimated duration of 6 min for the in vitro degradation of one molecule of the 436-amino acid enolase by one yeast 20S proteasome is similar to that for other 20S proteasomes, but the rate is far lower than the expected physiological rates (Nussbaum et al., 1998).

The exact determinant of size and type of peptide products generated by 20S proteasomes remains unclear. Originally it was noted that the average length of products synthesized during hydrolysis of proteins by the T. acidophilum 20S proteasome was between seven and nine amino acids, which is in good agreement with the distance between the proteolytic active sites of 2.8 nm (Wenzel et al., 1994). Thus, the spacing between the multiple active sites was proposed to act as a molecular ruler in determining the length of the peptide product (Löwe et al., 1995; Baumeister and Lupas, 1997; Groll et al., 1997). However, when 20S proteasomes are incubated with a molar excess of substrate protein and the relative amounts of peptide products are determined at a constant rate of degradation, the individual products range from 3 to 30 amino acids and their sizes fit a log-normal distribution (Kisselev et al., 1998, 1999a). A comparison of yeast 20S proteasomes with six, four, or two active sites reveals that the average fragment lengths for peptide products does not depend on the number of active sites in the chamber. These results provide strong evidence that the distance between the active sites does not influence the peptide product length (Nussbaum *et al.*, 1998). Thus, the dimensions of the axial gates formed by the α-ring may play a role in determining the size of products released from 20S proteasomes or, alternatively, the active β subunits preferentially bind certain lengths of polypeptides to generate the observed size distribution of products. Interestingly, regulatory components that associate with 20S proteasomes appear to influence product size. *In vitro* digestion of β-casein by human 20S and 26S proteasomes reveals that overlapping but distinct sets of peptide products are generated, with the 26S proteasome producing shorter cleavage products than the 20S proteasome (Emmerich *et al.*, 2000).

There is substantial evidence that 20S proteasomes do not cleave at random; instead cleavage is highly specific and allosterically regulated. In general, 20S proteasomes appear to cleave protein substrates with a preference for proline at position 4 (P4) and leucine at P1 upstream from the cleavage site. A preference for P1' amino acids that promote β-turns is also observed (Nussbaum *et al.*, 1998; Dick *et al.*, 1998; Emmerich *et al.*, 2000). There may also be an ordered, cyclical "bite–chew" mechanism of protein degradation that determines the type of products generated based on studies demonstrating allosteric control of the peptidase sites of rabbit muscle 20S proteasomes. The chymotrypsin-like site appears to be involved in the initial cleavage or "bite" of proteins followed by a stimulation of caspase-like activity which appears to facilitate "chewing" of the polypeptide fragments (Kisselev *et al.*, 1999b). This β- to β-subunit allostery, coupled with the demonstrated interdependence of the substrate binding and cleaving processes, may be linked to the allosteric transition of the axial gates formed by the α subunits from closed to open conformation after the addition of peptide substrates to uninhibited yeast 20S proteasomes (Dorn *et al.*, 1999; Osmulski and Gaczynska, 2000). However, recent data based on the complex kinetic effects of Ritonavir on 20S proteasomes challenge the bite–chew model and suggest a kinetic "two-site modifier" model. The model suggests that substrates or effectors may bind the active sites as well as a second noncatalytic modifier site, and this binding would influence the types of products generated by proteasomes (Schmidtke *et al.*, 2000).

D. Assembly and Processing of 20S Proteasomes

The assembly pathway of 20S proteasomes is not fully elucidated; however, current models suggest some distinct differences among the three domains (Gerards *et al.,* 1998a). One major difference is the number of β-type subunits cleaved at the amino terminus during maturation of the proteolytic complex. All of the β-type subunits of 20S proteasomes characterized from prokaryotes are processed to expose an active-site Thr1 residue. An exception may be the crenarchaeon *Aeropyrum*, which appears to encode both active and inactive β-type subunits based on the genome sequence (Kawarabayasi *et al.,* 1999). In contrast, the yeast 20S proteasome contains two unprocessed subunits, β3 and β4, and five processed subunits, including the "inactive" β6 and β7 as well as the "active" β1, β2, and β5 (Kimura *et al.,* 2000). The processing sites of these β subunits (β1, β2, and β5–β7) are conserved among yeast, plants, and mammals (Fu *et al.,* 1999). An additional difference is the length and sequence of the β propeptide. Some β-type proteins do not even have propeptides as seen for many eubacterial HslV proteases, whereas others have rather large propeptides, i.e., the 75 residue propeptide of the yeast β5 protein. Interestingly, higher eukaryotes use alternative leader exons to influence the length of some propeptides including those of the β1i and β5i proteins (Früh *et al.,* 1992; Martinez and Monaco, 1993).

Proteolytic exposure of the amino terminus of β-type proteasomal subunits is autocatalytic and most commonly occurs at a conserved Gly–Thr bond (Chen and Hochstrasser, 1996; Seemüller *et al.,* 1996; Schmidtke *et al.,* 1996; Zühl *et al.,* 1997b; Maupin-Furlow *et al.,* 1998; Ditzel *et al.,* 1998). Exceptions include the autocatalytic processing of β6 and β7 proteins of yeast at His–Gln and Asn–Thr bonds, respectively (Kimura *et al.,* 2000). The common fold of the Ntn hydrolase family (Brannigan *et al.,* 1995) suggests that β-type proteins are processed by an intramolecular mechanism due to the conformational strain at the cleavage site created by an unusual β-sheet structure. In yeast, maturation of β proteins to active subunits was examined by mutagenesis, which suggested that it is independent of the presence of other active β subunits and occurs by intrasubunit autolysis (Groll *et al.,* 1999). During removal of the β propeptide, the active-site amino group of Thr1 of the mature protein is not yet exposed in the precursor to accept protons for peptide bond hydrolysis. A catalytic water molecule may instead abstract a proton from the hydroxyl group of the Thr residue of the precursor (Bochtler *et al.,* 1999), consistent with crystal structures of Ntn hydrolases (Duggleby *et al.,* 1995; Groll *et al.,* 1997). This would promote nucleophilic attack by the Thr γ-oxygen on the carbonyl carbon of the preceding Gly–Thr bond of the β precursor. This bond is ultimately hydrolyzed to release the propeptide and expose the active-site Thr as the amino-terminal residue of the mature protein. In contrast, the inactive β6 and β7 subunits that are intermediately processed are likely cleaved by an intermolecular mechanism mediated by the active β subunits (Nussbaum *et al.*, 1998). *In vitro* studies with the *T. acidophilum* 20S proteasome suggest that archaeal β proteins may also mature by an intermolecular event (Seemüller *et al.*, 1996). Mixing wild-type β proteins with β mutants, which alone are not processed, results in processing of up to 90% of the mutant β protein in recombinant 20S proteasomes synthesized in *E. coli* (Seemüller *et al.*, 1996). Interestingly, the conserved Gly residue amino to the cleavage site of the propeptide appears to be more important for propeptide removal than the Thr residue carboxyl to this site (Seemüller *et al.*, 1996).

The exact role(s) of the propeptide at the amino terminus of the β-type precursor proteins of 20S proteasomes is not fully understood. Although no direct evidence is available, the β propeptide is widely believed to protect the cell from unregulated hydrolysis of intracellular proteins that would otherwise occur if 20S proteasome assembly intermediates were exposed in the cytosol. The β propeptide is apparently needed for proper folding of the β precursor and may act as a molecular chaperone that protects the β protein against degradation or premature multimerization similar to the propeptides of elastase, α-toxin, and lipase (Braun and Tommassen, 1998). The yeast β propeptides also protect the active-site amino-terminal Thr from modification prior to incorporation into 20S proteasomes as described below.

Many proteasome-like HslV proteases do not even have a propeptide and instead rely upon a methionine aminopeptidase to expose an active-site Thr1 residue. However, unlike genuine 20S proteasomes, HslV displays only weak peptidase activity and weak binding of inhibitors in the absence of the ATPase regulatory component, HslU (Yoo *et al.*, 1996; Rohrwild *et al.*, 1996; Huang and Goldberg, 1997). HslU and ATP are apparently needed to modify allosterically the geometry of the HslV active site to increase the binding affinity for peptides and proteins (Bogyo *et al.*, 1997). Thus, in the absence of the regulatory HslU ATPase, assembly intermediates of HslV probably have limited activity in the nonspecific, unregulated hydrolysis of proteins in the cell.

The eubacterial β proteasomal propeptides are relatively large, ranging from 49 to 65 residues. They are needed in *cis* or *trans* for productive assembly of 20S proteasomes *in vitro* as well as in recombinant *E. coli* (Zühl *et al.*, 1997b). Likewise, the yeast β5 propeptide of 75 amino acid residues is essential for incorporation of the β5 subunit into active 20S proteasomes and can be supplied in *trans* (Chen and Hochstrasser, 1995, 1996). Thus, many β propeptides appear to facilitate folding of the precursor or are necessary for appropriate contacts with other subunits or proteins for efficient 20S proteasome assembly. Removal of the propeptide from the yeast β1, β2, and β5 residues by mutagenesis prior to

20S proteasome assembly suggests that another critical function of the β propeptide is protection of the amino-terminal catalytic Thr residues against cotranslational amino-α-acetylation (Arendt and Hochstrasser, 1999). Surprisingly, many archaeal β propeptides are relatively short, ranging from 6 to 11 residues, and are dispensable in the assembly of 20S proteasomes *in vitro* as well as in recombinant *E. coli* (Zwickl *et al.,* 1994; Seemüller *et al.,* 1996; Maupin-Furlow *et al.,* 1998; Wilson *et al.,* 2000). Even the 49-residue β propeptide of *H. volcanii* is not essential for *in vitro* assembly of 20S proteasomes from individual $\alpha 1$ and β subunits dissociated in low salt (Wilson *et al.,* 1999). Thus, the archaeal β propeptide appears to have a minimal, if any, role in the actual assembly mechanism of individual subunits into active 20S proteasomes outside the archaeal cell. Instead, the archaeal β propeptide may be needed for proper folding, stabilization, and/or protection of the amino terminus of the β subunit when synthesized *in vivo* as suggested by a recent study in *H. volcanii* (Kaczowka and Maupin-Furlow, unpublished).

The order and mechanism of incorporation of individual subunits into active 20S proteasomes are still unclear. In the eubacteria, the α- and β-type proteins alone remain monomeric but, when mixed together, form half ($\alpha_7\beta_7$) and full ($\alpha_7\beta_7\beta_7\alpha_7$) 20S proteasomes (Zühl *et al.,* 1997b). Mature, correctly processed β subunits are not detected in these half-proteasomes, suggesting that cleavage of the β propeptide occurs during the joining of the half-proteasomes to form active 20S proteasomes (Zühl *et al.,* 1997b). The α and β proteins of the eubacteria must interact to initiate half-complex formation, and the rate-limiting step of assembly appears to be the subsequent processing of the β propeptide after dimerization of the two half-proteasomes (Zühl *et al.,* 1997b). Unlike the archaea and eukarya, the genes encoding the eubacterial 20S proteasome α and β proteins are in tandem and appear to be cotranscribed from the same operon, which may ensure equimolar synthesis of these subunits for 20S proteasome assembly (Tamura *et al.,* 1995).

Like the eubacteria, the archaeal β proteins alone are inactive even in the absence of propeptides and form only monomers, dimers, trimers, or aggregates (Zwickl *et al.,* 1994; Maupin-Furlow *et al.,* 1998; Wilson *et al.,* 2000). However, in contrast, archaeal α subunits spontaneously self-assemble into single and/or double-stacked heptameric rings when produced in recombinant *E. coli* (Zwickl *et al.,* 1994; Maupin-Furlow *et al.,* 1998; Wilson *et al.,* 2000). Mutagenesis of archaeal α subunits reveals that the amino-terminal α-helix, which is absent in β-type proteins, is important for α-ring and 20S proteasome complex formation (Zwickl *et al.,* 1994). The archaeal α-ring appears to provide the scaffolding for autoassembly of the β-ring in the formation of active 20S proteasomes (Zwickl *et al.,* 1994; Maupin-Furlow *et al.,* 1998). Whether additional maturation factors are necessary to stimulate either eubacterial or archaeal 20S proteasome assembly *in vivo* is unclear.

Assembly of eukaryal 20S proteasomes is more complicated, with 14 different subunits located at specific positions in either the α- or the β-ring. Some eukaryal α subunits including the human α7 (HsC8) and *Trypanosoma brucei* α5 proteins self-assemble into single, double, and even four-stacked protein rings when produced in recombinant *E. coli* (Gerards *et al.*, 1997; Yao *et al.*, 1999). The human α7 protein also induces heteroligomeric ring formation when synthesized with α1 (PROS27) and α6 (PROS30) that are adjacent to α7 in the axial ring of 20S proteasomes (Gerards *et al.*, 1998b). Similar to archaea, the amino terminus of eukaryal α-type subunits is essential for incorporation of these subunits into 20S proteasomes (Seelig *et al.*, 1993). It remains to be determined whether all seven eukaryal α subunits spontaneously assemble into a distinct ring or whether the α subunits require interaction with β-type proteins or additional factors for proper oligomerization. *In vivo*, eukaryal assembly intermediates of 13S to 16S have been identified which contain at least six α-type and several β-type proteins (Frentzel *et al.*, 1994; Patel *et al.*, 1994; Yang *et al.*, 1995; Schmidtke *et al.*, 1997). Thus, α-ring formation may precede β protein incorporation, although further studies are needed. Unprocessed and partially processed forms of β-type proteins have been observed in 13S to 16S half-proteasome complexes including the β1i, β5i, and β6 proteins (Frentzel *et al.*, 1994; Yang *et al.*, 1995; Schmidtke *et al.*, 1996; Nandi *et al.*, 1997). In addition, putative maturation factors are associated with half-proteasomes including Ump1 and Hsc73, an Hsp70 family member (Schmidtke *et al.*, 1997; Schmidt *et al.*, 1997; Ramos *et al.*, 1998). Ump1 appears to influence the conformation of β propeptides and to ensure proper autocatalytic processing of the β-type subunits in yeast. Ump1 then becomes trapped inside the assembled, active 20S proteasome for destruction (Ramos *et al.*, 1998). The human Ump1 homologue appears to serve a similar role and is IFN-γ inducible (Burri *et al.*, 2000). Proteasome maturation requires the supply of newly translated proteins, but it is unknown whether this is a need for newly synthesized subunits or a maturation factor(s) that are otherwise rate limiting (Frentzel *et al.*, 1994).

E. CHEMICAL INHIBITORS OF PROTEASOME ACTIVITY

Pharmacological inhibitors of proteasomes have been available since 1994 and are used extensively for *in vivo* studies. Several types of proteasome inhibitors have been analyzed and include peptide aldehydes, vinyl sulfones, lactacystin and its derivative *clasto*-lactacystin

β-lactone, and peptide boronates. Many of these inhibitors are used *in vivo* since they readily enter many cell types. However, selective inhibitors of lysosomal and Ca^{2+}-activated proteases are needed as controls in eukaryotic cell studies, since many proteasome inhibitors also influence cathepsin and calpain activities (Lee and Goldberg, 1998). *In vivo* application of inhibitors has only recently been initiated in the archaea and may provide some insight into archaeal proteasome function (Ruepp *et al.*, 1998).

Peptide aldehydes [e.g., Cbz-ile-glu(O-tBu)-ala-leucinal (PSI), Cbz-leu-leu-leucinal (MG132), Cbz-leu-leu-norvalinal (MG115), and acetyl-leu-leu-norleucinal (MG101), where Cbz is N-benzyloxycarbonyl] are substrate analogues which inhibit the transition state during hydrolysis of peptide bonds carboxyl to hydrophobic or acidic residues (Rock *et al.*, 1994; Figueiredo-Pereira *et al.*, 1994; Lee and Goldberg, 1996). The inhibitors form reversible hemiacetyl complexes with the active-site amino-terminal threonine hydroxyl groups, thus inhibiting the proteolytic activity of the proteasome without influencing its ATPase or isopeptidase activities. Lactacystin and its derivative *clasto*-lactacystin β-lactone are natural products that were originally isolated from the actinomycetes (Omura *et al.*, 1991). Lactacystin forms a β-lactone in aqueous solutions. The β-lactone is the active form of the inhibitor and functions as a pseudosubstrate that is covalently linked to the active site Thr1 hydroxyl groups of the proteasome (Fenteany *et al.*, 1995; Dick *et al.*, 1996). Recently peptides containing either a carboxyl-terminal vinyl sulfone (e.g., Cbz-leu-leu-leu-vinyl sulfone) or boronate groups (e.g., Cbz-leu-leu-leu-boronate) have been developed and are potent inhibitors that covalently modify the Thr1 active sites of the proteasome (Bogyo *et al.*, 1997; Adams *et al.*, 1998b).

Potential therapeutic applications of proteasome inhibitors include reducing the production of inflammatory mediators through their ability to block the activation of NF-κB by the proteasome. Furthermore, the use of proteasome inhibitors for cancer therapy appears promising since these compounds induce apoptosis in rapidly dividing transformed cells while seeming to reduce apoptosis in other cells (Lee and Goldberg, 1998; Adams *et al.*, 2000).

III. Protease-Associated Regulators

A. ENERGY-DEPENDENT REGULATORS

Self-compartmentalized proteases alone typically catalyze only the hydrolysis of unfolded polypeptides and short peptides and are unable to degrade most native or aggregated proteins. Instead, separate

ATPase (nucleotide triphosphatase, NTPase) domains or complexes are needed to activate the hydrolysis of folded and aggregated proteins by these proteases (Schmidt et al., 1999). The ATPase component is likely to function as a reverse chaperone to unfold and disaggregate protein substrates and to facilitate their entry into the chamber harboring the proteolytic active sites.

1. The AAA+ Superfamily

The majority of NTPases involved in energy-dependent proteolysis classify to the AAA family (ATPases associated with various cellular activities) or AAA+ superfamily (Neuwald et al., 1999). The Rpt (regulatory particle triple-A) subunits of the 19S cap of the eukaryal 26S proteasome, archaeal PAN (proteasome-activating nucleotidase), eubacterial ARC (ATPase ring-forming complex), ClpX, ClpA, HslU, Lon, and FtsH proteins are all proposed to be involved in proteolysis and are included in this superfamily. Molecular chaperones such as the S. cerevisiae Hsp104 protein and other enzymes mediating protein:protein interactions are also related. The AAA proteins are P-loop nucleotidases with at least one characteristic 250 residue ATP-binding domain which includes a Walker A motif ($G-X_2-GXGKT$), predicted to be involved in coordination of Mg^{2+} and formation of hydrogen bonds with nucleotide triphosphates including the β- and γ-phosphates. A modified Walker B motif or "DEAD" box is also located in this domain and is predicted to be involved in Mg^{2+} binding and ATP hydrolysis (Walker et al., 1982; Saraste et al., 1990). A second region of homology or SRH motif [(T/S)–(N/S)–X_5–DXA–X_2–R–X_2–RX–(D/E)] that distinguishes AAA+ proteins from the broader family of Walker-type ATPases is conserved. The SRH motif is part of the interface between neighboring subunits in the crystal structure of the NEM-sensitive fusion protein (NSF) (Lenzen et al., 1998; Yu et al., 1998). Mutations of the SRH motif of FtsH suggest that it may play an intermolecular catalytic role that is ATP hydrolysis dependent (Karata et al., 1999).

The substrate-binding domain(s) of the unfoldase components of proteases is (are) not fully elucidated. AAA+ proteins contain a carboxyl-terminal helical extension, distal to the ATPase domain(s), which has been implicated in substrate selection and discrimination (SSD) (Levchenko et al., 1997a; Smith et al., 1999). Carboxyl-terminal portions of the AAA domains of eubacterial Lon and Clp proteases are necessary for binding and degradation of substrates in vitro (Smith et al., 1999). However, deletion of this region in Yme1p (FtsH orthologue) does not influence the binding of unfolded substrates in vivo (Leonhard et al., 1999). A recent crystal structure of the HslUV protease suggests that the carboxyl terminus of the HslU ATPase may not be substrate accessible and may instead bind between the subunits of the HslV protease (Sousa *et al.*, 2000). This would be consistent with the recent finding that the addition of an epitope tag to the carboxyl terminus of an Rpt (ATPase) subunit of the 19S cap regulator inhibits assembly of the 26S proteasome in yeast (Verma *et al.*, 2000). Modification of the amino-terminal coiled-coil domain found in many AAA^+ proteins influences binding of substrate proteins to the yeast Yme1-AAA (FtsH-like) protease (Leonhard *et al.*, 1999). However, amino-terminal modification also influences a variety of other activities including nucleotide hydrolysis by Rpt2 of the 26S proteasome and related archaeal PAN (Lucero *et al.*, 1995; Wilson *et al.*, 2000), subunit interaction of the 26S proteasome (Richmond *et al.*, 1997), and autodegradation of ClpA by the ClpAP protease (Seol *et al.*, 1995). Furthermore, the amino-terminal coiled-coil domain is not needed for stimulation of β-casein hydrolysis by the archaeal PAN-20S proteasome (Wilson *et al.*, 2000).

2. Proteasome-Associated AAA Proteins

In the eukaryotic cell, complexes containing AAA proteins are known to associate with 20S proteasomes and include the 19S cap (PA700) and a 300-kDa modulator. Of these, the 19S cap, composed of at least 17 or 18 subunits, associates with 20S proteasomes to form the energy-dependent 26S proteasome involved in the degradation of folded "native" proteins often covalently tagged with ubiquitin. The 19S regulator is thought to perform a number of functions in this process including recognition of polyubiquitinated substrates, recycling of ubiquitin, opening of the axial gates of the 20S proteasomes, and unfolding and translocation of substrate proteins into the proteolytic compartment of the 20S proteasome. Biochemical and genetic studies have been used to analyze the subunit organization of the asymmetrical 19S caps (Ferrell *et al.*, 2000). Six of the 19S subunits, Rpt1 to Rpt6, are AAA proteins and the remaining non-ATPase subunits are designated Rpn1 to Rpn12 (regulatory particle non-ATPase) in yeast (Finley *et al.*, 1998).

Recently, the 19S cap was separated into Lid and Base domains and characterized *in vitro* (Glickman *et al.*, 1998). The Lid is needed for recognition of ubiquitinated proteins and does not directly interact with the 20S core. The Base is the energy-dependent component that recognizes, unfolds, and feeds model "native" protein substrates such as casein into the proteolytic chamber of 20S proteasomes. The Base directly associates with the outer α-rings of the 20S core and is composed of eight subunits including the six AAA proteins (Rpt1 to Rpt6). Thus, it is speculated that these six Rpt proteins form a hexameric ring that interfaces the heptameric α-ring of the 20S core to form a symmetry mismatch similar to ClpAP protease (Beuron et al., 1998). Three-dimensional electron microscopy suggests that there is a flexible linkage or "wagging-type movement" between the 19S caps and the 20S core that may be a consequence of this symmetry mismatch (Walz et al., 1998).

The assembly of 19S caps with 20S proteasomes is not well understood in comparison with the biogenesis of 20S proteasomes. In a variety of eukaryotes, a 300-kDa modulator is involved in stimulating assembly of the 26S proteasome from 19S and 20S particles via an ATP-dependent mechanism. The modulator is composed of the Rpt4 and Rpt5 ATPase subunits of the 19S cap and an additional protein, p27 (DeMartino et al., 1996; Adams et al., 1997; Watanabe et al., 1998; Hastings et al., 1999). The exact role of the modulator in this process is unknown but may be an intermediate involved in insertion of the Rpt subunits to complete the maturation of immature or damaged 19S complexes.

Orthologs of eukaryal Base domain AAA$^+$ (Rpt) proteins are predicted from the archaeal genome sequence and include the PAN protein (proteasome-activating nucleotidase). Interestingly, not all archaea encode PAN proteins including T. acidophilum (Ruepp et al., 2000) and Pyrobaculum aerophilum (Zwickl et al., 1999). It is speculated that these organisms may use alternative AAA$^+$ proteins with closer relationships to the FtsH/CDC48 and Lon families for activation of proteasome-mediated protein degradation in the cell. In addition, the halophilic archaea including Halobacterium NRC-1 (Ng et al., 2000) and Haloferax volcanii (Maupin-Furlow, unpublished results) encode two PAN paralogs with over 70% identical amino acid residues. It is possible that this adds flexibility to the repertoire of substrates that are degraded by the proteasome similar to the related ClpA and ClpX ATPases that associate with the ClpP protease in eubacteria (see discussion below).

The M. jannaschii PAN protein has been purified and is an irregular ring-shaped ATPase and CTPase complex with a diameter similar to that of 20S proteasomes (Zwickl et al., 1999; Wilson et al., 2000). In vitro, PAN associates with the ends of 20S proteasomes in the presence of ATP much like the 19S cap complexes of eukaryotes (Wilson et al., 2000) and activates the hydrolysis of substrate proteins such as bovine β-casein in the presence of either ATP or CTP (Zwickl et al., 1999; Wilson et al., 2000). The amino-terminal 73 amino acids modify the specificity of nucleotides hydrolyzed by PAN and are not necessary for activation of β-casein degradation (Wilson et al., 2000). Interestingly, 5'-adenylyl β,γ-imidophosphate (AMP-PNP) is not hydrolyzed by PAN yet supports PAN-dependent activation of β-casein degradation as determined by the generation of new α-amino groups using fluorescamine

(Wilson *et al.*, 2000). AMP-PNP does not support the conversion of β-[^{14}C]-casein into acid-soluble products (Zwickl *et al.*, 1999). This suggests that the binding of a nonhydrolyzable nucleotide analogue to PAN supports a single or limited number of cleavages by 20S proteasomes of substrate protein into large, acid-precipitable products much like the Lon protease of *E. coli* (Edmunds and Goldberg, 1986).

The ARC protein of the actinomycetes is predicted to activate proteolysis by 20S proteasomes similar to the archaeal PAN protein. ARC is a divergent member of the AAA$^+$ family, which in *R. erythropolis* is encoded by a separate operon but spatially linked to genes encoding the α2 and β2 subunits of the eubacterial 20S proteasome (Lupas *et al.*, 1997b; Wolf *et al.*, 1998). The ARC protein purifies as a complex of single and double-stacked hexameric rings that hydrolyzes nucleotides including ATP, CTP, and ADP (Wolf *et al.*, 1998). ARC does not appear to activate protein degradation or associate with the 20S proteasome *in vitro*; however, the association constant for these isolated proteins may be too low to detect specific interactions outside the cell (Wolf *et al.*, 1998).

3. AAA Protein Structure and Mechanism

Recent biochemical advances are providing detailed insight into the structure and mechanism of the AAA$^+$ proteins. Many of these proteins form ring-like structures that are likely to be physiologically advantageous in the catalysis of protein folding and degradation. The ring may provide a compartment to isolate the substrate protein in a central cavity away from other proteins in the cytosol. This architecture may also facilitate cooperative interactions between the substrate and a high local concentration of binding sites (Horwich *et al.*, 1999).

X-ray crystal structures of several members of the AAA$^+$ family have been determined including the δ' subunit of DNA polymerase III clamp loader (Guenther *et al.*, 1997), the hexamerization domain D2 of the NSF vesicle-fusion protein (Lenzen *et al.*, 1998; Yu *et al.*, 1998), and the HslU nucleotide triphosphatase in the presence and absence of the HslV protease (Bochtler *et al.*, 2000; Sousa *et al.*, 2000). The 3.4-Å crystal structure of the *H. influenzae* HslUV protease is consistent with the solution structure of the active complex derived from small-angle X-ray scattering data (Sousa *et al.*, 2000). Cryo-electron micrograph three-dimensional reconstructions of the 26S proteasome (Walz *et al.*, 1998), ClpAP (Beuron *et al.*, 1998), and VAT (CDC48) (Rockel *et al.*, 1999) as well as structural analysis of the amino-terminal domain of VAT via nuclear magnetic resonance (Coles *et al.*, 1999) also provide insight into the overall AAA$^+$ protein structure. This, in combination with biochemical studies on Walker-type NTPases, has led to the development of models to explain how AAA⁺ proteins function (Schmidt *et al.*, 1999; Vale, 2000; Sousa *et al.*, 2000).

It appears that the AAA⁺ proteins classify to a broad group of mechanoenzymes that use similar nucleotide-dependent conformational changes to mediate many biological processes (e.g., proteolysis, protein folding, RNA unwinding). Based on current structural models, it appears that the nucleotide providing energy binds in regions that promote significant conformational changes in the AAA⁺ proteins. The nucleotide-binding site for the NSF D2 domain is at the interface of neighboring subunits. This arrangement is likely to promote the observed ATP-dependent association of the subunits into hexameric rings and facilitate intersubunit cooperativity (Lenzen *et al.*, 1998; Yu *et al.*, 1998). In HslUV the nucleotide binds at an interface formed by the carboxyl and amino domains of HslU, which are joined by an intermediate hinge. This nucleotide-binding region also associates with the apical helices of HslV. Thus, ATP-binding to HslU appears to induce a shift in the apical helices of HslV that propagates to the threonine active site, consistent with the ATP-dependent assembly and activation of HslUV (Sousa *et al.*, 2000).

The conformational changes of AAA⁺ proteins are most likely mediated by the presence or absence of the γ-phosphate of the NTP (Lorsch and Herschlag, 1998). In general, conformational changes in Walker-type NTPases are mediated by contacts between residues of the enzyme and the NTP γ-phosphate as well as the Mg^{2+} chelated to the β- and γ-phosphates (Smith and Rayment, 1996; Vale, 1996). A key residue involved in the conformational changes is probably the invariant aspartate (D) that is the first residue of the DEAD box (the Walker B motif) which binds Mg^{2+} via a water molecule. After hydrolysis of the NTP and release of the inorganic phosphate, the Mg^{2+} moves, causing conformational changes in the enzyme. In addition, the conserved lysine of the Walker A motif as well as other residues in direct contact with the β- and γ-phosphates of the bound NTP becomes disrupted after nucleotide hydrolysis. The β- and γ-phosphates of the nucleotide interact with an arginine finger motif, similar to the Ras–RasGAP complex (Ahmadian *et al.*, 1997). This may provide a mechanism for transducing structural alterations through ATP binding/hydrolysis to mediate intersubunit cooperativity and communication within the regulatory nucleotidase ring and protease for complex assembly, substrate affinity, protein unfolding, and protein translocation.

When AAA ATPases associate with proteases such as 20S proteasomes and ClpP, they appear to have an internal cavity and a central pore in the hexameric ring which forms a passage that is continuous

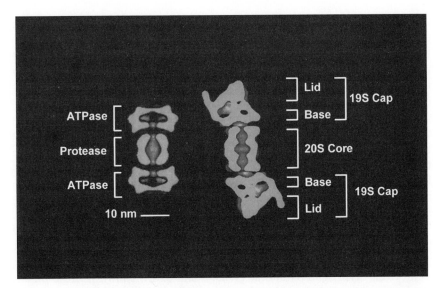

FIG. 3. Comparison of the energy-dependent ClpAP protease (left) to the 26S proteasome (right). The self-compartmentalized proteolytic chambers (20S proteasome and ClpP protease) and energy-dependent AAA⁺-regulatory complexes (19S cap or PA700 regulator and ClpA ATPase) are indicated. The 19S cap Lid and Base domains are also labeled. The Base domain is composed of eight subunits including the six AAA⁺ proteins, which are proposed to form a hexmeric ring similar to ClpA. (Figure adapted with permission from S. Wickner, M. R. Maurizi, and S. Gottesman.)

with the central channel of the protease (Fig. 3). This suggests that as the substrate protein becomes unfolded it passes through the central cavity of the AAA⁺ complex. Then it travels through a passage formed by a combination of the pore of the AAA⁺ ring and the narrow openings located at either end of the protease cylinder.

Similarly to molecular chaperones such as GroEL/ES, a first approximation for energy-dependent proteolysis incorporates a minimum two-state model with ATP- versus ADP-bound enzyme. The ATP-bound state may involve association of the regulatory unfoldase with the protease as well as an increased affinity of the unfoldase for the polypeptide ligand. Denaturation of substrate protein, however, may require the enzyme to cycle through at least one additional conformational state that is readily accessed by hydrolysis of ATP to the ADP-bound enzyme. The unfolding of substrate is likely to occur in the ring-like structure of the reverse chaperone. The conformational switch due to ATP hydrolysis may be similar to opening a trap door, which joins the unfoldase to the proteolytic chamber and allows the substrate to enter. Alternatively, the unfoldase may actively thread the extended polypeptide chain through the narrow passage of the protease while the protease is actively pulling the substrate into the proteolytic chamber. In either case, it is likely that multiple turnovers of the nucleotide are required for the unfoldase both to denature and to translocate the protein substrate into the proteolytic cylinder (Horwich *et al.*, 1999).

The protein-unfolding activities of a variety of AAA$^+$ proteins associated with proteases have been demonstrated *in vitro* by monitoring the fluorescence of derivatives of green fluorescent protein (GFP). Initial studies tested the action of ClpA on GFP-SsrA, a stable monomeric form of GFP with an 11-amino acid nonpolar carboxyl-terminal recognition peptide (SsrA-tag; see section on targeting substrates for degradation). Unfolding was examined in the presence and absence of a GroEL "trap" mutant that captures but does not release nonnative forms of GFP. These studies demonstrate that ClpA catalyzes ATP-dependent protein unfolding and that this unfolding is global as determined using deuterium–hydrogen exchange (Weber-Ban *et al.*, 1999). A more recent study demonstrates that ClpA binds specifically to a GFP-RepA protein fusion in a reaction that requires ATP binding, not hydrolysis, and that ATP hydrolysis is needed to unfold the fusion protein. In addition, ClpA binds globally unfolded proteins and releases these substrates only in the presence of hydrolyzable ATP, even when ClpA is in association with ClpP. These results reveal that there are two ATP-requiring steps for proteolysis: (1) the initial protein unfolding step and (2) the translocation of unfolded substrate into the proteolytic chamber (Hoskins *et al.*, 2000a). Interestingly, ClpXP with translocated substrate trapped within the proteolytic chamber retains the ability to unfold additional substrate proteins (Singh *et al.*, 2000). ClpX also catalyzes the release of substrates that are trapped in ClpP complexes in which the active site serine is inactivated by chemical modification (Kim *et al.*, 2000). More recent studies with the archaeal PAN demonstrate that it too catalyzes the ATP-dependent unfolding of GFP-SsrA and can be linked to degradation by 20S proteasomes (Benaroudj and Goldberg, 2000).

4. Chaperone Activities of AAA Proteins

The same energy-dependent AAA$^+$ complexes promoting the unfolding and translocation of proteins for degradation may also function as molecular chaperones that promote protein folding in the cell. This dual activity would provide a proofreading step following the initial binding of substrate proteins, enabling the cell to distinguish among proteins destined for refolding, disaggregation, or destruction (Wickner *et al.*, 1999). The number of ATPase complexes participating in proteolysis and protein folding in the cell is unclear (Schmidt *et al.*, 1999).

Often when putative "unfoldases" are separated from the protease component *in vitro*, they catalyze the reverse reaction by promoting protein folding and preventing protein aggregation similar to molecular chaperones (Glickman *et al.*, 1998; Hoskins *et al.*, 1998; Leonhard *et al.*, 1999; Pak *et al.*, 1999; Braun *et al.*, 1999).

It is currently unknown whether these *in vitro* results are consistent with reactions that occur in the cell; however, there is some evidence for chaperone activities mediated by these AAA$^+$ proteins *in vivo*. In eubacteria, the ClpX ATPase acts in the specific disassociation of proteins into active monomers, such as the phage MuA transposase tetramer that is disassembled from the DNA after recombination (Mhammedi-Alaoui *et al.*, 1994). Likewise the ClpA ATPase may also act as a chaperone in the transition of inactive bacteriophage RepA dimers to monomers that reinitiate P1 phage DNA replication in *E. coli* (Wickner *et al.*, 1994; Hoskins *et al.*, 1998; Pak *et al.*, 1999). Furthermore, mutations in *ftsH* cause improper orientation of some proteins in the cytoplasmic membrane that are overcome by high-level synthesis of molecular chaperones in *E. coli* (Shirai *et al.*, 1996). Abnormal assembly of penicillin binding protein 3 in the membrane has also been reported for *ftsH* mutants (Ogura *et al.*, 1991). Proper assembly of membrane proteins including the F_o domain of the proton ATPase and cytochrome *c* oxidase also requires active FtsH in *E. coli* and yeast (Arlt *et al.*, 1996; Akiyama *et al.*, 1996a; Leonhard *et al.*, 1999).

The 19S cap of the eukaryal 26S proteasome, which contains six Rpt (AAA$^+$) subunits, may also participate in maintaining the quality control of proteins by functioning as an unfoldase and a molecular chaperone. Recently, the 19S cap and the eight-subunit Base domain, which contains the six Rpt (AAA$^+$) proteins, have been shown to inhibit the aggregation of incompletely folded, nonubiquitinated proteins and to promote the refolding of denatured proteins (e.g., citrate synthase) to their native state *in vitro* (Braun *et al.*, 1999; Schmidt *et al.*, 1999; Strickland *et al.*, 2000). This chaperone activity appears also to operate *in vivo* by playing a nonproteolytic role in nucleotide excision repair to disassemble or rearrange the repair complex in yeast (Russell *et al.*, 1999). The exact mechanism of the refolding and disaggregation activity of the 19S cap ATPases is unclear. The stoichiometric ratios of enzyme-to-substrate protein needed for activity, ATP dependence, and detection of unfolded substrate:enzyme intermediates differ somewhat among current studies (Braun *et al.*, 1999; Strickland *et al.*, 2000).

Recent *in vitro* evidence suggests that the archaeal PAN protein may also reduce the aggregation of denatured proteins and refold certain denatured enzymes. These processes appear to require ATP binding but not ATP hydrolysis (Benaroudj and Goldberg, 2000). However, studies have not been performed to confirm the chaperone activity of the PAN protein *in vivo.*

5. Additional Nonproteolytic Roles for Proteasome-Associated AAA Proteins

ATPase (Rpt) proteins of the 19S cap are also found in other complexes that may regulate transcription and other cellular processes. It is possible that these proteins function independently of the proteasome as ATP-dependent molecular chaperones and mediate conformational changes in substrate proteins. Interestingly, Rpt6 (SUG1, Trip1, S8) rescues mutations in the GAL4 transcriptional activating domain (Swaffield *et al.,* 1992). Rpt6 also copurifies with a RNA polymerase II holoenzyme complex that responds to the GAL-VP16 and GCN4 transcriptional activators and interacts with the activating domains of these transcription factors *in vitro* (Kim *et al.,* 1994; Hengartner *et al.,* 1995). Furthermore, Rpt6 alone interacts with at least two components of the transcriptional machinery that belong to the TFIID complex (i.e., TBP and hTAF$_{II}$30) (Swaffield *et al.,* 1995; vom Baur *et al.,* 1996). Rpt6 also appears to mediate ligand-dependent transcriptional activity of several nuclear receptors including those which bind thyroid hormone, estrogen, and vitamin D3 (vom Baur *et al.,* 1996). Thus, Rpt6 may function as a linker between the transcriptional enhancer binding proteins and the basal transcriptional complex. Whether Rpt6 is present in the cell in free form and/or functions with other complexes in addition to the 26S proteasome remains to be determined.

B. NON-AAA⁺ REGULATORS OF PROTEASOMES

There are several non-AAA$^+$ complexes that associate with 20S proteasomes including 11S regulators (PA28, PA26, or REG), various inhibitors, and several heat shock protein (Hsp) complexes reflecting the central role of proteasomes in a variety of pathways.

The 11S regulators are IFN-γ inducible (Realini *et al.,* 1994; Ahn *et al.,* 1995) heptamers that exist in ring-like particles including heterooligomers of REGα and REGβ as well as a homooligomer of REGγ (Realini *et al.,* 1997; Zhang *et al.,* 1999). In the absence of ATP, the 11S regulators can bind to either one or both ends of 20S proteasomes and stimulate peptidase activity up to 60-fold, with little effect on protein degradation (Ma *et al.,* 1992; Dubiel *et al.,* 1992; Realini *et al.,* 1997). They behave as positive allosteric activators that activate certain peptidase activities by decreasing the K_m and reducing the positive cooperativity of substrate binding, with minimal changes in the V_{max}. These kinetic findings suggest that the rate of peptide cleavage in 20S proteasomes is limited by peptide substrate uptake or release, and association of 11S regulators improves this rate of peptide translocation (Stohwasser et al., 2000). The structural basis for activation was recently investigated by cocrystallization of the yeast 20S proteasome associated with two trypanosome 11S regulators (Whitby et al., 2000). At 3.4-Å resolution, the β subunits of the 20S proteasome were essentially unchanged after binding the regulators, suggesting that there was not an alteration in conformation at the catalytic centers. Instead, the 11S regulators opened the 20S axial gates, consistent with activation of peptidase activities. Therefore, the 11S regulator may act to stimulate release of peptide products, while the 19S caps recognize, unfold, and translocate substrate proteins into the 20S proteasome catalytic chamber. This model is consistent with the observation that 11S and 19S regulators can bind simultaneously to separate ends of the same 20S proteasome (Hendil et al., 1998). The 11S regulators may also facilitate the generation of longer peptide ligands for MHC class I molecules by reducing the processivity of 20S proteasomes.

Proteasome inhibitors include the eukaryal CF-2, which affects both peptidase and proteinase activities. Inhibition appears to be linked to the ubiquitination state of CF-2. Surprisingly, CF-2 is identical to δ-aminolevulinic acid dehydratase (ALAD), the second enzyme in the pathway of heme biosynthesis. The domains for each enzymatic activity (i.e., proteasome inhibition and ALAD) are separate based on differences in Zn^{2+} dependence. This suggests that these activities either are physiologically related or represent an example of gene sharing similar to the lens crystalline family (Volker et al., 1993). The eukaryal PI31 (proteasome inhibitor of 31,000 Da) also inhibits the peptidase and proteinase activities of the 20S proteasome. The PI31 protein has a proline-rich carboxyl terminus that forms a random coil and is required for proteasome inhibition (McCutchen-Maloney et al., 2000). A 200-kDa homooligomer of 50-kDa subunits has also been described that noncompetitively inhibits proteolysis and chymotrypsin-like peptidase activity (Li et al., 1991). The β-amyloid protein is also an inhibitor that has been shown to enter the central channel of mammalian 20S proteasomes (Gregori et al., 1997). Hsp90 often copurifies with the proteasome; it inhibits the chymotrypsin-like peptidase activity of the proteasome but also protects proteasome peptidase activities against oxidative damage (Tsubuki et al., 1994; Conconi et al., 1996). Interestingly, in screening for molecules interacting with archaeal 20S proteasomes, an inhibitor of Ca^{2+}-dependent proteinase activity was partially purified but not characterized (Ehlers et al., 1997). In general, the mechanisms of inhibition for all of these proteins remain unclear. It is possible that the inhibitors cap the axial pores of the proteasome or that a segment of the inhibitor may enter the proteasomal channel similar to a plug. Alternatively, the inhibitor may bind to the proteasome at sites distant from the terminal rings and promote conformational changes that influence proteasome–substrate interactions prior to or following substrate entry into the proteolytic chamber.

IV. Targeting Substrates for Degradation

Many questions still remain as to how the cell regulates the sudden and specific targeting of a fully functional protein for degradation by energy-dependent proteases such as the proteasome. The known mechanisms of protein targeting can be categorized into ubiquitin-dependent and -independent regulatory pathways.

A. Ubiquitin-Dependent Pathways

In eukaryotes, the majority of short-lived proteins are targeted for degradation by conjugation with ubiquitin, a 76-residue protein, which provides a signal for recognition by regulatory 19S caps of 26S proteasomes (Laney and Hochstrasser, 1999; Ciechanover et al., 2000). Thus, the specificity of substrate protein recognition is primarily at the level of the ubiquitin conjugation system. Ubiquitination is an energy-dependent process that occurs through the covalent attachment of the carboxyl terminus of ubiquitin to ε-amino groups of Lys residues of the target protein. The process is initiated by the ubiquitin-activating enzyme (E1), which hydrolyzes ATP, forming an adenylated enzyme intermediate that forms a thioester-linked complex with ubiquitin. Then, through a transthiolation reaction, ubiquitin is transferred to one of many distinct ubiquitin carrier/conjugating enzymes (E2). Ubiquitin may be linked to the substrate protein either directly by an E2 enzyme or, more often, requiring assistance from one of many substrate-specific ubiquitin protein ligases (E3) (Jackson et al., 2000). Multiubiquitin chains are then formed on the target proteins by the conjugation of additional ubiquitin moieties to one of seven lysine residues in the ubiquitin molecules already attached to the target protein. Substrates ligated to multiubiquitin chains in which the carboxyl terminus of one ubiquitin is linked to the Lys48 of the previously attached ubiquitin are preferentially degraded by 26S proteasomes (Chau et al., 1989; Thrower et al., 2000). Often an additional factor, E4, assists in the elongation of these chains for recognition by the proteasome (Koegl et al., 1999). Numerous deubiquitination enzymes (DUBs) have been identified, and it is likely that the pathways of ubiquitination and deubiquitination compete in the cell. Thus, selectivity of proteolysis by the proteasome would depend strongly on the exact combination of ubiquitination and deubiquitination enzymes present in the cell at any given time (D'Andrea and Pellman, 1998; Wilkinson, 2000).

There is an extensive array of E3 ligases involved in recognizing substrate degrons, specific sequence motifs associated with protein hydrolysis by 26S proteasomes. Each degron has its own E3, thus allowing numerous independent pathways for regulation of protein degradation. Degrons include proline–glutamate/aspartate–serine–threonine (PEST) sequences (Rechsteiner and Rogers, 1996), destabilizing amino-terminal residues (Varshavsky, 1997), hydrophobic regions (Beal et al., 1998), glycine-rich regions (GRR) (Lin and Ghosh, 1996), destruction boxes of mitotic cyclins (Glotzer et al., 1991; Luca et al., 1991), and Deg1 degradation signals (Johnson et al., 1998). In some instances the destruction of degron-containing substrates is facilitated by the presence of functional nuclear localization sequences (NLS). Presumably, the NLS concentrates some substrates to the vicinity of the endoplasmic reticulum–nuclear envelope where specific E2/E3 proteins such as Ubc6 and Ubc7 are located (Johnson et al., 1998; Hochstrasser et al., 1999). This is consistent with the finding that rapid degradation of a Deg1-containing fusion protein is observed only when the reporter is imported into the nucleus (Lenk and Sommer, 2000).

Recognition of the degron by the E3 enzyme can be regulated by several other mechanisms (Pickart, 2000). Interaction of E3 with the substrate protein can be altered through covalent modifications of the substrate including phosphorylation (Fuchs et al., 1998) and glycosylation (Su et al., 1999; Chung et al., 2000) as well as noncovalent modifications such as interaction with other proteins. Unfolding of a protein, which exposes sequences or nonnative structures recognized by E3, may activate ubiquitin conjugation. Changes in the levels of E3 expression can also regulate substrate recognition. In addition, the activity of E3's is regulated by external signals through allosteric activation by small ligands. For example, the E3 known as Ubr1p or E3α, which is involved in the N-end rule pathway, is activated by specific dipeptides that accelerate the proteasome-mediated degradation of a transcriptional repressor of a di-/tripeptide transporter in yeast (Turner et al., 2000). Allosteric regulation may also be involved in the proteasome-mediated turnover of methylglutaryl–CoA reductase and fructose bisphosphatase, which is stimulated by metabolites (Gardner and Hampton, 1999; Schule et al., 2000). Interestingly, E3's may also facilitate hydrolysis of the target protein by directly interacting as an E3:substrate complex with the 26S proteasome (Xie and Varshavsky, 2000).

Ubiquitin-like modifier proteins (Ubls) and ubiquitin-domain proteins (UbDs) are emerging as further controls that regulate the ubiquitin–proteasome system (Liakopoulos et al., 1998; Hochstrasser et al., 1999). Ubl proteins resemble ubiquitin in their amino acid sequences and are covalently attached to target proteins similar to ubiquitin; however, their physiological roles differ (Hochstrasser, 2000a). After Ubl modification, a target protein can change its subcellular location or its interactions with other proteins/DNA. Ubls such as the small ubiquitin-related modifier (SUMO) are conjugated to proteins that are also targets for ubiquitination (e.g., I-κBα, c-Jun, p53) and appear to modulate the stability and/or activity of these proteins (Desterro, et al., 1998; Muller et al., 2000). Several E3 ligases are also modified by SUMO or the related-to-ubiquitin 1 (RUB1) protein (Buschmann et al., 2000; Read et al., 2000). This modification appears to trigger a conformational change and/or limit self-ubiquitination of E3's, which is necessary for optimal ubiquitination of the target protein.

UbD proteins bear protein domains that are related to ubiquitin but are otherwise unrelated to each other in sequence (Jentsch and Pyrowolakis, 2000; Hochstrasser, 2000a). About 30 UbD proteins have been identified, and many play an important role in protein folding/degradation by linking pathways to functions of the proteasome. For example, RAD23 is a UbD that is required for nucleotide excision repair and interacts with the ubiquitin-binding Rpn10 subunit of the 19S cap of the 26S proteasome (Schauber et al., 1998; Hiyama et al., 1999). This interaction appears to target RAD23 for either remodeling by the 19S ATPase subunits or degradation by the 26S proteasome. Several UbDs including BAG-1 have affinities for the Hsp70 chaperones and may recruit these chaperones to the proteasome to couple protein folding with proteolysis (Lüders et al., 2000).

Not all ubiquitinated proteins are targeted for degradation by the 26S proteasome. Many plasma membrane proteins such as cell surface receptors, transporters, and channels are ubiquitinated in response to ligand binding (Hicke 1999). This often triggers endocytosis and subsequent degradation of these proteins in the vacuole or lysosome. In yeast, L28, which is a component of the large subunit of the ribosome, is the most abundant ubiquitinated protein. Modification of L28 with Lys63-linked ubiquitin chains is strongly dependent on the phase of the yeast cell cycle. Removal of ubiquitin from L28 reduces the activity and stability of the ribosome, suggesting that ubiquitination has a role in facilitating protein translation (Spence et al., 2000). Likewise several histones modified by ubiquitin are stable in vivo; their incorporation into nucleosomes is proposed to alter the local chromatin structure, resulting in increased transcriptional activity, replication, and meiosis (Nickel *et al.,* 1989; Pham and Sauer, 2000; Robzyk *et al.,* 2000). Interestingly, ligand binding also appears to determine the fate of some ubiquitinated proteins as demonstrated by an experiment with a modified dihydrofolate reductase (mDHFR) bearing a destabilizing amino-terminal residue. Ligand binding to mDHFR was found to inhibit its degradation but not ubiquitination, suggesting that the folded state of the enzyme is stabilized in the ligand-bound form (Johnston *et al.,* 1995).

To date, ubiquitin targeting systems have not been identified in prokaryotes including those whose genomes have been completely sequenced; however, distant cousins of the ubiquitin system do exist. There are several parallels between the activation of ubiquitin and the biosynthesis of sulfur-containing coenzymes including thiamin and molybdopterin (Rajagopalan, 1997; Taylor *et al.,* 1998; Hochstrasser, 2000b). For example, the synthesis of the thiazole moiety of thiamin requires ThiS, a short polypeptide with a carboxyl-terminal Gly–Gly dipeptide similar to ubiquitin. Like the ubiquitin system, the ThiF sulfurtransferase protein is related to E1 enzymes and catalyzes the adenylation of ThiS. The ThiS–COAMP then reacts with cysteine or a cysteine-derived sulfur donor to generate ThiS–COSH, which is involved in subsequent sulfur transfer reactions in thiamine biosynthesis. Other prokaryotic relationships to the ubiquitin system include the E1-like MccB enzyme, which catalyzes the phosphoramide linkage of a modified adenylate to the carboxyl terminus of an antibiotic protein, microcin C7, found in many enteric bacteria (Gonzalez-Pastor *et al.,* 1995). In addition, cyanobacteria synthesize an E1-like HesA protein required for nitrogen fixation that is encoded in the same operon as the ferredoxin HesB, which has structural similarities to ubiquitin (Borthakur *et al.,* 1990). HesB-like proteins are also required for iron–sulfur cluster assembly in many other prokaryotic organisms. Thus, it has been speculated that ubiquitin and ubiquitin-like proteins with their activating enzymes may have evolved from protein-based sulfide-donor systems or systems involved in iron–sulfur cluster assembly. The E3 ligases and other components of ubiquitin systems may be more recent evolutionary developments.

Interestingly, many eubacterial and viral systems appear to interact with and regulate the ubiquitin systems of eukaryotic cells. For example, the virulence effector YopJ of the eubacterium *Yersinia* is translocated via a type III secretory system into animal host cells. YopJ then disrupts the universal signaling pathways of animal hosts and, thus, prevents phagocytosis and induction of the host immune response. YopJ has recently been identified as a cysteine protease that is envisioned to cleave the isopeptide bond that links the carboxyl termini of Ubls to the Lys residues of host proteins (Orth *et al.,* 2000). YopJ may also process ubiquitin-like protein precursors so that the carboxyl-terminal Gly residue is available for covalent modification of target proteins. Likewise, evidence suggests that enteric organisms suppress the immune response of host cells by blocking the ubiquitination of IκBα and its subsequent degradation by the proteasome (Neish et al., 2000).

B. UBIQUITIN-INDEPENDENT PATHWAYS

1. Ubiquitin-Independent Systems in Eukaryotes

There are now several examples of the ubiquitin-independent degradation of substrate proteins mediated by the eukaryal proteasome (Pickart, 1997). The prototype is the turnover of ornithine decarboxylase (ODC), which catalyzes the rate-limiting step in polyamine biosynthesis, a pathway directly linked to cell proliferation (Hayashi and Murakami, 1995). The noncovalent interaction of ODC with the protein factor antizyme triggers ODC degradation by the proteasome (Murakami et al., 1992; Li and Coffino, 1993; Hayashi et al., 1996). Some proteasome substrates, such as IκBα and c-Jun, are subject to bimodal targeting in which the signal-induced turnover is mediated by ubiquitin and basal turnover is ubiquitin independent (Treier et al., 1994; Jariel-Encontre et al., 1995; Krappmann et al., 1996). Also, several nonubiquitinated proteins bind directly to proteasome subunits including the viral Tax protein, which binds to two 20S core subunits (Rousset et al., 1996); the HBx protein of hepatitis B virus, which interacts with a proteasomal α-type subunit (Fischer et al., 1995); the p55 tumor necrosis factor receptor, which binds to a non-ATPase 19S cap protein (Boldin et al., 1995); and an ankyrin repeat protein which binds to an Rpt (AAA) 19S cap protein (Dawson et al., 1997). The binding of scavenger proteins (e.g., ankyrin-repeat proteins) to substrate proteins appears to signal proteasome-mediated degradation independent of the ubiquitin pathway (Dawson et al., 1997). Some Rpt subunits of the proteasome also form alternative complexes that do not contain the ubiquitin recognition proteins of the 19S cap and appear to recognize and modulate the folding state of protein substrates in the absence of ubiquitin (Hastings et al., 1999). Whether the above examples of ubiquitin-independent interaction with proteasome proteins are exceptions or commonly used in eukaryotic cells for proteasome-mediated hydrolysis is unclear.

2. Ubiquitin-Independent Systems in Prokaryotes

Similar to eukarya, the compartmentalized proteolytic active sites of prokaryotic energy-dependent proteases are rather nonspecific. However, in contrast to the vbiguitin system, the unfoldase that associates with the energy-dependent protease appears to be directly involved in the recognition of substrate and may also decide between protein destruction and refolding. Some evidence suggests that stimulation of proteolysis may also require accessory proteins to modify covalently the substrate or substrate-associated protein, dock the substrate to the unfoldase/protease complex, or mediate conformational changes in the substrate or unfoldase. Examples of these mechanisms include phosphorylation of RssR as well as the SsrA-tagging system stimulated by SspB (see discussion below).

The eubacterial unfoldases or their accessory proteins recognize different motifs in their substrates including residues that are not usually surface-exposed in native proteins. For example, improperly folded or unfolded proteins generally have an increased exposure of hydrophobic residues on their surface that are often recognized for degradation by proteases. This is demonstrated by the addition of amino acid analogues or puromycin, which increases the intracellular concentration of abnormal proteins and stimulates energy-dependent proteolysis (Goff and Goldberg, 1985). Likewise, the synthesis of certain "foreign" or modified proteins that do not properly fold in the cell stimulates their degradation (Huang *et al.,* 2000). Thermal stability of a protein can also influence its turnover rate; those that are most thermally stable have the longest intracellular half-lives, and vice versa (Parsell and Sauer, 1989). Energy-dependent proteases in the membrane, such as *E. coli* FtsH, are also able to sense the folding state of protein substrates typically located in the membrane (Akiyama *et al.,* 1998). FtsH catalyzes the selective degradation of subunits of the Sec translocase and F_0 sector of the proton ATPase when they fail to associate with other subunits of their respective complexes (Kihara *et al.,* 1995; Akiyama *et al.,* 1996a,b). Thus, it is likely that residues that are normally hidden from the surface of native proteins, which subsequently become exposed, can then be recognized as substrate by energy-dependent proteases for degradation.

Although ubiquitin-like proteins are not associated with tagging substrate proteins in the prokaryotes, there are specific substrate motifs and even tags recognized by unfoldases and/or accessory proteins that activate degradation pathways in the eubacteria. Often these recognition sequences are located at the amino or carboxyl terminus of a protein substrate. The addition of amino-terminal arginine, lysine, leucine, phenylalanine, tyrosine, or tryptophan residues to β-galactosidase confers short half-lives (<2 min) that are dramatically increased in a *clpA* mutant. The same test protein modified with other amino-terminal residues is highly stable, with a half-life of more than 10 h. This suggests that ClpAP protease is necessary for the recognition of substrates to be degraded by the N-end rule in eubacteria (Varshavsky, 1992). Components of the N-end rule have not been reconstituted *in vitro*, suggesting that additional factors are necessary for function. Furthermore, the pathway has not been extended to physiological substrates.

There are, however, examples of amino-terminal sequences playing a critical role in facilitating proteolysis *in vivo*. ClpA recognizes N-terminal signals in RepA in the vicinity of amino acids 10 to 15 (–LYADIE–) (Hoskins *et al.*, 2000b). ClpA also recognizes HemA, glutamyl-tRNA reductase, through sites in the amino-terminal 18 amino acids (MTLLALGINHKTAPVSLR–) (Wang *et al.*, 1999b). Residues located between 15 and 18 (–FPLF–) and 26 and 29 (–FPSP–) of the amino terminus of UmuD, a component of the SOS mutagenesis pathway, are recognized by Lon for degradation (Gonzalez *et al.*, 1998). Furthermore, deletion of the first 18 amino acids of a bacteriophage λO replication protein derivative (MTNTAKILNFGRGNFAGQ–) blocks its degradation by the ClpXP protease, whereas deletions in the carboxyl terminus do not influence protein stability (Gonciarz-Swiatek *et al.*, 1999). Interestingly, ClpX still retains its capacity to bind the amino-terminal deletion derivative of λO, apparently through an internal motif, without efficiently promoting its energy-dependent binding to ClpP protease. Thus stable binding of substrate polypeptides to AAA$^+$ unfoldases does not appear to be sufficient to insure proteolysis.

The most elaborate system identified in eubacteria is the SsrA-tagging system, which is used to mark an incomplete nascent polypeptide chain for destruction when it becomes stalled on the ribosome due to the generation of 3′-truncated mRNA, rare codons within the mRNA, or under conditions which promote tRNA scarcity (e.g., starvation) (Tu *et al.*, 1995; Keiler *et al.*, 1996; Roche and Sauer, 1999). Recent evidence suggests that premature transcription termination resulting from a strong DNA-binding protein that binds within an open reading frame may also be linked to the SsrA-tagging system (Abo *et al.*, 2000). The *ssrA* gene encodes a hybrid tmRNA that resembles tRNA-ala at the 5′-end and mRNA encoding a 10-residue peptide (ANDENYALAA) with a stop codon at the 3′-end. This tmRNA enters the unoccupied P site of a stalled ribosome complex and adds the alanine from the SsrA-tRNA structure; it then switches to translation of the SsrA-mRNA, resulting in the addition of a total of 11 nonpolar residues to the nascent chain. *In vitro* studies suggest that SsrA-tagged proteins are recognized for destruction by the ClpAP and ClpXP proteases and, to a lesser extent, by the FtsH protease (Gottesman *et al.*, 1998; Herman *et al.*, 1998). However, in vivo the ClpAP and FtsH proteases do not appear to have a role in degrading SsrA-tagged substrates (Levchenko *et al.*, 2000). Instead, a ribosome-associated SspB protein is required to bind SsrA-tagged proteins and enhance recognition of these proteins for destruction by the ClpXP protease. Modification of the terminal Ala–Ala residues of the SsrA tag to Asp–Asp significantly reduces this substrate recognition and degradation (Levchenko *et al.*, 2000). An additional component of the system is a 160-residue SmpB protein that specifically binds SsrA-RNA (Karzai *et al.*, 2000). SmpB appears to be critical for a step(s) after the charging of SsrA-RNA with alanine but prior to the transpeptidation reaction that couples this alanine to the growing peptide chain. Together these studies reveal that the elaborate SsrA system provides a means for the cell rapidly to identify damaged proteins for destruction while still associated with the ribosome.

The extreme carboxyl terminus of a protein itself can also serve as a recognition signal for substrate protein degradation; the determinants are often nonpolar residues similar to the SsrA tag. Examples include the carboxyl-terminal 10-residue sequence of MuA (–LEQNRRKKAI), which is required for recognition and disassembly of tetrameric MuA by ClpX (Levchenko *et al.*, 1997b). The carboxyl-terminal eight amino acids of SulA (–KIHSNLYH) is sufficient for degradation by Lon protease (Zaiss and Belote, 1997). The Mu *vir* repressor protein is recognized by ClpXP protease for degradation by a carboxyl-terminal motif of seven amino acids (–FMNRKVL). A single amino acid substitution (N to D) in this motif is sufficient to prevent degradation by the Clp protease. Fusion of the Mu *vir* carboxyl-terminus to CcdA, a substrate normally degraded only by Lon, targets the protein for degradation by both Clp and Lon proteases. However, similar experiments with the naturally stable CcdB protein had no effect, suggesting that certain carboxyl-terminal motifs are not sufficient to target all proteins for degradation (Laachouch *et al.*, 1996).

Another pathway for signaling protein degradation in enterics involves a two-component response regulator, RssB. In its phosphorylated form RssB binds σ^S (RpoS) and specifically targets this general stress response σ factor for degradation by ClpXP. Stress apparently influences the affinity of RssB for σ^S by modulating the rate of RssB phosphorylation (Becker *et al.*, 1999).

Molecular chaperones and cochaperones such as the DnaK/DnaJ/Hsp70 and GroEL/GroES families are essential for the selective breakdown of some protein substrates by energy-dependent proteases. For example, in *E. coli* a nonsecreted form of alkaline phosphatase that lacks a signal sequence fails to fold in the cytosol and is rapidly degraded by ClpAP and Lon proteases. DnaJ is required for the degradation of this modified phosphatase and appears to promote the formation of protease–substrate complexes (Huang *et al.*, 2000). Likewise DnaK may sequester σ^{32} from the RNA polymerase core ($\alpha\beta\beta'$) and make it accessible for FtsH protease degradation, whereas σ^{32} in association with the core polymerase is stabilized (Blaszczak et al., 1999). Furthermore, repeated cycles of substrate protein binding and release from GroEL appear to facilitate degradation by the Clp proteases, which is presumed to solubilize protein aggregates for access to the Clp ATPases (Kandror et al., 1999).

Little is known about how substrate proteins are recognized for degradation by proteasomes in the archaea or actinomycetes. However, in analogy to the ubiguitin-independent Clp, Lon, and FtsH, it is likely that unfoldase components of proteasomes (e.g., PAN and ARC) are involved in this recognition. In vitro studies suggest that the archaeal PAN regulator is able to recognize substrates such as β-casein that lack any appreciable secondary structure and to stimulate degradation of these substrates by the proteasome in the presence of either ATP or AMP-PNP, a nonhydrolyzable analogue of ATP (Zwickl et al., 1999; Wilson et al., 2000). Based on genome sequence, the SsrA-tag system is not present in the archaea; however, this nonpolar tag is recognized in vitro by the archaeal PAN protein to stimulate ATP-dependent degradation by the 20S proteasome (Benaroudj and Goldberg, 2000).

V. Proteasomes in the Cell

A. ROLE OF THE PROTEASOME IN EUKARYOTES

Protein degradation is an essential process involved in many aspects of cell growth, development, and environmental responses. In eukaryotes, the ubiquitin–proteasome pathway is linked to numerous cell processes including cell division, class I antigen processing, signal transduction, development, DNA repair, heat shock responses, heavy metal resistance, and degradation of abnormal proteins (Teichert et al., 1989; Jentsch, 1992; Lee and Goldberg, 1996). The central role of proteasomes is highlighted by yeast mutagenesis studies, which demonstrate that many subunits of the 26S proteasome are essential for life (Hilt and Wolf, 1996; Rubin et al., 1998). The ubiquitin–proteasome system is required for the targeted destruction of specific proteins as well as the bulk turnover of proteins. In mammalian cells, the 26S proteasome has the tremendous responsibility of degrading the majority of cytosolic proteins (Lee and Goldberg, 1998) and about 30% of newly synthesized proteins (Schubert et al., 2000). Likewise in yeast more than 50% of nascent proteins with an amino-terminal degron are cotranslationally degraded (Turner and Varshavsky, 2000). Thus, the folding of nascent proteins appears to be in kinetic competition with cotranslational degradation mediated by the ubiquitin–proteasome pathway.

The rate-limiting step in the ubiquitin–proteasome pathway is assumed to be the elaborate and highly regulated conjugation of polyubiquitin chains to the substrate as discussed above. However, the subsequent degradation of ubiquitin conjugates by the proteasome is not simply a rapid and unregulated process. The cell appears to modulate the intracellular distribution, content, composition, and activity of 20S proteasomes as well as their associated regulatory components (e.g., the 19S cap).

1. The Subcellular Location of Proteasomes

In eukaryotic cells, proteasomes are abundant and evenly distributed in the cytoplasm, throughout the nuclear matrix, and around the nucleolus (Peters *et al.*, 1994; Reits *et al.*, 1997; Rivett, 1998). Populations of proteasomes are also found associated with the endoplasmic reticulum, intermediate filaments, and centrosome (or microtubule-organizing centers) (Grossi de Sa *et al.*, 1988; Rivett *et al.*, 1992; Palmer *et al.*, 1994, 1996; Arcangeletti *et al.*, 1997; Wigley *et al.*, 1999). They also can be found concentrated at the nuclear envelope–ER network (Wilkinson *et al.*, 1998; Enenkel *et al.*, 1998). There are several mechanisms that may govern the intracellular distribution of 20S proteasomes. Adaptor proteins such as the 19S and 11S complexes may guide 20S proteasomes to different locations. The incorporation of different α- or β-type subunit isoenzymes and/or covalent modification of proteasomal subunits may also modify the distribution. Proteasomes may also be tethered to various substrate proteins located in different subcellular compartments (e.g., misfolded proteins associated with ER membranes) (Rivett, 1998; Hirsch and Ploegh, 2000).

2. Phosphorylation of Proteasome Subunits

Many subunits of the proteasome 20S core and regulatory complexes possess putative nuclear localization signals as well as tyrosine and serine/threonine phosphorylation sites that may regulate the activity and subcellular location of the proteasome. Casein kinase II appears to mediate the phosphorylation of some of these subunits. The kinase copurifies with human erythrocyte proteasomes and mediates *in vitro* phosphorylation of a 30-kDa proteasomal subunit (Ludemann *et al.*, 1993). In addition, casein kinase II phosphorylates the α6 subunit from rice proteasomes; however, the phosphorylation site of α6 is not conserved among different organisms (Umeda *et al.*, 1997). Purified human placental and rat liver proteasomes contain α2 subunits with phosphotyrosine and phosphothreonine residues, while the α3, α5, α7, and β6 subunits have phosphoserine residues (Wehren *et al.*, 1996). The phosphorylation of α3 and α7 has been confirmed by immunoprecipitation of proteasomes labeled with [32]P *in vivo*. A serine residue within a potential cGMP-dependent phosphorylation site (–RRYDSR–) was labeled within the $\alpha 3$ subunit. Phosphoserine residues were identified in the $\alpha 7$ subunit (Ser-243 and Ser-250) which were no longer phosphorylated by casein kinase II after mutation to Ala (Castaño *et al.*, 1996; Mason *et al.*, 1996). The phosphorylation sites of $\alpha 3$ and $\alpha 7$ are located at the ends of the proteasome cylinder, suggesting that covalent modification of these subunits may influence interaction of 20S proteasomes with regulatory components (Bose *et al.*, 1999). Both NLS and tyrosine phosphorylation sites are important for nuclear translocation of the $\alpha 2$ subunit *in vivo* (Benedict and Clawson, 1996). Subunits of the 11S regulator α/β complex and γ homooligomer are all phosphorylated at serine residues. Dephosphorylation of these subunits eliminates the ability of the 11S regulators to stimulate and modify the peptidase activities of the proteasome (Li *et al.*, 1996). Several subunits of the 19S caps are also phosphorylated *in vivo* including the Rpt2 (S4) ATPase (Mason *et al.*, 1998), which is highly related to the archaeal PAN protein.

3. Protein and mRNA Levels of Eukaryal Proteasomes

The overall content and individual populations of 20S core and regulatory subunits of the proteasome differ based on growth and environmental conditions. These changes are likely to reflect the adaptations of the cell to particular environmental conditions. The most fluid and dynamic changes appear in the subunit composition for non-ATPase components of the 26S proteasome, which has made these proteins difficult to characterize. There are also transient associations of 26S proteasome subpopulations with other proteins, which may shuttle substrates to the 26S proteasome (Seeger *et al.*, 1997). It should be noted that an increase in the intracellular levels of proteasome mRNAs does not always parallel an increase in the levels of the corresponding subunits. This suggests that the production of proteasome protein may be regulated posttranscriptionally (e.g., unassembled subunits are rapidly degraded and/or translation is regulated).

Typically ubiquitin and components of ubiquitin-dependent proteolysis are induced in response to various stresses including desiccation, heat shock, heavy metals, and infection (Vierstra, 1996). The most established changes in proteasome composition are after IFN-γ induction. The levels of the $\beta 1$i, $\beta 2$i, and $\beta 5$i subunits increase and replace their constitutive counterparts in the 20S core, resulting in modification of the peptidase activities of the proteasome. The levels of 11S regulator proteins are also increased in the presence of IFN-γ, resulting in the stimulation of peptide hydrolysis. There are also general differences in the β-subunit composition of 20S proteasomes in different subcellular compartments. The β2 protein is enriched in nuclear proteasomes but low in microsomes, whereas the β1i is relatively low in the nuclei and somewhat enriched in the microsomes (Palmer et al., 1996).

In yeast, the heavy metal cadmium appears to increase the levels of β4, α5, Rpn11, and Rpt5 proteins over three- to fivefold based on proteomics (Vido et al., 2000). Yeast DNA chip analysis suggests that the mRNA transcripts of the Rpn1 (non-ATPase) and β1 subunits of the 26S proteasome are increased 5- to 10-fold after exposure to DNA-damaging agents (Jelinsky and Samson, 1999). Growth-dependent changes in the levels of 26S proteasome protein content and activity are also observed in yeast, with increases occurring in the stationary phase (Fujimuro et al., 1998). Recent analysis using mild heat shock conditions in yeast suggests that there is a rapid reorganization of the cellular distribution of proteasomes and possible covalent modification of the α1 and α6 subunits. However, no changes in the levels of 20S/26S proteasome proteins were evident after this mild heat shock (Kuckelkorn et al., 2000).

Changes in the levels, subunit composition, and tissue distribution of 20S/26S proteasome proteins also occur during development and programmed cell death (Klein et al., 1990; Hashimoto et al., 1996). This is observed for Drosophila species that express genes encoding alternative isoforms of the α4 subunit exclusively in the germline of the testes during spermatogenesis (Yuan et al., 1996; Belote et al., 1998). Likewise, in Manduca an eightfold increase in the absolute amounts of proteasome proteins is detected and parallels the synthesis of four new proteasome subunits in intersegmental muscles during programmed cell death (Jones et al., 1995; Dawson et al., 1995). Increases in the levels of a molting hormone prevent the occurrence of programmed cell death and also prevent the rise in the levels of an Rpt ATPase subunit in this organism (Low et al., 1997). During fetal development in rats, heterogeneous proteasome populations are found throughout the cell, and subpopulations of proteasomes are found preferentially in the apical domain of hepatocytes (Briane et al., 1992).

Glucose starvation and hypoxia appear to induce the nuclear accumulation of proteasome proteins and increase peptidase activities in tumor cells (Ogiso et al., 1999). Similarly, the levels of proteasome proteins are increased in breast cancer cells and in neighboring normal cells (Bhui-Kaur et al., 1998). This response may enhance the ability of cancer cells to respond to stress under adverse conditions. In contrast, carbon starvation does not appear to change the amount of 20S proteasome protein levels or location in plants. Instead, an increase in vacuolar endoprotease activity is responsible for the increased degradation of bulk proteins in these cells (Brouquisse et al., 1998).

Proteasome levels also appear to influence how a cell ages. Typically aged cells accumulate increased levels of oxidized and glycated proteins as well as proteins modified with the lipid peroxidation product 4-hydroxy-2-nonenal. Cell aging and replicative senescence decrease proteasome activity and subunit levels, which may explain the observed accumulation of oxidized and abnormal proteins in these cells (Petropoulos *et al.*, 2000). Complementing this observation, the gene expression of 20S, 26S, and 11S subunits was found to be significantly elevated in cells from centenarians compared to the normal aging population (Chondrogianni *et al.*, 2000). Analysis of the subunit composition of 20S proteasomes from normal and aging cells by 2D-PAGE suggests that four of these subunits (α3, α4, α5, and β4) may be altered by post-translational modification after aging (Bulteau *et al.*, 2000). This may influence the observed changes in activity and/or subunit levels of the proteasome.

B. PROTEASOME FUNCTION IN EUBACTERIA AND ARCHAEA

The proteasome-related HslUV protease, found in most eubacteria, is encoded by the heat shock-inducible *hslUV* operon (Chuang *et al.*, 1993). *In vitro* HslUV functions as an ATP-dependent protease and protein-activated ATPase toward denatured substrates (Seol *et al.*, 1997). *In vivo* HslUV is involved in the overall proteolysis of misfolded proteins in *E. coli* (Missiakas *et al.*, 1996) and regulates the heat shock response by controlling the hydrolysis of σ^{32} (RpoH) in a synergistic mechanism with other ATP-dependent proteases (Kanemori *et al.*, 1997). A transient increase in σ^{32} then activates transcription of heat shock genes in *E. coli*. The HslUV protease also suppresses the SOS-mediated inhibition of cell division by its ability to degrade SulA, an inhibitor of the FtsZ cell division protein (Khattar, 1997).

Little is known about the biological significance of 20S proteasomes or AAA^+ proteins in the archaea or actinomycetes. The role of 20S proteasomes in protein degradation has not been demonstrated in prokaryotic cells, and substrate proteins have not been identified. In analogy to the eukaryal proteasome it is possible that the prokaryotic enzyme plays an active role in mediating stress responses of the cell. In general, most purified proteasomes are stable at temperatures above the optimum for growth and are resistant to damage by oxidizing agents (Wagner and Margolis, 1993; Wilson *et al.*, 1999; Andersson *et al.*, 1999). This suggests that the enzyme complex is stable and may function after exposure to stress in the cell.

In the actinomycetes, the levels of the 20S proteasome subunits do not increase after cells are exposed to heat shock (De Mot *et al.*, 1999). In addition, two-dimensional electrophoresis reveals that the ARC (AAA$^+$) protein is expressed in a relatively small amount, and its levels are independent of the growth temperature of the organism from 4 to 34°C (Wolf *et al.*, 1998). Knockout mutations of α- and β-type 20S proteasomal genes (*prcAB*) in the eubacterium *Mycobacterium smegmatis* result in no discernible phenotypic changes including response to stresses such as heat shock, nutrient depletion, ethanol shock, heavy metals, amino acid analogues, and osmotic shock (Knipfer and Shrader, 1997). Even *M. smegmatis prcAB lon* double mutants display wild-type growth rates and stress tolerances (Knipfer *et al.*, 1999). Based on genome sequences, the actinomycetes do not synthesize HslUV proteases; however, it is possible that other proteases such as Clp or distantly related Lon homologues compensate for the loss of Lon and 20S proteasome function in these mutant strains. This compensation for loss of protease function has been observed for eukaryotic cells adapted for growth in the presence of proteasome inhibitors. These adapted cells exhibited enhanced tripeptidyl peptidase II (TPPII) activity catalyzed by a large rod-shaped protease that appears to substitute for some metabolic functions of the proteasome (Geier *et al.*, 1999). In *E. coli*, the HslUV protease is able to hydrolyze substrates typically degraded by the Lon protease (Wu *et al.*, 1999).

Many archaea grow optimally in what are considered to be "extreme" environments such as high/low temperature, high/low pH, high salinity, high pressure, and absence of oxygen. However, even extremophiles exhibit stress responses and acquire thermotolerance after brief exposure to heat shock temperatures (Trent *et al.*, 1994; Macario *et al.*, 1999). To investigate the role of the archaeal proteasome in stress responses, *T. acidophilum* cells were incubated with a tripeptide vinyl sulfone inhibitor that inactivated about 80% of the 20S proteasome β subunits through covalent modification of the amino-terminal Thr active-site nucleophile (Ruepp *et al.*, 1998). This partial inhibition of 20S proteasome activity significantly reduced the ability of cells to overcome heat shock. These results suggest that as active 20S proteasomes become limiting, *T. acidophilum* is unable to overcome an increased level of heat-damaged proteins. It is likely that these proteins formed intracellular aggregates or inclusion bodies that inhibited cell growth. Analysis of *H. volcanii* reveals that during normal growth the levels of the α1 proteasome protein are constitutive, whereas the levels of the α2 protein increase during the stationary phase. After heat shock or addition of canavanine, the levels of 20S proteasome protein and mRNA transcripts are increased (Kaczowka, Wilson, and Maupin-Furlow, unpublished). Although further studies are needed, these results suggest that 20S proteasomes may play a role in the stress responses of the archaeal cell.

Genome analysis of numerous archaea reveals that the majority of genes predicted to be cotranscribed with the genes encoding the α and β subunits of the 20S proteasome have a high similarity to proteins involved in RNA processing and binding. Thus, the archaeal proteasome may associate with or regulate the turnover of proteins that bind to and modify RNA. In eukaryotes, the proteasome has been implicated in repression of mRNA translation (Schmid et al., 1984), pre-tRNA 5′-processing endonuclease activity (Castaño et al., 1986), amino acyl transferase activity (Shelton et al., 1970), and RNase activity (Petit et al., 1997a, b). Small cytoplasmic RNA molecules have been found associated with eukaryal proteasomes (Arrigo et al., 1988; Falkenburg et al., 1988). Interestingly, the CCT-TriC (Hsp60) chaperone of the archaeon Sulfolobus solfataricus is also an RNA-binding protein that participates in ribosomal RNA processing (Ruggero et al., 1998). Whether specific RNA molecules are an integral part of the proteasome or reflect any physiological role is unclear (Pamnani et al., 1994).

Evidence suggests that the archaeal proteasome may be regulated by processing and/or covalent modification of α subunits. Amino acid sequence analysis of 20S proteasomes purified from M. thermophila revealed two amino termini for α subunits encoded from the same gene. One subunit is four amino acids shorter than the other, suggesting the possibility of in vivo processing; however, nonspecific hydrolysis during purification cannot be discounted (Maupin-Furlow and Ferry, 1995). Primary sequence analysis reveals that many archaeal α subunits contain highly conserved NLS/cNLS and phosphorylation sites with similarity to the analogous eukaryal proteins. Surprisingly, the NLS/cNLS sites of the α subunit from T. acidophilum are able to localize proteins to the nucleus in eukaryotic cells; the functional significance of this in the archaea, which do not have a nucleus, remains to be determined (Nederlof et al., 1995). Furthermore, isoelectric focusing suggests the α subunit of T. acidophilum proteasomes is covalently modified when produced in recombinant E. coli; however, this has not been demonstrated in vivo (Zwickl et al., 1992b).

VI. Additional Energy-Dependent Proteases of the Archaea

Archaeal genome sequence analysis has aided in understanding of the types of energy-dependent proteases that may be synthesized in the archaea and how the proteasome may interact with these pathways (Maupin-Furlow et al., 2000; Ruepp et al., 2000). It appears that a Lon-related energy-dependent serine protease is membrane associated and universally distributed among the archaea. It is possible that this archaeal Lon-like protease enables the cell to fold and/or degrade membrane-associated proteins much like the FtsH protein of eubacteria.

Various AAA$^+$ ATPases lacking conserved protease domains also appear to be synthesized in the archaea and are predicted to assist in protein folding and/or unfolding for protein degradation. These include a second cytosolic Lon-related ATPase found in some archaea (e.g., *T. acidophilum*) and a cytosolic FtsH/CDC48-related ATPase encoded by most archaea. Although the Clp family is not common among the archaea, a putative Clp-like ATPase protein appears to be found in the methanogenic archaeon, *M. thermoautotrophicum* ΔH (Smith *et al.*, 1997).

Recently the periplasmic HtrA(DegP) protease of *E. coli* was found to form a self-compartmentalized structure similar to other energy-dependent proteases. Although the openings at the ends of its cylindrical structure were narrower than the openings of GroEL, they were larger than those of the 20S proteasome, ClpP, and HslV (Kim *et al.*, 1999). Because the periplasmic space of *E. coli* is probably depleted of ATP, it is speculated that HtrA may not associate with energy-dependent unfoldases and instead hydrolyzes proteins denatured by stress. Interestingly, *Halobacterium* NRC-1 and *M. thermoautotrophicum* both encode HtrA(DegP)-like proteins with conserved residues required for proteolytic activity but without the conserved signal sequences required for protein export/secretion. It remains to be determined whether the archaeal HtrA forms a self-compartmentalized protease in the cytosol or is exported.

VII. Summary and Perspective

Future research in the field of archaeal proteasomes is focused on understanding its physiological role. We are encouraged by the recent genetic and biochemical analysis of proteasomes in the haloarchaeon *H. volcanii* and methanoarchaeon *Methanococcus maripaludis* (Kaczowka, Wilson, and Maupin-Furlow, unpublished results). Both of these organisms have established genetic systems and are currently being used for analysis of proteasome function *in vivo*. It will be interesting to integrate the detailed information of the biochemical and biophysical properties of 20S proteasome and PAN proteins with studies focused on understanding how these elaborate complexes function in the archaeal cell.

ACKNOWLEDGMENTS

We are greatly indebted to Henry C. Aldrich and Donna Williams for their help with transmission electron microscopy of archaeal proteasomes. The work was supported in part by National Institutes of Health Award R01GM57498 and the Florida Agricultural Experiment Station (Journal Series R-08275).

References

Abo, T., Inada, T., Ogawa, K., and Aiba, H. (2000). *EMBO J.* **19**, 3762–3769.
Adams, G. M., Falke, S., Goldberg, A. L., Slaughter, C. A., DeMartino, G. N., and Gogol, E. P. (1997). *J. Mol. Biol.* **273**, 646–657.
Adams, G. M., Crotchett, B., Slaughter, C. A., DeMartino, G. N., and Gogol, E. P. (1998a). *Biochemistry* **37**, 12927–12932.
Adams, J., Behnke, M., Chen, S., Cruickshank, A. A., Dick, L. R., Grenier, L., Klunder, J. M., Ma, Y. T., Plamondon, L., and Stein, R. L. (1998b). *Bioorg. Med. Chem. Lett.* **8**, 333–338.
Adams, J., Palombella, V. J., and Elliott, P. J. (2000). *Invest. New Drugs* **18**, 109–121.
Ahmadian, M. R., Stege, P., Scheffzek, K., and Wittinghofer, A. (1997). *Nature Struct. Biol.* **4**, 686–689.
Ahn, J. Y., Tanahashi, N., Akiyama, K., Hisamatsu, H., Noda, C., Tanaka, K., Chung, C. H., Shimbara, N., Willy, P. J., Mott, J. D., Slaughter, C. A., and DeMartino, G. N. (1995). *FEBS Lett.* **366**, 37–42.
Akiyama, Y., Kihara, A., and Ito, K. (1996a). *FEBS Lett.* **399**, 26–28.
Akiyama, Y., Kihara, A., Tokuda, H., and Ito, K. (1996b). *J. Biol. Chem.* **271**, 31196–31201.
Akiyama, Y., Ehrmann, M., Kihara, A., and Ito, K. (1998). *Mol. Microbiol.* **28**, 803–812.
Akopian, T. N., Kisselev, A. F., and Goldberg, A. L. (1997). *J. Biol. Chem.* **272**, 1791–1798.
Andersson, M., Sjostrand, J., and Karlsson, J. O. (1999). *Exp. Eye Res.* **69**, 129–138.
Aravind, L., and Ponting, C. P. (1998). *Protein Sci.* **7**, 1250–1254.
Arcangeletti, C., Sutterlin, R., Aebi, U., Deconto, F., Missorini, S., Chezzi, C., and Scherrer, K. (1997). *J. Struct. Biol.* **119**, 35–58.
Arendt, C. S., and Hochstrasser, M. (1997). *Proc. Natl. Acad. Sci. USA* **94**, 7156–7161.
Arendt, C. S., and Hochstrasser, M. (1999). *EMBO J.* **18**, 3575–3585.
Arlt, H., Tauer, R., Feldmann, H., Neupert, W., and Langer, T. (1996). *Cell* **85**, 875–885.
Arribas, J., and Castaño, J. G. (1990). *J. Biol. Chem.* **265**, 13969–13973.
Arribas, J., Arizti, P., and Castaño, J. G. (1994). *J. Biol. Chem.* **269**, 12858–12864.
Arrigo, A.-P., Tanaka, K., Goldberg, A. L., and Welch, W. J. (1988). *Nature* **331**, 192–194.
Bauer, M. W., Halio, S. B., and Kelly, R. M. (1997). *Appl. Environ. Microbiol.* **63**, 1160–1164.
Baumeister, W., and Lupas, A. (1997). *Curr. Opin. Struct. Biol.* **7**, 273–278.
Baumeister, W., Walz, J., Zühl, F., and Seemüller, E. (1998). *Cell* **92**, 367–380.
Beal, R. E., Toscano-Cantaffa, D., Young, P., Rechsteiner, M., and Pickart, C. M. (1998). *Biochemistry* **37**, 2925–2934.
Becker, G., Klauck, E., and Hengge-Aronis, R. (1999). *Proc. Natl. Acad. Sci. USA* **96**, 6439–6444.
Belote, J. M., Miller, M., and Smyth, K. A. (1998). *Gene* **215**, 93–100.
Benaroudj, N., and Goldberg, A. L. (2000). *Nature Cell Biol.* **2**, 833–839.
Benedict, C. M., and Clawson, G. A. (1996). *Biochemistry* **35**, 11612–11621.
Beuron, F., Maurizi, M. R., Belnap, D. M., Kocsis, E., Booy, F. P., Kessel, M., and Steven, A. C. (1998). *J. Struct. Biol.* **123**, 248–259.
Bhui-Kaur, A., Therwath, A., Henry, L., Chiesa, J., Kurkure, A., Scherrer, K., and Bureau, J. P. (1998). *J. Cancer Res. Clin. Oncol.* **124**, 117–126.
Blaszczak, A., Georgopoulos, C., and Liberek, K. (1999). *Mol. Microbiol.* **31**, 157–166.
Bochtler, M., Ditzel, L., Groll, M., and Huber, R. (1997). *Proc. Natl. Acad. Sci. USA* **94**, 6070–6074.
Bochtler, M., Ditzel, L., Groll, M., Hartmann, C., and Huber, R. (1999). *Annu. Rev. Biophys. Biomol. Struct.* **28**, 295–317.
Bochtler, M., Hartmann, C., Song, H. K., Bourenkov, G. P., Bartunik, H. D., and Huber, R. (2000). *Nature* **403**, 800–805.

Bogyo, M., McMaster, J. S., Gaczynska, M., Tortorella, D., Goldberg, A. L., and Ploegh, H. (1997). *Proc. Natl. Acad. Sci. USA* **94,** 6629–6634.

Boisvert, D. C., Wang, J., Otwinowski, Z., Horwich, A. L., and Sigler, P. B. (1996). *Nat. Struct. Biol.* **3,** 170–177.

Boldin, M. P., Mett, I. L., and Wallach, D. (1995). *FEBS Lett.* **367,** 39–44.

Borthakur, D., Basche, M., Buikema, W. J., Borthakur, P. B., and Haselkorn, R. (1990). *Mol. Gen. Genet.* **221,** 227–234.

Bose, S., Mason, G. G., and Rivett, A. J. (1999). *Mol. Biol. Rep.* **26,** 11–14.

Braig, K., Otwinowski, Z., Hegde, R., Boisvert, D. C., Joachimiak, A., Horwich, A. L., and Sigler, P. B. (1994). *Nature* **371,** 578–586.

Brannigan, J. A., Dodson, G., Duggleby, H. J., Moody, P. C. E., Smith, J. L., Tomchick, D. R., and Murzin, A. G. (1995). *Nature* **378,** 416–419.

Braun, P., and Tommassen, J. (1998). *Trends. Microbiol.* **6,** 6–8.

Braun, B. C., Glickman, M., Kraft, R., Dahlmann, B., Kloetzel, P. M., Finley, D., and Schmidt, M. (1999). *Nature Cell Biol.* **1,** 221–226.

Briane, D., Olink-Coux, M., Vassy, J., Oudar, O., Huesca, M., Scherrer, K., and Foucrier, J. (1992). *Eur. J. Cell Biol.* **57,** 30–39.

Brouquisse, R., Gaudillere, J. P., and Raymond, P. (1998). *Plant Physiol.* **117,** 1281–1291.

Brown, M. G., Driscoll, J., and Monaco, J. J. (1993). *J. Immunol.* **151,** 1193–1204.

Bulteau, A., Petropoulos, I., and Friguet, B. (2000). *Exp. Gerontol.* **35,** 767–777.

Burri, L., Hockendorff, J., Boehm, U., Klamp, T., Dohmen, R. J., and Levy, F. (2000). *Proc. Natl. Acad. Sci. USA* **97,** 10348–10353.

Buschmann, T., Fuchs, S. Y., Lee, C. G., Pan, Z. Q., and Ronai, Z. (2000). *Cell* **101,** 753–762.

Cardozo, C. (1993). *Enzyme Protein* **47,** 296–305.

Cardozo, C., and Kohanski, R. A. (1998). *J. Biol. Chem.* **273,** 16764–16770.

Cardozo, C., Vinitsky, A., Hidalgo, M. C., Michaud, C., and Orlowski, M. (1992). *Biochemistry* **31,** 7373–7380.

Cardozo, C., Michaud, C., and Orlowski, M. (1999). *Biochemistry* **38,** 9768–9777.

Castaño, J. G., Ornberg, R., Koster, J. G., Tobian, J. A., and Zasloff, M. (1986). *Cell* **46,** 377–385.

Castaño, J. G., Mahillo, E., Arizti, P., and Arribas, J. (1996). *Biochemistry* **35,** 3782–3789.

Cerundolo, V., Kelly, A., Elliott, T., Trowsdale, J., and Townsend, A. (1995). *Eur. J. Immunol.* **25,** 554–562.

Chau, V., Tobias, J. W., Bachmair, A., Marriott, D., Ecker, D. J., Gonda, D. K., and Varshavsky, A. (1989). *Science* **243,** 1576–1583.

Chen, P., and Hochstrasser, M. (1995). *EMBO J.* **14,** 2620–2630.

Chen, P., and Hochstrasser, M. (1996). *Cell* **86,** 961–972.

Chondrogianni, N., Petropoulos, I., Franceschi, C., Friguet, B., and Gonos, E. S. (2000). *Exp. Gerontol.* **35,** 721–728.

Chuang, S. E., Burland, V., Plunkett, G., Daniels, D. L., and Blattner, F. R. (1993). *Gene* **134,** 1–6.

Chung, D. H., Ohashi, K., Watanabe, M., Miyasaka, N., and Hirosawa, S. (2000). *J. Biol. Chem.* **275,** 4981–4987.

Ciechanover, A., Orian, A., and Schwartz, A. L. (2000). *BioEssays* **22,** 442–451.

Coles, M., Diercks, T., Liermann, J., Groger, A., Rockel, B., Baumeister, W., Koretke, K. K., Lupas, A., Peters, J., and Kessler, H. (1999). *Curr. Biol.* **9,** 1158–1168.

Conconi, M., Szweda, L. I., Levine, R. L., Stadtman, E. R., and Friguet, B. (1996). *Arch. Biochem. Biophys.* **331,** 232–240.

Coux, O., and Goldberg, A. L. (1998). *J. Biol. Chem.* **273,** 8820–8828.

Coux, O., Nothwang, H. G., Pereira, I. S., Targa, F. R., Bey, F., and Scherrer, K. (1994). *Mol. Gen. Genet.* **245,** 769–780.

Coux, O., Tanaka, K., and Goldberg, A. L. (1996). *Annu. Rev. Biochem.* **65,** 801–847.

Dahlmann, B., Kopp, F., Kuehn, L., Niedel, B., Pfeifer, G., Hegerl, R., and Baumeister, W. (1989). *FEBS Lett.* **251**, 125–131.

Dahlmann, B., Kuehn, L., Grizwa, A., Zwickl, P., and Baumeister, W. (1992). *Eur. J. Biochem.* **208**, 789–797.

D'Andrea, A., and Pellman, D. (1998). *Crit. Rev. Biochem. Mol. Biol.* **33**, 337–352.

Dawson, S. P., Arnold, J. E., Mayer, N. J., Reynolds, S. E., Billett, M. A., Gordon, C., Colleaux, L., Kloetzel, P. M., Tanaka, K., and Mayer, R. J. (1995). *J. Biol. Chem.* **270**, 1850–1858.

Dawson, S., Hastings, R., Takayanagi, K., Reynolds, S., Low, P., Billett, M., and Mayer, R. J. (1997). *Mol. Biol. Rep.* **24**, 39–44.

DeMartino, G. N., Proske, R. J., Moomaw, C. R., Strong, A. A., Song, X., Hisamatsu, H., Tanaka, K., and Slaughter, C. A. (1996). *J. Biol. Chem.* **271**, 3112–3118.

De Mot, R., Nagy, I., Walz, J., and Baumeister, W. (1999). *Trends Microbiol.* **7**, 88–92.

Desterro, J. M., Rodriguez, M. S., and Hay, R. T. (1998). *Mol. Cell* **2**, 233–239.

Dick, L. R., Moomaw, C. R., DeMartino, G. N., and Slaughter, C. A. (1991). *Biochemistry* **30**, 2725–2734.

Dick, L. R., Aldrich, C., Jameson, S. C., Moomaw, C. R., Pramanik, B. C., Doyle, C. K., DeMartino, G. N., Bevan, M. J., Forman, J. M., and Slaughter, C. A. (1994). *J. Immunol.* **152**, 3884–3894.

Dick, L. R., Cruikshank, A. A., Grenier, L., Melandri, F. D., Nunes, S. L., Stein, R., and Stein, R. L. (1996). *J. Biol. Chem.* **271**, 7273–7276.

Dick, T. P., Nussbaum, A. K., Deeg, M., Heinemeyer, W., Groll, M., Schirle, M., Keilholz, W., Stevanovic, S., Wolf, D. H., Huber, R., Rammensee, H. G., and Schild, H. (1998). *J. Biol. Chem.* **273**, 25637–25646.

Ditzel, L., Huber, R., Mann, K., Heinemeyer, W., Wolf, D. H., and Groll, M. (1998). *J. Mol. Biol.* **279**, 1187–1191.

Djaballah, H., and Rivett, A. J. (1992). *Biochemistry* **31**, 4133–4141.

Djaballah, H., Harness, J. A., Savory, P. J., and Rivett, A. J. (1992). *Eur. J. Biochem.* **209**, 629–634.

Dorn, I. T., Eschrich, R., Seemüller, E., Guckenberger, R., and Tampe, R. (1999). *J. Mol. Biol.* **288**, 1027–1036.

Driscoll, J., Brown, M. G., Finley, D., and Monaco, J. J. (1993). *Nature* **365**, 262–264.

Dubiel, W., Pratt, G., Ferrell, K., and Rechsteiner, M. (1992). *J. Biol. Chem.* **267**, 22369–22377.

Duggleby, H. J., Tolley, S. P., Hill, C. P., Dodson, E. J., Dodson, G., and Moody, P. C. (1995). *Nature* **373**, 264–268.

Edmunds, T., and Goldberg, A. L. (1986). *J. Cell Biochem.* **32**, 187–191.

Ehlers, C., Kopp, F., and Dahlmann, B. (1997). *Biol. Chem.* **378**, 249–253.

Ehring, B., Meyer, T. H., Eckerskorn, C., Lottspeich, F., and Tampe, R. (1996). *Eur. J. Biochem.* **235**, 404–415.

Eleuteri, A. M., Kohanski, R. A., Cardozo, C., and Orlowski, M. (1997). *J. Biol. Chem.* **272**, 11824–11831.

Emmerich, N. P., Nussbaum, A. K., Stevanovic, S., Priemer, M., Toes, R. E., Rammensee, H. G., and Schild, H. (2000). *J. Biol. Chem.* **275**, 21140–21148.

Enenkel, C., Lehmann, H., Kipper, J., Guckel, R., Hilt, W., and Wolf, D. H. (1994). *FEBS Lett.* **341**, 193–196.

Enenkel, C., Lehmann, A., and Kloetzel, P. M. (1998). *EMBO J.* **17**, 6144–6154.

Fagan, J. M., Waxman, L., and Goldberg, A. L. (1986). *J. Biol. Chem.* **261**, 5705–5713.

Falkenburg, P. E., Haass, C., Kloetzel, P. M., Niedel, B., Kopp, F., Kuehn, L., and Dahlmann, B. (1988). *Nature* **331**, 190–192.

Fehling, H. J., Swat, W., Laplace, C., Kuhn, R., Rajewsky, K., Muller, U., and von Boehmer, H. (1994). *Science* **265**, 1234–1237.

Fenteany, G., Standaert, R. F., Lane, W. S., Choi, S., Corey, E. J., and Schreiber, S. L. (1995). *Science* **268**, 726–731.

Ferrell, K., Wilkinson, C. R., Dubiel, W., and Gordon, C. (2000). *Trends Biochem. Sci.* **25**, 83–88.

Figueiredo-Pereira, M. E., Berg, K. A., and Wilk, S. (1994). *J. Neurochem.* **63**, 1578–1581.

Finley, D., Tanaka, K., Mann, C., Feldmann, H., Hochstrasser, M., Vierstra, R., Johnston, S., Hampton, R., Haber, J., Mccusker, J., Silver, P., Frontali, L., Thorsness, P., Varshavsky, A., Byers, B., Madura, K., Reed, S. I., Wolf, D., Jentsch, S. (1998). *Trends Biochem. Sci.* **23**, 244–245.

Fischer, M., Runkel, L., and Schaller, H. (1995). *Virus Genes* **10**, 99–102.

Frentzel, S., Pesold-Hurt, B., Seelig, A., Kloetzel, P.-M., and Kloetzel, P. M. (1994). *J. Mol. Biol.* **236**, 975–981.

Früh, K., Yang, Y., Arnold, D., Chambers, J., Wu, L., Waters, J. B., Spies, T., and Peterson, P. A. (1992). *J. Biol. Chem.* **267**, 22131–22140.

Früh, K., Gossen, M., Wang, K., Bujard, H., Peterson, P. A., and Yang, Y. (1994). *EMBO J.* **13**, 3236–3244.

Fu, H., Girod, P. A., Doelling, J. H., van Nocker, S., Hochstrasser, M., Finley, D., and Vierstra, R. D. *Mol. Biol. Rep.* **26**, 137–146.

Fuchs, S. Y., Fried, V. A., and Ronai, Z. (1998). *Oncogene* **17**, 1483–1490.

Fujimuro, M., Takada, H., Saeki, Y., Toh-e, A., Tanaka, K., and Yokosawa, H. (1998). *Biochem. Biophys. Res. Commun.* **251**, 818–823.

Gaczynska, M., Rock, K. L., and Goldberg, A. L. (1993). *Nature* **365**, 264–267.

Gaczynska, M., Goldberg, A. L., Tanaka, K., Hendil, K., Rock, K. L., and Hendil, K. B. (1996). *J. Biol. Chem.* **271**, 17275–17280.

Gardner, R. G., and Hampton, R. Y. (1999). *J. Biol. Chem.* **274**, 31671–31678.

Gardrat, F., Fraigneau, B., Montel, V., Raymond, J., and Azanza, J. L. (1999). *Eur. J. Biochem.* **262**, 900–906.

Geier, E., Pfeifer, G., Wilm, M., Lucchiari-Hartz, M., Baumeister, W., Eichmann, K., and Niedermann, G. (1999). *Science* **283**, 978–981.

Gerards, W. L., Enzlin, J., Haner, M., Hendriks, I. L., Aebi, U., Bloemendal, H., and Boelens, W. (1997). *J. Biol. Chem.* **272**, 10080–10086.

Gerards, W. L., de Jong, W. W., Boelens, W., and Bloemendal, H. (1998a). *Cell Mol. Life Sci.* **54**, 253–262.

Gerards, W. L., de Jong, W. W., Bloemendal, H., and Boelens, W. (1998b). *J. Mol. Biol.* **275**, 113–121.

Glickman, M. H., Rubin, D. M., Coux, O., Wefes, I., Pfeifer, G., Cjeka, Z., Baumeister, W., Fried, V. A., and Finley, D. (1998). *Cell* **94**, 615–623.

Glotzer, M., Murray, A. W., and Kirschner, M. W. (1991). *Nature* **349**, 132–138.

Glynne, R., Powis, S. H., Beck, S., Kelly, A., Kerr, L. A., and Trowsdale, J. (1991). *Nature* **353**, 357–360.

Goff, S. A., and Goldberg, A. L. (1985). *Cell* **41**, 587–595.

Goldberg, A. L. (1992). *Eur. J. Biochem.* **203**, 9–23.

Gonciarz-Swiatek, M., Wawrzynow, A., Um, S. J., Learn, B. A., McMacken, R., Kelley, W. L., Georgopoulos, C., Sliekers, O., and Zylicz, M. (1999). *J. Biol. Chem.* **274**, 13999–14005.

Gonzalez, M., Frank, E. G., Levine, A. S., and Woodgate, R. (1998). *Genes Dev.* **12**, 3889–3899.

Gonzalez-Pastor, J. E., San Millan, J. L., Castilla, M. A., and Moreno, F. (1995). *J. Bacteriol.* **177**, 7131–7140.

Gottesman, S. (1999). *Curr. Opin. Microbiol.* **2**, 142–147.

Gottesman, S., Roche, E., Zhou, Y., and Sauer, R. T. (1998). *Genes Dev.* **12**, 1338–1347.

Gregori, L., Hainfeld, J. F., Simon, M. N., and Goldgaber, D. (1997). *J. Biol. Chem.* **272**,

58–62.

Grizwa, A., Baumeister, W., Dahlmann, B., and Kopp, F. (1991). *FEBS Lett.* **290**, 186–190.

Groettrup, M., Soza, A., Eggers, M., Kuehn, L., Dick, T. P., Schild, H., Rammensee, H. G., Koszinowski, U. H., and Kloetzel, P. M. (1996). *Nature* **381**, 166–168.

Groll, M., Ditzel, L., Löwe, J., Stock, D., Bochtler, M., Bartunik, H. D., and Huber, R. (1997). *Nature* **386**, 463–471.

Groll, M., Heinemeyer, W., Jäger, S., Ullrich, T., Bochtler, M., Wolf, D. H., and Huber, R. (1999). *Proc. Natl. Acad. Sci. USA* **96**, 10976–10983.

Grossi de Sa, M. F., Martins de Sa, C., Harper, F., Olink-Coux, M., Huesca, M., and Scherrer, K. (1988). *J. Cell Biol.* **107**, 1517–1530.

Gueckel, R., Enenkel, C., Wolf, D. H., and Hilt, W. (1998). *J. Biol. Chem.* **273**, 19443–19452.

Guenther, B., Onrust, R., Sali, A., O'Donnell, M., and Kuriyan, J. (1997). *Cell* **91**, 335–345.

Hashimoto, M. K., Mykles, D. L., Schwartz, L. M., and Fahrbach, S. E. (1996). *J. Comp. Neurol.* **365**, 329–341.

Hastings, R., Walker, G., Eyheralde, I., Dawson, S., Billett, M., and Mayer, R. J. (1999). *Mol. Biol. Rep.* **26**, 35–38.

Hayashi, S., and Murakami, Y. (1995). *Biochem. J.* **306**, 1–10.

Hayashi, S., Murakami, Y., and Matsufuji, S. (1996). *Trends Biochem. Sci.* **21**, 27–30.

Hegerl, R., Pfeifer, G., Pühler, G., Dahlmann, B., and Baumeister, W. (1991). *FEBS Lett.* **283**, 117–121.

Heilig, R. (2000). Genbank accession AJ248283 to AJ248288.

Heinemeyer, W., Kleinschmidt, J. A., Saidowsky, J., Escher, C., and Wolf, D. H. (1991). *EMBO J.* **10**, 555–562.

Heinemeyer, W., Gruhler, A., Möhrle, V., Mahé, Y., and Wolf, D. H. (1993). *J. Biol. Chem.* **268**, 5115–5120.

Heinemeyer, W., Trondle, N., Albrecht, G., and Wolf, D. H. (1994). *Biochemistry* **33**, 12229–12237.

Hendil, K. B., Khan, S., and Tanaka, K. (1998). *Biochem. J.* **332**, 749–754.

Hengartner, C. J., Thompson, C. M., Zhang, J., Chao, D. M., Liao, S. M., Koleske, A. J., Okamura, S., and Young, R. A. (1995). *Genes Dev.* **9**, 897–910.

Herman, C., Thevenet, D., Bouloc, P., Walker, G. C., and D'Ari, R. (1998). *Genes Dev.* **12**, 1348–1355.

Heusch, M., Lin, L., Geleziunas, R., and Greene, W. C. (1999). *Oncogene* **18**, 6201–6208.

Hicke, L. (1999). *Trends Cell Biol.* **9**, 107–112.

Hilt, W., and Wolf, D. H. (1996). *Trends Biochem. Sci.* **21**, 96–102.

Hilt, W., Enenkel, C., Gruhler, A., Singer, T., and Wolf, D. H. (1993). *J. Biol. Chem.* **268**, 3479–3486.

Hirsch, C., and Ploegh, H. L. (2000). *Trends Cell Biol.* **10**, 268–272.

Hisamatsu, H., Shimbara, N., Saito, Y., Kristensen, P., Hendil, K. B., Fujiwara, T., Takahashi, E., Tanahashi, N., Tamura, T., Ichihara, A., and Tanaka, K. (1996). *J. Exp. Med.* **183**, 1807–1816.

Hiyama, H., Yokoi, M., Masutani, C., Sugasawa, K., Maekawa, T., Tanaka, K., Hoeijmakers, J. H., and Hanaoka, F. (1999). *J. Biol. Chem.* **274**, 28019–28025.

Hochstrasser, M. (2000a). *Science* **289**, 563–564.

Hochstrasser, M. (2000b). *Nature Cell Biol.* **2**, E153–E157.

Hochstrasser, M., Johnson, P. R., Arendt, C. S., Amerik, A. Y., Swaminathan, S., Swanson, R., Li, S. J., Laney, J., Pals-Rylaarsdam, R., Nowak, J., and Connerly, P. L. (1999). *Philos. Trans. R. Soc. Lond. B Biol. Sci.* **354**, 1513–1522.

Horwich, A. L., Weber-Ban, E. U., and Finley, D. (1999). *Proc. Natl. Acad. Sci. USA* **96**, 11033–11040.

Hoskins, J. R., Pak, M., Maurizi, M. R., and Wickner, S. (1998). *Proc. Natl. Acad. Sci. USA*

95, 12135–12140.
Hoskins, J. R., Singh, S. K., Maurizi, M. R., and Wickner, S. (2000a). *Proc. Natl. Acad. Sci. USA* **97,** 8892–8897.
Hoskins, J. R., Kim, S.-Y., and Wickner, S. (2000b). *J. Biol. Chem.* **275,** 35361–35367.
Huang, H., and Goldberg, A. L. (1997). *J. Biol. Chem.* **272,** 21364–21372.
Huang, H., Sherman, M. Y., Kandror, O., and Goldberg, A. L. (2001). *J. Biol. Chem.* **276,** 3920–3928.
Isupov, M. N., Obmolova, G., Butterworth, S., Badet-Denisot, M. A., Badet, B., Polikarpov, I., Littlechild, J. A., and Teplyakov, A. (1996). *Structure* **4,** 801–810.
Jackson, P. K., Eldridge, A. G., Freed, E., Furstenthal, L., Hsu, J. Y., Kaiser, B. K., and Reimann, J. D. (2000). *Trends Cell Biol.* **10,** 429–439.
Jariel-Encontre, I., Pariat, M., Martin, F., Carillo, S., Salvat, C., and Piechaczyk, M. (1995). *J. Biol. Chem.* **270,** 11623–11627.
Jelinsky, S. A., and Samson, L. D. (1999). *Proc. Natl. Acad. Sci. USA* **96,** 1486–1491.
Jentsch, S. (1992). *Annu. Rev. Genet.* **26,** 179–207.
Jentsch, S., and Pyrowolakis, G. (2000). *Trends Cell Biol.* **10,** 335–342.
Johnson, P. R., Swanson, R., Rakhilina, L., and Hochstrasser, M. (1998). *Cell* **94,** 217–227.
Johnston, J. A., Johnson, E. S., Waller, P. R., and Varshavsky, A. (1995). *J. Biol. Chem.* **270,** 8172–8178.
Jones, M. E., Haire, M. F., Kloetzel, P. M., Mykles, D. L., and Schwartz, L. M. (1995). *Dev. Biol.* **169,** 436–447.
Kandror, O., Sherman, M., and Goldberg, A. (1999). *J. Biol. Chem.* **274,** 37743–37749.
Kanemori, M., Nishihara, K., Yanagi, H., and Yura, T. (1997). *J. Bacteriol.* **179,** 7219–7225.
Karata, K., Inagawa, T., Wilkinson, A. J., Tatsuta, T., and Ogura, T. (1999). *J. Biol. Chem.* **274,** 26225–26232.
Karzai, A. W., Roche, E. D., and Sauer, R. T. (2000). *Nature Struct. Biol.* **7,** 449–455.
Kawarabayasi, Y., Sawada, M., Horikawa, H., Haidawa, Y., Hino, Y., Yamamoto, S., Sekine, M., Baba, S., Kosugi, H., Hosoyama, A., Nagai, Y., Sakai, M., Ogura, K., Otsuka, R., Nakazawa, H., Takamiya, M., Ohfuku, Y., Funahasi, T., Tanaka, T. (1998). *DNA Res.* **5,** 147–155.
Kawarabayasi, Y., Hino, Y., Horikawa, H., Yamazaki, S., Haikawa, Y., Jin-no, K., Takahashi, M., Sekine, M., Baba, S., Ankai, A., Kosugi, H., Hosoyama, A., Fukui, S., Nagai, Y., Nishijima, K., Nakazawa, H., Takamiya, M., Masuda, S., Funahashi, T. (1999). *DNA Res.* **6,** 83–52.
Keiler, K. C., Waller, P. R. H., Sauer, R. T., and Waller, P. R. (1996). *Science* **271,** 990–993.
Khattar, M. M. (1997). *FEBS Lett.* **414,** 402–404.
Kihara, A., Akiyama, Y., and Ito, K. (1995). *Proc. Natl. Acad. Sci. USA* **92,** 4532–4536.
Kim, K. I., Park, S. C., Kang, S. H., Cheong, G. W., and Chung, C. H. (1999). *J. Mol. Biol.* **294,** 1363–1374.
Kim, Y. J., Bjorklund, S., Li, Y., Sayre, M. H., and Kornberg, R. D. (1994). *Cell* **77,** 599–608.
Kim, Y. I., Burton, R. E., Burton, B. M., Sauer, R. T., and Baker, T. A. (2000). *Mol. Cell* **5,** 639–648.
Kimura, Y., Takaoka, M., Tanaka, S., Sassa, H., Tanaka, K., Polevoda, B., Sherman, F., and Hirano, H. (2000). *J. Biol. Chem.* **275,** 4635–4639.
Kisselev, A. F., Akopian, T. N., and Goldberg, A. L. (1998). *J. Biol. Chem.* **273,** 1982–1989.
Kisselev, A. F., Akopian, T. N., Woo, K. M., and Goldberg, A. L. (1999a). *J. Biol. Chem.* **274,** 3363–3371.
Kisselev, A. F., Akopian, T. N., Castillo, V., and Goldberg, A. L. (1999b). *Mol. Cell* **4,** 395–402.
Kisselev, A. F., Songyang, Z., and Goldberg, A. L. (2000). *J. Biol. Chem.* **275,** 14831–14837.
Klein, U., Gernold, M., and Kloetzel, P. M. (1990). *J. Cell Biol.* **111,** 2275–2282.

Knipfer, N., Seth, A., Roudiak, S. G., and Shrader, T. E. (1999). *Gene* **231**, 95–104.
Knipfer, N., and Shrader, T. E. (1997). *Mol. Microbiol.* **25**, 375–383.
Koegl, M., Hoppe, T., Schlenker, S., Ulrich, H. D., Mayer, T. U., and Jentsch, S. (1999). *Cell* **96**, 635–644.
Kopp, F., Dahlmann, B., and Hendil, K. B. (1993). *J. Mol. Biol.* **229**, 14–19.
Kopp, F., Hendil, K. B., Dahlmann, B., Kristensen, P., Sobek, A., and Uerkvitz, W. (1997). *Proc. Natl. Acad. Sci. USA* **94**, 2939–2944.
Krappmann, D., Wulczyn, F. G., and Scheidereit, C. (1996). *EMBO J.* **15**, 6716–6726.
Kuckelkorn, U., Knuehl, C., Boes-Fabian, B., Drung, I., and Kloetzel, P. M. (2000). *Biol. Chem.* **381**, 1017–1023.
Laachouch, J. E., Desmet, L., Geuskens, V., Grimaud, R., and Toussiant, A. (1996). *EMBO J.* **15**, 437–444.
Laney, J. D., and Hochstrasser, M. (1999). *Cell* **97**, 427–430.
Larsen, T. M., Boehlein, S. K., Schuster, S. M., Richards, N. G., Thoden, J. B., Holden, H. M., and Rayment, I. (1999). *Biochemistry* **38**, 16146–16157.
Lee, D. H., and Goldberg, A. L. (1996). *J. Biol. Chem.* **271**, 27280–27284.
Lee, D. H., and Goldberg, A. L. (1998). *Trends Cell Biol.* **8**, 397–403.
Leibovitz, D., Koch, Y., Pitzer, F., Fridkin, M., Dantes, A., Baumeister, W., and Amsterdam, A. (1994). *FEBS Lett.* **346**, 203–206.
Leibovitz, D., Koch, Y., Fridkin, M., Pitzer, F., Zwickl, P., Dantes, A., Baumeister, W., and Amsterdam, A. (1995). *J. Biol. Chem.* **270**, 11029–11032.
Lenk, U., and Sommer, T. (2000). *J. Biol. Chem.* **275**, 39403–39410.
Lenzen, C. U., Steinmann, D., Whiteheart, S. W., and Weis, W. I. (1998). *Cell* **94**, 525–536.
Leonhard, K., Stiegler, A., Neupert, W., and Langer, T. (1999). *Nature* **398**, 348–351.
Levchenko, I., Smith, C. K., Walsh, N. P., Sauer, R. T., and Baker, T. A. (1997a). *Cell* **91**, 939–947.
Levchenko, I., Yamauchi, M., and Baker, T. A. (1997b). *Genes Dev.* **11**, 1561–1572.
Levchenko, I., Seidel, M., Sauer, R. T., and Baker, T. A. (2000). *Science* **289**, 2354–2356.
Li, N., Lerea, K. M., and Etlinger, J. D. (1996). *Biochem. Biophys. Res. Commun.* **225**, 855–860.
Li, X., and Coffino, P. (1993). *Mol. Cell Biol.* **13**, 2377–2383.
Li, X., Gu, M., and Etlinger, J. D. (1991). *Biochemistry* **30**, 9709–9715.
Liakopoulos, D., Doenges, G., Matuschewski, K., and Jentsch, S. (1998). *EMBO J.* **17**, 2208–2214.
Lin, L., and Ghosh, S. (1996). *Mol. Cell Biol.* **16**, 2248–2254.
Lin, L., DeMartino, G. N., and Greene, W. C. (1998). *Cell* **92**, 819–828.
Lorsch, J. R., and Herschlag, D. (1998). *Biochemistry* **37**, 2194–2206.
Low, P., Bussell, K., Dawson, S. P., Billett, M. A., Mayer, R. J., and Reynolds, S. E. (1997). *FEBS Lett.* **400**, 345–349.
Löwe, J., Stock, D., Jap, B., Zwickl, P., Baumeister, W., and Huber, R. (1995). *Science* **268**, 533–539.
Luca, F. C., Shibuya, E. K., Dohrmann, C. E., and Ruderman, J. V. (1991). *EMBO J.* **10**, 4311–4320.
Lucero, H. A., Chojnicki, E. W., Mandiyan, S., Nelson, H., and Nelson, N. (1995). *J. Biol. Chem.* **270**, 9178–9184.
Ludemann, R., Lerea, K. M., and Etlinger, J. D. (1993). *J. Biol. Chem.* **268**, 17413–17417.
Lüders, J., Demand, J., and Hohfeld, J. (2000). *J. Biol. Chem.* **275**, 4613–4617.
Lupas, A., Koster, A. J., Walz, J., and Baumeister, W. (1994). *FEBS Lett.* **354**, 45–49.
Lupas, A., Flanagan, J. M., Tamura, T., and Baumeister, W. (1997a). *Trends Biochem. Sci.* **22**, 399–404.
Lupas, A., Zühl, F., Tamura, T., Wolf, S., Nagy, I., De Mot, R., and Baumeister, W. (1997b).

Mol. Biol. Rep. **24,** 125–131.

Ma, C. P., Slaughter, C. A., and DeMartino, G. N. (1992). *J. Biol. Chem.* **267,** 10515–10523.

Macario, A. J., Lange, M., Ahring, B. K., and de Macario, E. C. (1999). *Microbiol. Mol. Biol. Rev.* **63,** 923–967.

Martinez, C. K., and Monaco, J. J. (1991). *Science* **353,** 664–667.

Martinez, C. K., and Monaco, J. J. (1993). *Mol. Immunol.* **30,** 1177–1183.

Mason, G. G. F., Hendil, K. B., and Rivett, A. J. (1996). *Eur. J. Biochem.* **238,** 453–462.

Mason, G. G., Murray, R. Z., Pappin, D., and Rivett, A. J. (1998). *FEBS Lett.* **430,** 269–274.

Mason, R. W. (1990). *Biochem. J.* **265,** 479–484.

Mathew, A., and Morimoto, R. I. (1998). *Ann. N.Y. Acad. Sci.* **851,** 99–111.

Maupin-Furlow, J. A., and Ferry, J. G. (1995). *J. Biol. Chem.* **270,** 28617–28622.

Maupin-Furlow, J. A., Aldrich, H. C., and Ferry, J. G. (1998). *J. Bacteriol.* **180,** 1480–1487.

Maupin-Furlow, J. A., Wilson, H. L., Kaczowka, S. J., and Ou, M. S. (2000). *Front. Biosci.* **5,** d837–d865.

Maurizi, M. R. (1992). *Experientia* **48,** 178–201.

McCutchen-Maloney, S. L., Matsuda, K., Shimbara, N., Binns, D. D., Tanaka, K., Slaughter, C. A., and DeMartino, G. N. (2000). *J. Biol. Chem.* **275,** 18557–18565.

McGuire, M. J., McCullough, M. L., Croall, D. E., and DeMartino, G. N. (1989). *Biochim. Biophys. Acta* **995,** 181–186.

Mhammedi-Alaoui, A., Pato, M., Gama, M. J., and Toussaint, A. (1994). *Mol. Microbiol.* **11,** 1109–1116.

Missiakas, D., Schwager, F., Betton, J.-M., Georgopoulos, C., Raina, S., and Betton, J. M. (1996). *EMBO J.* **15,** 6899–6909.

Muller, S., Berger, M., Lehembre, F., Seeler, J. S., Haupt, Y., and Dejean, A. (2000). *J. Biol. Chem.* **275,** 13321–13329.

Murakami, Y., Matsufuji, S., Kameji, T., Hayashi, S.-I., Igarashi, K., Tamura, T., Tanaka, K., and Ichihara, A. (1992). *Nature* **360,** 597–599.

Murray, B. W., Sültmann, H., and Klein, J. (1999). *J. Immunol.* **163,** 2657–2666.

Mykles, D. L. (1996). *Arch. Biochem. Biophys.* **325,** 77–81.

Mykles, D. L., and Haire, M. F. (1991). *Arch. Biochem. Biophys.* **288,** 543–551.

Nagy, I., Tamura, T., Vanderleyden, J., Baumeister, W., and De Mot, R. (1998). *J. Bacteriol.* **180,** 5448–5453.

Nandi, D., Jiang, H., and Monaco, J. J. (1996). *J. Immunol.* **156,** 2361–2364.

Nandi, D., Woodward, E., Ginsburg, D. B., and Monaco, J. J. (1997). *EMBO J.* **16,** 5363–5375.

Nederlof, P. M., Wang, H.-R., and Baumeister, W. (1995). *Proc. Natl. Acad. Sci. USA* **92,** 12060–12064.

Neish, A. S., Gewirtz, A. T., Zeng, H., Young, A. N., Hobert, M. E., Karmali, V., Rao, A. S., and Madara, J. L. (2000). *Science* **289,** 1560–1563.

Neuwald, A. F., Aravind, L., Spouge, J. L., and Koonin, E. V. (1999). *Genome Res.* **9,** 27–43.

Ng, W. V., Kennedy, S. P., Mahairas, G. G., Berquist, B., Pan, M., Shukla, H. D., Lasky, S. R., Baliga, N. S., Thorsson, V., Sbrogna, J., Swartzell, S., Weir, D., Hall, J., Dahl, T. A., Welti, R., Goo, Y. A., Leithauser, B., Keller, K., Cruz, R. (2000). *Proc. Natl. Acad. Sci. USA* **97,** 12176–12181.

Nickel, B. E., Allis, C. D., and Davie, J. R. (1989). *Biochemistry* **28,** 958–963.

Nussbaum, A. K., Dick, T. P., Keilholz, W., Schirle, M., Stevanovic, S., Dietz, K., Heinemeyer, W., Groll, M., Wolf, D. H., Huber, R., Rammensee, H. G., and Schild, H. (1998). *Proc. Natl. Acad. Sci. USA* **95,** 12504–12509.

Ogiso, Y., Tomida, A., Kim, H. D., and Tsuruo, T. (1999). *Biochem. Biophys. Res. Commun.* **258,** 448–452.

Ogura, T., Tomoyasu, T., Yuki, T., Morimura, S., Begg, K. J., Donachie, W. D., Mori, H., Niki, H., and Hiraga, S. (1991). *Res. Microbiol.* **142**, 279–282.

Oinonen, C., Tikkanen, R., Rouvinen, J., and Peltonen, L. (1995). *Nature Struct. Biol.* **2**, 1102–1108.

Omura, S., Matsuzaki, K., Fujimoto, T., Kosuge, K., Furuya, T., Fujita, S., and Nakagawa, A. (1991). *J. Antibiot. (Tokyo)* **44**, 117–118.

Orian, A., Schwartz, A. L., Israel, A., Whiteside, S., Kahana, C., and Ciechanover, A. (1999). *Mol. Cell. Biol.* **19**, 3664–3673.

Orlowski, M., Cardozo, C., Hidalgo, M. C., and Michaud, C. (1991). *Biochemistry* **30**, 5999–6005.

Orlowski, M., Cardozo, C., and Michaud, C. (1993). *Biochemistry* **32**, 1563–1572.

Orlowski, M., Cardozo, C., Eleuteri, A. M., Kohanski, R., Kam, C. M., and Powers, J. C. (1997). *Biochemistry* **36**, 13946–13953.

Orth, K., Xu, Z., Mudgett, M. B., Bao, Z. Q., Palmer, L. E., Bliska, J. B., Mangel, W. F., Staskawicz, B., and Dixon, J. E. (2000). *Science* **290**, 1594–1597.

Ortiz-Navarrete, V., Seelig, A., Gernold, M., Frentzel, S., Kloetzel, P. M., and Hammerling, G. J. (1991). *Nature* **353**, 662–664.

Osmulski, P. A., and Gaczynska, M. (2000). *J. Biol. Chem.* **275**, 13171–13174.

Pacifici, R. E., Salo, D. C., and Davies, K. J. (1989). *Free Radic. Biol. Med.* **7**, 521–536.

Pacifici, R. E., Kono, Y., and Davies, K. J. (1993). *J. Biol. Chem.* **268**, 15405–15411.

Pak, M., Hoskins, J. R., Singh, S. K., Maurizi, M. R., and Wickner, S. (1999). *J. Biol. Chem.* **274**, 19316–19322.

Palmer, A., Mason, G. G., Paramio, J. M., Knecht, E., and Rivett, A. J. (1994). *Eur. J. Cell Biol.* **64**, 163–175.

Palmer, A., Rivett, A. J., Thomson, S., Hendil, K. B., Butcher, G. W., Fuertes, G., and Knecht, E. (1996). *Biochem. J.* **316**, 401–407.

Palombella, V. J., Rando, O. J., Goldberg, A. L., and Maniatis, T. (1994). *Cell* **78**, 773–785.

Pamnani, V., Haas, B., Puhler, G., Sanger, H. L., and Baumeister, W. (1994). *Eur. J. Biochem.* **225**, 511–519.

Parsell, D. A., and Sauer, R. T. (1989). *J. Biol. Chem.* **264**, 7590–7595.

Patel, S. D., Monaco, J. J., and McDevitt, H. O. (1994). *Proc. Natl. Acad. Sci. USA* **91**, 296–300.

Peters, J.-M. (1994). *Trends Biochem. Sci.* **19**, 377–382.

Peters, J. M., Franke, W. W., and Kleinschmidt, J. A. (1994). *J. Biol. Chem.* **269**, 7709–7718.

Petit, F., Jarrousse, A. S., Dahlmann, B., Sobek, A., Hendil, K. B., Buri, J., Briand, Y., and Schmid, H. P. (1997a). *Biochem. J.* **326**, 93–98.

Petit, F., Jarrousse, A. S., Boissonnet, G., Dadet, M. H., Buri, J., Briand, Y., and Schmid, H. P. (1997b). *Mol. Biol. Rep.* **24**, 113–117.

Petropoulos, I., Conconi, M., Wang, X., Hoenel, B., Bregegere, F., Milner, Y., and Friguet, B. (2000). *J. Gerontol. A Biol. Sci. Med. Sci.* **55**, B220–B227.

Pham, A. D., and Sauer, F. (2000). *Science* **289**, 2357–2360.

Pickart, C. M. (1997). *FASEB J.* **11**, 1055–1066.

Pickart, C. M. (2000). *Nature Cell Biol.* **2**, E139–E141.

Pouch, M. N., Cournoyer, B., and Baumeister, W. (2000). *Mol. Microbiol.* **35**, 368–377.

Rajagopalan, K. V. (1997). *Biochem Soc. Trans.* **25**, 757–761.

Ramos, P. C., Hockendorff, J., Johnson, E. S., Varshavsky, A., and Dohmen, R. J. (1998). *Cell* **92**, 489–499.

Read, M. A., Brownell, J. E., Gladysheva, T. B., Hottelet, M., Parent, L. A., Coggins, M. B., Pierce, J. W., Podust, V. N., Luo, R. S., Chau, V., and Palombella, V. J. (2000). *Mol. Cell. Biol.* **20**, 2326–2333.

Realini, C., Dubiel, W., Pratt, G., Ferrell, K., and Rechsteiner, M. (1994). *J. Biol. Chem.* **269**, 20727–20732.

Realini, C., Jensen, C. C., Zhang Zg, Johnston, S. C., Knowlton, J. R., Hill, C. P., and Rechsteiner, M. (1997). *J. Biol. Chem.* **272**, 25483–25492.

Rechsteiner, M., and Rogers, S. W. (1996). *Trends Biochem. Sci.* **21**, 267–271.

Reidlinger, J., Pike, A. M., Savory, P. J., Murray, R. Z., and Rivett, A. J. (1997). *J. Biol. Chem.* **272**, 24899–24905.

Reits, E. A. J., Benham, A. M., Plougastel, B., Neefjes, J., and Trowsdale, J. (1997). *EMBO J.* **16**, 6087–6094.

Richmond, C., Gorbea, C., and Rechsteiner, M. (1997). *J. Biol. Chem.* **272**, 13403–13411.

Rivett, A. J. (1993). *Biochem. J.* **291**, 1–10.

Rivett, A. J. (1998). *Curr. Opin. Immunol.* **10**, 110–114.

Rivett, A. J., Palmer, A., and Knecht, E. (1992). *J. Histochem. Cytochem.* **40**, 1165–1172.

Robzyk, K., Recht, J., and Osley, M. A. (2000). *Science* **287**, 501–504.

Roche, E. D., and Sauer, R. T. (1999). *EMBO J.* **18**, 4579–4589.

Rock, K. L., Gramm, C., Rothstein, L., Clark, K., Stein, R., Dick, L., Hwang, D., and Goldberg, A. L. (1994). *Cell* **78**, 761–771.

Rockel, B., Walz, J., Hegerl, R., Peters, J., Typke, D., and Baumeister, W. (1999). *FEBS Lett.* **451**, 27–32.

Rohrwild, M., Coux, O., Huang, H.-C., Moerschell, R. P., Yoo, S. J., Seol, J. H., Chung, C. H., Goldberg, A. L., and Huang, H. C. (1996). *Proc. Natl. Acad. Sci. USA* **93**, 5808–5813.

Rousset, R., Desbois, C., Bantignies, F., and Jalinot, P. (1996). *Nature* **381**, 328–331.

Rubin, D. M., Glickman, M. H., Larsen, C. N., Dhruvakumar, S., and Finley, D. (1998). *EMBO J.* **17**, 4909–4919.

Ruepp, A., Eckerskorn, C., Bogyo, M., and Baumeister, W. (1998). *FEBS Lett.* **425**, 87–90.

Ruepp, A., Graml, W., Santos-Martinez, M. L., Koretke, K. K., Volker, C., Mewes, H. W., Frishman, D., Stocker, S., Lupas, A. N., and Baumeister, W. (2000). *Nature* **407**, 508–513.

Ruggero, D., Ciammaruconi, A., and Londei, P. (1998). *EMBO J.* **17**, 3471–3477.

Russell, S. J., Reed, S. H., Huang, W., Friedberg, E. C., and Johnston, S. A. (1999). *Mol. Cell* **3**, 687–695.

Sacchetta, P., Battista, P., Santarone, S., and Di Cola, D. (1990). *Biochim. Biophys. Acta* **1037**, 337–343.

Saraste, M., Sibbald, P. R., and Wittinghofer, A. (1990). *Trends Biochem. Sci.* **15**, 430–434.

Schauber, C., Chen, L., Tongaonkar, P., Vega, I., Lambertson, D., Potts, W., and Madura, K. (1998). *Nature* **391**, 715–718.

Schmid, H. P., Akhayat, O., de Sa, C. M., Puvion, F., Koehler, K., and Scherrer, K. (1984). *EMBO J.* **3**, 29–34.

Schmidt, M., Schmidtke, G., and Kloetzel, P. M. (1997). *Mol. Biol. Rep.* **24**, 103–112.

Schmidt, M., Lupas, A. N., and Finley, D. (1999). *Curr. Opin. Chem. Biol.* **3**, 584–591.

Schmidtke, G., Kraft, R., Kostka, S., Henklein, P., Frommel, C., Löwe, J., Huber, R., Kloetzel, P. M., and Schmidt, M. (1996). *EMBO J.* **15**, 6887–6898.

Schmidtke, G., Schmidt, M., and Kloetzel, P.-M. (1997). *J. Mol. Biol.* **268**, 95–106.

Schmidtke, G., Emch, S., Groettrup, M., and Holzhutter, H. G. (2000). *J. Biol. Chem.* **275**, 22056–22063.

Schubert, U., Anton, L. C., Gibbs, J., Norbury, C. C., Yewdell, J. W., and Bennink, J. R. (2000). *Nature* **404**, 770–774.

Schule, T., Rose, M., Entian, K. D., Thumm, M., and Wolf, D. H. (2000). *EMBO J.* **19**, 2161–2167.

Sears, C., Olesen, J., Rubin, D., Finley, D., and Maniatis, T. (1998). *J. Biol. Chem.* **273**, 1409–1419.

Seeger, M., Ferrell, K., and Dubiel, W. (1997). *Mol. Biol. Rep.* **24**, 83–88.

Seelig, A., Multhaup, G., Pesold-Hurt, B., Beyreuther, K., and Kloetzel, P. M. (1993). *J. Biol. Chem.* **268**, 25561–25567.

Seemüller, E., Lupas, A., Stock, D., Löwe, J., Huber, R., and Baumeister, W. (1995a). *Science* **268,** 579–582.

Seemüller, E., Lupas, A., Zühl, F., Zwickl, P., and Baumeister, W. (1995b). *FEBS Lett.* **359,** 173–178.

Seemüller, E., Lupas, A., and Baumeister, W. (1996). *Nature* **382,** 468–470.

Seol, J. H., Yoo, S. J., Kang, M. S., Ha, D. B., and Chung, C. H. (1995). *FEBS Lett.* **377,** 41–43.

Seol, J. H., Yoo, S. J., Shin, D. H., Shim, Y. K., Kang, M. S., Goldberg, A. L., and Chung, C. H. (1997). *Eur. J. Biochem.* **247,** 1143–1150.

Shelton, E., Kuff, E. L., Maxwell, E. S., and Harrington, J. T. (1970). *J. Cell Biol.* **45,** 1–8.

Shirai, Y., Akiyama, Y., and Ito, K. (1996). *J. Bacteriol.* **178,** 1141–1145.

Shotland, Y., Koby, S., Teff, D., Mansur, N., Oren, D. A., Tatematsu, K., Tomoyasu, T., Kessel, M., Bukau, B., Ogura, T., and Oppenheim, A. B. (1997). *Mol. Microbiol.* **24,** 1303–1310.

Sibille, C., Gould, K. G., Willard-Gallo, K., Thomson, S., Rivett, A. J., Powis, S., Butcher, G. W., and De Baetselier, P. (1995). *Curr. Biol.* **5,** 923–930.

Singh, S. K., Grimaud, R., Hoskins, J. R., Wickner, S., and Maurizi, M. R. (2000). *Proc. Natl. Acad. Sci. USA* **97,** 8898–8903.

Smith, C. A., and Rayment, I. (1996). *Biophys. J.* **70,** 1590–1602.

Smith, C. K., Baker, T. A., and Sauer, R. T. (1999). *Proc. Natl. Acad. Sci. USA* **96,** 6678–6682.

Smith, D. R., Doucette-Stamm, L. A., Deloughery, C., Lee, H., Dubois, J., Aldredge, T., Bashirzadeh, R., Blakely, D., Cook, R., Gilbert, K., Harrison, D., Hoang, L., Keagle, P., Lumm, W., Pothier, B., Qiu, D., Spadafora, R., Vicaire, R., Wang, Y. (1997). *J. Bacteriol.* **179,** 7135–7155.

Smith, J. L., Zaluzec, E. J., Wery, J. P., Niu, L., Switzer, R. L., Zalkin, H., and Satow, Y. (1994). *Science* **264,** 1427–1433.

Sousa, M. C., Trame, C. B., Tsuruta, H., Wilbanks, S. M., Reddy, V. S., and McKay, D. B. (2000). *Cell* **103,** 633–643.

Spence, J., Gali, R. R., Dittmar, G., Sherman, F., Karin, M., and Finley, D. (2000). *Cell* **102,** 67–76.

Stohwasser, R., Salzmann, U., Giesebrecht, J., Kloetzel, P. M., and Holzhutter, H. G. (2000). *Eur. J. Biochem.* **267,** 6221–6230.

Strickland, E., Hakala, K., Thomas, P. J., and DeMartino, G. N. (2000). *J. Biol. Chem.* **275,** 5565–5572.

Su, K., Roos, M. D., Yang, X., Han, I., Paterson, A. J., and Kudlow, J. E. (1999). *J. Biol. Chem.* **274,** 15194–15202.

Swaffield, J. C., Bromberg, J. F., and Johnston, S. A. (1992). *Nature* **360,** 768–768.

Swaffield, J. C., Melcher, K., and Johnston, S. A. (1995). *Nature* **374,** 88–91.

Tamura, T., Nagy, I., Lupas, A., Lottspeich, F., Cejka, Z., Schoofs, G., Tanaka, K., De Mot, R., and Baumeister, W. (1995). *Curr. Biol.* **5,** 766–774.

Tanaka, K., and Kasahara, M. (1998). *Immunol. Rev.* **163,** 161–176.

Taylor, S. V., Kelleher, N. L., Kinsland, C., Chiu, H. J., Costello, C. A., Backstrom, A. D., McLafferty, F. W., and Begley, T. P. (1998). *J. Biol. Chem.* **273,** 16555–16560.

Teichert, U., Mechler, B., Muller, H., and Wolf, D. H. (1989). *J. Biol. Chem.* **264,** 16037–16045.

Thompson, M. W., Singh, S. K., and Maurizi, M. R. (1994). *J. Biol. Chem.* **269,** 18209–18215.

Thrower, J. A., Hoffman, L., Rechsteiner, M., and Pickart, C. M. (2000). *EMBO J.* **19,** 94–102.

Treier, M., Staszewski, L. M., and Bohmann, D. (1994). *Cell* **78,** 787–798.

Trent, J. D., Gabrielsen, M., Jensen, B., Neuhard, J., and Olsen, J. (1994). *J. Bacteriol.* **176,** 6148–6152.

Tsubuki, S., Saito, Y., and Kawashima, S. (1994). *FEBS Lett.* **344,** 229–233.

Tu, G. F., Reid, G. E., Zhang, J. G., Moritz, R. L., and Simpson, R. J. (1995). *J. Biol. Chem.* **270,** 9322–9326.

Turner, G. C., and Varshavsky, A. (2000). *Science* **289,** 2117–2120.

Turner, G. C., Du, F., and Varshavsky, A. (2000). *Nature* **405,** 579–583.

Umeda, M., Manabe, Y., and Uchimiya, H. (1997). *FEBS Lett.* **403,** 313–317.

Ustrell, V., Realini, C., Pratt, G., and Rechsteiner, M. (1995). *FEBS Lett.* **376,** 155–158.

Vale, R. D. (1996). *J. Cell Biol.* **135,** 291–302.

Vale, R. D. (2000). *J. Cell Biol.* **150,** F13–F19.

Van Kaer, L., Ashton-Rickardt, P. G., Eichelberger, M., Gaczynska, M., Nagashima, K., Rock, K. L., Goldberg, A. L., Doherty, P. C., and Tonegawa, S. (1994). *Immunity* **1,** 533–541.

Varshavsky, A. (1992). *Cell* **69,** 725–735.

Varshavsky, A. (1997). *Genes Cells* **2,** 13–28.

Verma, R., Chen, S., Feldman, R., Schieltz, D., Yates, J., Dohmen, J., and Deshaies, R. J. (2000). *Mol. Biol. Cell* **11,** 3425–3439.

Vido, K., Spector, D., Lagniel, G., Lopez, S., Toledano, M. B., and Labarre, J. (2001). *J. Biol. Chem.* **267,** 8469–8474.

Vierstra, R. D. (1996). *Plant Mol. Biol.* **32,** 275–302.

Volker, U., Riethdorf, S., Winkler, A., Weigend, B., Fortnagel, P., and Hecker, M. (1993). *FEMS Microbiol. Lett.* **106,** 287–293.

vom Baur, E., Zechel, C., Heery, D., Heine, M. J., Garnier, J. M., Vivat, V., Le Douarin, B., Gronemeyer, H., Chambon, P., and Losson, R. (1996). *EMBO J.* **15,** 110–124.

Wagner, B. J., and Margolis, J. W. (1993). *Arch. Biochem. Biophys.* **307,** 146–152.

Walker, J. E., Saraste, M., Runswick, M. J., and Gay, N. J. (1982). *EMBO J.* **1,** 945–951.

Walz, J., Erdmann, A., Kania, M., Typke, D., Koster, A. J., and Baumeister, W. (1998). *J. Struct. Biol.* **121,** 19–29.

Wang, J., Hartling, J. A., and Flanagan, J. M. (1997). *Cell* **91,** 447–456.

Wang, L., Elliott, M., and Elliott, T. (1999b). *J. Bacteriol.* **181,** 1211–1219.

Wang, R., Chait, B. T., Wolf, I., Kohanski, R. A., and Cardozo, C. (1999a). *Biochemistry* **38,** 14573–14581.

Watanabe, T. K., Saito, A., Suzuki, M., Fujiwara, T., Takahashi, E., Slaughter, C. A., DeMartino, G. N., Hendil, K. B., Chung, C. H., Tanahashi, N., and Tanaka, K. (1998). *Genomics* **50,** 241–250.

Weber-Ban, E. U., Reid, B. G., Miranker, A. D., and Horwich, A. L. (1999). *Nature* **401,** 90–93.

Wehren, A., Meyer, H. E., Sobek, A., Kloetzel, P.-M., and Dahlmann, B. (1996). *Biol. Chem.* **377,** 497–503.

Weissman, J. S., Sigler, P. B., and Horwich, A. L. (1995). *Science* **268,** 523–524.

Weitman, D., and Etlinger, J. D. (1992). *J. Biol. Chem.* **267,** 6977–6982.

Wenzel, T., and Baumeister, W. (1993). *FEBS Lett.* **326,** 215–218.

Wenzel, T., and Baumeister, W. (1995). *Nature Struct. Biol.* **2,** 199–204.

Wenzel, T., Eckerskorn, C., Lottspeich, F., and Baumeister, W. (1994). *FEBS Lett.* **349,** 205–209.

Whitby, F. G., Masters, E. I., Kramer, L., Knowlton, J. R., Yao, Y., Wang, C. C., and Hill, C. P. (2000). *Nature* **408,** 115–120.

Wickner, S., Gottesman, S., Skowyra, D., Hoskins, J., McKenney, K., and Maurizi, M. R. (1994). *Proc. Natl. Acad. Sci. USA* **91,** 12218–12222.

Wickner, S., Maurizi, M. R., and Gottesman, S. (1999). *Science* **286,** 1888–1893.

Wigley, W. C., Fabunmi, R. P., Lee, M. G., Marino, C. R., Muallem, S., DeMartino, G. N., and Thomas, P. J. (1999). *J. Cell Biol.* **145,** 481–490.

Wilkinson, C. R., Wallace, M., Morphew, M., Perry, P., Allshire, R., Javerzat, J. P., McIntosh, J. R., and Gordon, C. (1998). *EMBO J.* **17,** 6465–6476.

Wilkinson, K. D. (2000). *Semin. Cell Dev. Biol.* **11,** 141–148.

Wilson, H. L., Aldrich, H. C., and Maupin-Furlow, J. A. (1999). *J. Bacteriol.* **181,** 5814–5824.

Wilson, H. L., Ou, M. S., Aldrich, H. C., and Maupin-Furlow, J. A. (2000). *J. Bacteriol.* **182,** 1680–1692.

Wolf, S., Nagy, I., Lupas, A., Pfeifer, G., Cejka, Z., Müller, S. A., Engel, A., De Mot, R., and Baumeister, W. (1998). *J. Mol. Biol.* **277,** 13–25.

Wu, W. F., Zhou, Y., and Gottesman, S. (1999). *J. Bacteriol.* **181,** 3681–3687.

Xie, Y., and Varshavsky, A. (2000). *Proc. Natl. Acad. Sci. USA* **97,** 2497–2502.

Xu, Z., Horwich, A. L., and Sigler, P. B. (1997). *Nature* **388,** 741–750.

Yang, Y., Früh, K., Ahn, K., and Peterson, P. A. (1995). *J. Biol. Chem.* **270,** 27687–27964.

Yao, Y., Toth, C. R., Huang, L., Wong, M. L., Dias, P., Burlingame, A. L., Coffino, P., and Wang, C. C. (1999). *Biochem. J.* **344,** 349–358.

Yoo, S. J., Seol, J. H., Shin, D. H., Rohrwild, M., Kang, M.-S., Tanaka, K., Goldberg, A. L., Chung, C. H., and Kang, M. S. (1996). *J. Biol. Chem.* **271,** 14035–14040.

Yoo, S. J., Shim, Y. K., Seong, I. S., Seol, J. H., Kang, M. S., and Chung, C. H. (1997). *FEBS Lett.* **412,** 57–60.

Yu, R. C., Hanson, P. I., Jahn, R., and Brunger, A. T. (1998). *Nature Struct. Biol.* **5,** 803–811.

Yuan, X., Miller, M., and Belote, J. M. (1996). *Genetics* **144,** 147–157.

Zaiss, D., and Belote, J. M. (1997). *Gene* **201,** 99–105.

Zhang, Z., Krutchinsky, A., Endicott, S., Realini, C., Rechsteiner, M., and Standing, K. G. (1999). *Biochemistry* **38,** 5651–5658.

Zühl, F., Tamura, T., Dolenc, I., Cejka, Z., Nagy, I., De Mot, R., Baumeister, W., and De Mot, R. (1997a). *FEBS Lett.* **400,** 83–90.

Zühl, F., Seemüller, E., Golbik, R., and Baumeister, W. (1997b). *FEBS Lett.* **418,** 189–194.

Zwickl, P., Pfeifer, G., Lottspeich, F., Kopp, F., Dahlmann, B., and Baumeister, W. (1990). *J. Struct. Biol.* **103,** 197–203.

Zwickl, P., Grizwa, A., Pühler, G., Dahlmann, B., Lottspeich, F., and Baumeister, W. (1992a). *Biochemistry* **31,** 964–972.

Zwickl, P., Lottspeich, F., and Baumeister, W. (1992b). *FEBS Lett.* **312,** 157–160.

Zwickl, P., Kleinz, J., and Baumeister, W. (1994). *Nature Struct. Biol.* **1,** 765–770.

Zwickl, P., Ng, D., Woo, K. M., Klenk, H.-P., and Goldberg, A. L. (1999). *J. Biol. Chem.* **274,** 26008–26014.

Archaeal Catabolite Repression: A Gene Regulatory Paradigm

ELISABETTA BINI AND PAUL BLUM

School of Biological Sciences
University of Nebraska
Lincoln, Nebraska 68588-0666

I. Catabolite Repression, Multiple Names for the Same Physiological Phenomenon

Over time, the terminology for catabolite repression has changed, reflecting improved understanding of the basic underlying mechanisms involved. For brevity we refer to this process as it occurred in bacteria. Adaptation and diauxic growth were first used to describe the sequential utilization of sugars when bacterial cultures were grown in the presence of two carbohydrates. Depending on the pair of sugars present, the

339

ADVANCES IN APPLIED MICROBIOLOGY, VOLUME 50

cultures showed an unusual growth curve, characterized by an initial cycle of growth on the preferred sugar. This was followed by a growth lag (adaptation) during which the enzymes for the utilization of the second sugar were synthesized. Finally, a second cycle of growth occurred. This two-cycle growth pattern was called diauxic growth or diauxie. For years diauxie was associated with glucose and termed the "glucose effect" because the presence of this sugar in most cases inhibits the metabolism of other carbon sources (Magasanik and Neidhardt, 1987). In time, Magasanik observed that not only glucose, but any intermediate metabolite in the pathway of glucose utilization, or any molecule that can be quickly converted to one of those intermediates, precipitates the same effect as glucose. Therefore, the term "catabolite repression" was invoked to indicate the glucose effect.

Further studies revealed that catabolite repression is composed of at least three components, which are activated depending on the organism or the conditions of growth. These include transient repression, permanent repression (or catabolite repression), and inducer exclusion. Transient repression produces a very strong repression, which lasts for not more than one generation. It occurs when a repressing carbon source is added to cells growing on a weak repressing carbon source. Then the activity of the affected enzyme will increase further up to the level of permanent repression (Tyler et al., 1967). Inducer exclusion occurs when the presence of glucose causes an inhibition of sugar permeases, preventing the transport of inducers into the cell (Saier, 1989).

A. ORGANIZATION OF THIS ARTICLE

This article presents an overview of the multiple mechanisms of catabolite repression employed by bacteria and eukaryotes as a basis for comparison with the recently discovered system of the archaeon *Sulfolobus solfataricus*. For more information the reader is directed to literature reviews available on this subject. For bacteria these include Deutscher et al. (1997), Hueck and Hillen (1995), Magasanik and Neidhardt (1987), Ramseier (1996), Saier (1989, 1996, 1998), Saier et al. (1996), and Stulke and Hillen (2000). For yeast and other fungi these include Ebbole (1998), Flores et al. (2000), Gancedo (1998), Ronne (1995), Trumbly (1992), and Wills (1996). For plants these include Farrar et al. (2000), Koch et al. (2000), and Roitsch (1999). For animals these include Conti (2000), Movesian (2000), Roesler (2000), and Skalhegg and Tasken (2000). The various mechanisms employed are classified according to their type of control, their targets, and their level of action. In contrast to bacteria and eukaryotes, the regulation of carbon metabolism in archaea is not as well studied. Most cultured archaea have evolved to live in extreme environments. Consequently, some consideration is necessary for certain metabolic characteristics, often unique, that allowed these organisms to adapt to environments once thought incompatible with life. Such information is used in this article to formulate hypotheses about mechanisms of catabolite repression in members of the archaeal domain.

II. How Bacterial Prokaryotes and Eukaryotes Accomplish Catabolite Repression

A. Multiple Mechanisms for Catabolite Repression in Bacteria and Eukaryotes

For historic reasons the mechanisms controlling catabolite repression in enteric bacteria, which rely on the use of the phospho*enol*pyruvate-dependent phosphotransferase system (PTS) to initiate signal transduction, have been regarded as a model for all other bacteria. More recently, observations on other bacterial species, notably gram-positive taxa, revealed that many different strategies have been developed to achieve catabolite repression. In eukaryotes, most data regarding catabolite repression derive from studies on budding yeast. In contrast, in multicellular eukaryotes, an interest on how sugars influence subcellular metabolism at the molecular level is relatively new. In higher animals, however, clinical implications related to cardiovascular disease, apoptosis, and learning are providing new stimuli for additional studies. The following section covers only the main systems of regulation, with the intent of demonstrating their mechanistic diversity, with an emphasis on those which act at the level of transcription. Table I provides a summary of catabolite repression protein factors which have been identified in the bacterial and eukaryotic domains. However, there are less common mechanisms of regulation that act instead at the posttranscriptional level, at the level of translation, or at the level of enzyme activity. Such systems are not discussed further here. The interested reader is encouraged to consult the reviews cited above for more information.

B. Mechanisms for Catabolite Repression in Bacteria

1. Escherichia coli

The catabolite repression systems known to operate in *E. coli* are thought to apply to other members of the enteric bacterial group, a subdivison within the proteobacterial kingdom of bacteria. This organism has two important mechanisms to accomplish catabolite repression. One employs the PTS. The PTS senses the presence of glucose and starts

TABLE I

BACTERIAL AND EUKARYOTIC FACTORS INVOLVED IN CATABOLITE REPRESSION

Organism	Protein	Accession No.	Gene family and function	Hits[a]
Bacteria				
E. coli	CAP	AAC76382	Transcription activator	b, a, f/b
	Cra		Transcription activator/repressor	
B. subtilis	Ccpa	S15318	Transcription repressor/activator	b
	Hpr		Kinase/phosphatase/activator of Ccpa	b
Eukaryotes				
Human	CREB	AAA52071	Basic protein/leucine zipper transcription activator	a, f, p
	CREM		Basic protein/leucine zipper transcription modulator	
	ATF		Basic protein/leucine zipper transcription activator	a, e
	HXK	P35557	Hexokinase	f, b
Yeast	HXK2	NP_011261	Hexokinase/sensor	f, pr, b
	SNF1[b]	NP_010765	sr/thr kinase/ inhibitor of MIG1	e
	SNF4		Inhibitor of MIG1	
	MIG1	S17248	Zn finger/transcription repressor	a, f
	CYC8		Activator of MIG1	
	TUP1		Activator of MIG1	
	GAL4	NP_015076	Zn finger	
Aspergillus	CREA		Related to MIG1	
Neurospora	CREA			
	CRE-1		Related to MIG1	
Plants	Q42525		Hexokinase	f, pr, b, a

[a] Searches against protein DB/against genomes at the NCBI. Single designations are as follows: b, bacteria; a, animals; f, fungi/yeast; p, plants; pr, protists; e, eukaryotes.

[b] Also called CAT1, and CCR1

a protein phosphorylation cascade terminating in the phosphorylation of glucose. The transport and linked phosphorylation of glucose determines the dephosphorylation of the PTS enzyme IIA. This has two consequences: a reduction in cAMP levels and inhibition of the uptake systems for other compounds by inducer exclusion. In the absence of glucose, elevated levels of cAMP activate the factor CRP (catabolite activator protein), also called CAP (catabolite receptor protein). CRP is a transcriptional activator for the genes affected by catabolite repression, which are therefore transcribed in the absence of glucose but not in its presence (Crasnier, 1996; Magasanik and Neidhardt, 1987).

The second catabolite repression mechanism employed by *E. coli* is cAMP independent. It is mediated by the pleiotropic transcriptional regulator Cra (catabolite repressor/activator). This factor acts at the level of transcription initiation, activates the Embden–Meyerhoff–Parnas (EMP) and Entner–Dourdoroff (ED) pathways, and exerts catabolite repression over the TCA cycle, the glyoxalate shunt, and gluconeogenesis (Magasanik and Neidhardt, 1987; Ramseier, 1996; Saier, 1996).

2. Bacillus Subtilis

The systems employed by *B. subtilis,* a member of the low-G+C group of gram-positive bacteria, are thought to apply to other genera classified within this grouping of gram-positive bacteria. *B. subtilis* uses at least three mechanisms to accomplish catabolite repression. All of them are activated by a signal generated by the PTS system. They include a global regulator of transcription, inducer exclusion, and induction prevention (Saier, 1989; Saier *et al.,* 1996; Stulke and Hillen, 2000). The PTS system in *B. subtilis* differs from that of *E. coli* because it is cAMP independent. In *B. subtilis,* the HPr component of the PTS system is phosphorylated in the presence of glucose, then binds and activates the catabolite control protein (CcpA) (Deutscher *et al.,* 1995; Jones *et al.,* 1997), and the resulting complex binds to the cis-acting catabolic responsive element (CRE). Complex binding inhibits transcription of target promoters (Fujita *et al.,* 1995; Miwa *et al.,* 2000). In some cases, CcpA can act as a transcriptional activator as in the case of the expression of genes involved in fermentative metabolism.

3. Examples of Reverse Catabolite Repression

TCA cycle intermediates repress enzymes involved in sugar metabolism in *Rhodospirillaceae.* Substrates such as succinate or malate are preferred carbon sources rather than sugars. This phenomenon is known as reverse catabolite repression (Inui *et al.,* 1996), presumably in recognition of the metabolic proximity of preferred carbon sources to central metabolism. *Rhizobium spp.* have no PTS for sugar transport. Succinate controls key enzymes of the ED and EMP pathways. Succinate also causes catabolite repression of lactose metabolism. Interestingly cAMP which is present may be involved in the regulation, acting as a signal molecule (O'Gara *et al.,* 1989). *Pseudomonas aeruginosa* is also characterized by reverse catabolite repression. In these organisms, also members of the proteobacteria but not the enteric group like *E. coli,* organic acids are preferred over carbohydrates. The existence of mutants that have lost catabolite repression suggests that catabolite repression acts by repression in the presence of the preferred carbon source. cAMP does not appear to be involved in this form of reverse catabolite repression (Collier *et al.,* 1996).

C. Mechanisms for Catabolite Repression in Eukaryotes

1. Yeast

In *Saccharomyces cerevisiae*, a member of the group of budding yeast and a subdivison of the fungi, catabolite repression is referred to as glucose repression and occurs mainly at the level of transcription. It consists of two types, a cAMP-independent form and a cAMP-dependent form. Glucose repression affects enzymes required for the metabolism of disaccharides and nonfermentable carbon sources, for the gluconeogenic pathway, and for respiration. Accordingly, multiple genes are involved in this regulatory system precipitated by glucose availability. In addition, the central pathway that controls glucose repressible genes interacts with secondary pathways that in turn regulate subsets of glucose repressible genes. *S. cerevisiae* should, however, be regarded as a eukaryotic model for catabolite repression with some caution. *S. cerevisiae* has evolved to ferment glucose exclusively. In the presence of nonfermentable carbon sources, carbon enters the TCA cycle and ATP is generated by respiration; at the same time, gluconegenesis is activated for production of sugars (hexoses). If instead glucose is present, it will be metabolized via the EMP pathway to form pyruvate. In the glycolysis of higher organisms, pyruvate enters the TCA cycle. In the case of *S. cerevisiae*, glucose instead represses the pathways related to respiration, resulting in ethanol formation. In spite of this peculiarity, catabolite repression in this organism has undergone extensive analysis and as yet represents the best-studied eukaryotic system for catabolite repression. The complexity of control of glucose repression in *S. cerevisiae* results from the overlap of various pathways presumably to avoid wasted energy.

a. cAMP-Independent Pathways. Regulation of expression of the SUC genes encoding the enzyme invertase has been among the most intensively studied of the cAMP-independent systems (del Castillo *et al.*, 1992; Lutfiyya *et al.*, 1998; Ronne, 1995; Trumbly, 1992). The hexokinase, HXK2, responds to the presence of glucose by generating a signal through the factors CID1, GRR1, and REG1. This signal inhibits the protein kinase SNF1–SNF4, which in turn inhibits the transcriptional repressor MIG1. MIG1 also requires the presence of the complex CYC8–TUP1 to repress the promoter. The end result of this regulatory cascade is that in the presence of glucose, SUC repression is relieved, and the level of SUC mRNA is 100 times higher than in the absence of glucose.

Regulation of expression of the GAL genes encoding enzymes for the catabolism of galactose has also received much attention as a cAMP-independent system (Trumbly, 1992). Two types of control have been observed. The first is induction by galactose, which results from the activation of transcription by GAL4, harboring a zinc finger domain. The second is repression by glucose, which results in the inhibition of induction using several pathways. Regulation is exerted primarily through the repression by MIG1 of GAL4 and secondarily by negative regulation of transcription of various target genes.

Regulation of expression of the MAL genes encoding enzymes for the catabolism of maltose is similar to that observed for the GAL genes (Lutfiyya et al., 1998; Trumbly, 1992). There is induction by maltose and repression by glucose. However, a third mechanism is also important for the occurrence of inducer exclusion.

Finally, genes involved in respiratory and gluconeogenic functions have also been examined. Genes involved in respiration are repressed by glucose by a pathway separate from the one used to repress genes for the metabolism of other carbohydrates [e.g., SUC, GAL, MAL (Ronne, 1995; Trumbly, 1992)]. Glucose represses the synthesis of the genes encoding the transcriptional activators HAP2 and HAP4, which in turn control the expression of the cytochrome c genes CYC1 and CYC7. Genes involved in gluconeogenesis including FBP1 and PCK1, which encode enzymes that catalyze the two irreversible steps of classical gluconeogenesis, are repressed by glucose (Ronne, 1995; Trumbly, 1992). Two overlapping mechanisms appear to be employed. One consists of the inhibition by the transcriptional repressor MIG1, and the other involves the inhibition of an activator complex with SNF1, which acts independently from MIG1, in contrast to its role in regulating expression of SUC.

b. The cAMP-Dependent Pathway. In this pathway, glucose generates a signal by inducing cAMP synthesis. It requires the presence of a G-protein-coupled receptor system [GPR1–GPR2 (Rolland et al., 2000)]. The G-protein system detects extracellular glucose and phosphorylates the incoming sugar through the action of a hexose kinase. Hexose transporters do not seem to be involved in the activation of this pathway, and the glucose sensors/carriers SNF3 and RGT2 (Ozcan et al., 1998), which are utilized in the cAMP-independent pathway, are also dispensable.

D. Plants

Sucrose acts as a signal molecule in plants. It is produced by photosynthesis and transported to the organs where it will be metabolized or stored. The effect of sucrose accumulation consists of the up-regulation of "sink metabolism," which is carbon fixation, and the down-regulation of photosynthesis metabolism and genes required for this process.

Sucrose signaling appears to exert this effect by regulation of gene expression and protein turnover. Other regulatory pathways, like the regulation of nitrogen metabolism, interact with the sucrose signaling system, with the end result that genes under sucrose control are also under nitrogen control. In contrast to efforts on lower eukaryotic organisms such as yeast, there is little information as yet on the mechanisms of regulation underlying sucrose signaling. A current hypothesis to explain sucrose signaling is that sugar sensing, and its consequent signal may be initiated by the action of hexokinases, or membrane sensors, by analogy with yeast (Farrar et al., 2000).

Two examples suffice to demonstrate this catabolite repression-like system in plants. The induction of invertase in maize occurs at the level of transcription and is mediated by sucrose and glucose. This seems somewhat curious, as glucose is produced by the invertase (sucrase) reaction. It is hypothesized that a second point of control occurs at the translational level because nonmetabolizable sugars cause an increase in mRNA but not in the level of protein or enzyme activity. The induction of the invertase is thought to be independent of the hexokinase pathway (Cheng et al., 1999). A second system which has been investigated in some depth is the hexokinase pathway (Farrar et al., 2000; Koch et al., 2000). In this system, a transduction cascade has been proposed which is similar to the one present in yeast and initiated by a hexokinase, glucokinase, or fructokinase. Plants have orthologues of the yeast glucose repression factor, SNF1, which is a protein kinase. This protein appears to play an analogous role in mediating catabolite control in plant cells (Sugden et al., 1999).

E. ANIMALS

In humans, catabolite repression is mediated by cAMP. This second messenger or signal molecule conveys the instructions of a diversity of endocrine-controlled systems at the intracellular level. cAMP levels regulate glucose metabolism and are in turn controlled by hormone–receptor interactions. Glucokinase is a sugar sensor which detects extracellular glucose for insulin secretion. It has an important role in the regulation of glucose homeostasis (Terauchi et al., 1995). cAMP acts primarily on the cAMP-dependent protein kinase or PKA (Skalhegg and Tasken, 2000). The signal transduction pathway ultimately targets a cAMP response element binding protein (CREB) which mediates changes in gene expression (Roesler, 2000). Changes in glucose metabolism reflect altered transcription of the phosphoenolpyruvate carboxykinase gene resulting from CREB action at conserved enhancer elements (Crosson and Roessler, 2000; Park et al., 1999).

III. A Comparative View of Bacterial Prokaryotic and Eukaryotic Systems

Bacterial and eukaryotic mechanisms can be compared at different levels. These include looking at the overall pathway, at individual steps within these pathways, and at the sequences of proteins through which regulation is exerted. Commonalities and distinctions are apparent at each level. The basic model employed by bacterial prokaryotes and eukaryotes for regulation of sugar metabolism begins with a sensor kinase. The initiating carbon source may be phosphorylated upon uptake, but in any case the kinase acts as sensor following activation by the carbon source.

A. Overview of Catabolite Repression Pathways

In enteric bacteria and bacilli the PTS system acts as the sensor and generates a signal in the presence of glucose; the signal is then transduced by phosphorylation (HPr components in *B. subtilis*) or dephosphorylation (enzyme IIA in *E. coli*), and cAMP may act as a mediator as in *E. coli*. The phosphorylation/dephosphorylation cascade ends in *E. coli* with the inhibition of activating transcription factors. In *B. subtilis* it ends with repression of activation of transcription by the same activated factor.

In some respects the bacterial mechanism appears analogous to that in budding yeast. In both cases the regulatory cascade begins at the level of the carbon uptake system and triggers a series of events ending with transcriptional control of target genes. In *S. cerevisiae* hexokinase (HXK2) or membrane proteins acting as sensors generate a signal in the presence of glucose. The signal initiates a phosphorylation cascade involving a kinase (SNF1) and ends with the inhibition of a transcription repressor (MIG1). Animals may use mechanisms similar to yeast. Plants, in contrast, do not use cAMP and resemble gram-positive bacteria in that regard. The overall regulatory eukaryotic pathway consists of the generation of a signal by a glucokinase (animals), a hexokinase, glucokinase, or fructokinase (plants), or membrane proteins functioning as sensors, followed by transduction of the signal (deduced by the presence of SNF1 orthologues), and action at the level of transcription.

B. Stepwise Comparison of Catabolite Repression Pathways

By comparing the single steps of the regulatory mechanisms, we can see differences between bacteria and yeast and, also, between, or even within, bacterial species. In *E. coli,* two systems for catabolite repression coexist: the PTS system cAMP-dependent mechanism and the Cra mechanism. They both act on the initiation of transcription but in different ways. The first mechanism represses transcription by inhibition of a transcriptional activator; the second mechanism acts by direct activation or repression of transcription, depending on the target genes. The *B. subtilis* mechanism is of a third type and different from those in *E. coli*. *B. subtilis* has a PTS system, but in contrast to *E. coli*, this one is cAMP independent. The *B. subtilis* signal transduction cascade also functions in an opposite manner to that of *E. coli;* the *B. subtilis* HPr component is phosphorylated, while the enzyme IIA of *E. coli* is dephosphorylated in the presence of glucose. The action on transcription is inhibitory (CcpA for *B. subtilis*). But for some genes there can be activation by CcpA, and in a sense this is similar to the Cra system of *E. coli*.

In yeast the role of cAMP in glucose repression remains somewhat unclear, with evidence both for and against a direct role. cAMP levels increase in the presence of glucose, which contrasts with the situation in enteric bacteria, where cAMP causes expression of glucose-repressible genes and its level decreases in the presence of glucose. Another difference between the *E. coli* PTS system and the *S. cerevisiae* hexokinase system is that glucose repression in *S. cerevisiae* involves direct negative control of the affected promoters by a repressor, rather than the inhibition of activation as in *E. coli*. In this sense the yeast system more closely resembles the *B. subtilis* PTS mechanism and *E. coli* Cra mechanism.

Despite the existence of regulatory similarities in the structure of the catabolite repression system found within members of these two domains, there is little protein sequence homology in the components which catalyze these events. This observation suggests that conservation of regulatory function may result from evolutionary convergence. How, then, can we understand the origin of the catabolite repression systems in members of either the bacterial or the eukaryotic domain? One approach relies on using the system as it exists in archaea as a source of clues. The archaeal system may reveal the origins of this regulatory system by providing a glimpse of a mechanism transitioning between bacteria and eukaryotes.

IV. Archaeal Catabolite Repression

A. Metabolic Considerations for Adaptation to Extreme Environments

Archaea comprise a heterogeneous group of organisms. The best-studied members consist of three prominent biotypes: hyperthermophiles, halophiles, and methanogens. Such organisms are viewed by many as extremophiles inhabiting conditions defining the limit of life. Remarkably, archaea employ pathways for carbohydrate metabolism in many ways analogous to those in bacteria and eukaryotes. Accordingly, the need to conduct catabolite repression should be conserved as well in these organisms.

Both obligate and facultative chemoheterotrophs and lithoautotrophs are found in the archaea. Some species appear to be fastidious and cannot grow heterotrophically on simple sugars as sole carbon sources, without the addition of complex medium components. Perhaps as an adaptation to extreme environments, hyperthermophilic archaea often use modified pathways for glucose catabolism consisting of variations of the classical Embden–Meyerhof–Parnas (EMP) and Entner–Doudoroff (ED) pathways. Selected examples of such metabolic variations particularly in the hyperthermophilic archaea are described in more detail below. Most archaeal species were identified only in recent years, and as yet little is known about the regulation of their metabolisms. However, a number of archaeal genomes have been sequenced recently. The resulting sequence information has been used to predict much of their primary and secondary metabolic characteristics. These characteristics suggest that archaea must accomplish similar goals as performed by bacteria and eukaryotes for reproduction and survival.

B. The Archaeal General Transcription Apparatus

Gene expression is an area of intense interest in the archaea, due partly to the extensive overlap between their general transcription apparatus and that of eukaryotes (and not bacteria) as well as the attraction of a less complex model system for structural and functional studies (Baumann *et al.*, 1995; Langer *et al.*, 1995; Zillig *et al.*, 1993). Examples of this overlap include homologous promoter structure (Hain *et al.*, 1992; Reiter *et al.*, 1990), TBP orthologues (Marsh *et al.*, 1994; Qureshi *et al.*, 1995a; Rowlands *et al.*, 1994), TFIIB (TFB) orthologues (Gohl *et al.*, 1995; Qureshi *et al.*, 1995a,b), TFIIE (TFE) orthologues (Hanzelka *et al.*, 2001), and an RNA polymerase II (Pol II) orthologue (Klenk *et al.*, 1992). Unlike bacteria, archaea have a highly complex RNA polymerase consisting of 14 subunits, and unlike eukaryotes, they appear to lack other components of the basal transcription apparatus including TFIIA and TFIIH. Many of these observations derive from studies on *Sulfolobus* (Hain *et al.*, 1992; Klenk *et al.*, 1992; Qureshi *et al.*, 1995a; Reiter *et al.*, 1988, 1990). Recent X-ray crystallographic and solution NMR studies indicate that specific protein–DNA contacts between the transcription components of archaea and those of eukaryotes are strongly conserved (DeDecker *et al.*, 1996; Zhu *et al.*, 1996). Such studies further strengthen the justification for selection of hyperthermophilic archaeal transcription proteins for additional studies. Remarkably, conservation of sequence and structural features extends to include functional interactions as well; the eukaryotic TFIID subunit, TBP, promotes specific transcription of an archaeal promoter by an archaeal RNA polymerase *in vitro* (Wettach *et al.,* 1995). Despite such intriguing evolutionary overlaps, the identity of the accessory regulatory components of archaea which must control rates of transcription initiation in response to metabolic and environmental stimuli are virtually unknown.

C. REGULATION OF GENE EXPRESSION IN THE ARCHAEA

The regulation of transcription in *S. solfataricus* is of particular interest in light of the deeply rooted phylogenetic nature of its 16S rRNA. Such rooting, which appears common to hyperthermophiles, may mean that these organisms were among the first to appear on earth (Pace, 1991). Therefore unique information on the origins of basic cellular processes may be most apparent in these organisms. Regulation of gene expression at temperature extremes also must overcome unique biophysical constraints, particularly the need for controlled DNA melting to ensure protein access to critical ssDNA regions. *S. solfataricus* presents additional challenges for such protein–nucleic acid interactions since it is has an AT-rich genome (38 G+C mol%); it is thus unlike most other hyperthermophilic organisms, which tend to have GC-rich genomes.

A number of examples of gene regulation have been described in the other major archaeal subdivision called the euryarchaea. In the halophilic archaeal group examples include the regulation of synthesis of bacteriorhodopsin [*bob* (Shand and Betlach, 1991)], halocins (Cheung *et al.,* 1997), gas vacuoles [*vac* (Roder and Pfeifer, 1996; Yang and DasSarma, 1990)], and heat shock [*cct* (Palmer and Daniels, 1995; Thompson and Daniels, 1998)]. In the methanogenic archaeal group examples include the regulation of methane biosynthesis (Hennigan and Reeve, 1994; Palmer and Reeve, 1993), histones (Sandman *et al.,* 1994), carbon monoxide dehydrogenase [*cdh* (Sowers *et al.,* 1993)], and nitrogen fixation [*nif* (Cohen-Kupiec *et al.,*1997)]. However, for the hyperthermophilic archaeal group, gene regulatory studies are less common, perhaps because most are obligate anaerobes with fastidious growth requirements (Danson, 1993); it is likely that the physiological manipulations necessary to study gene regulation may be particularly daunting. In comparison to the euryarchaea, studies on gene regulation among the crenarchaea such as *S. solfataricus* are even more unusual. In any case, only rarely have accessory regulatory factors been identified (Roder and Pfeifer, 1996; Yang and DasSarma, 1990) and little mechanistic information is yet available on how they might function.

D. Catabolite Regulation in the Archaeon *S. solfataricus*

1. Sulfolobus solfataricus

Sulfolobus solfataricus is a hyperthermophilic microbe which inhabits acidic thermal hot springs. It has been assigned to one of the two archaeal subdivisions termed the crenarchaea by 16S rRNA analysis (Woese *et al.*, 1990). The relationship between cultured members of the crenarchaea, including *S. solfataricus,* is provided here using a phylogenetic distance tree based on 16S rRNA sequence comparisons (Fig. 1).

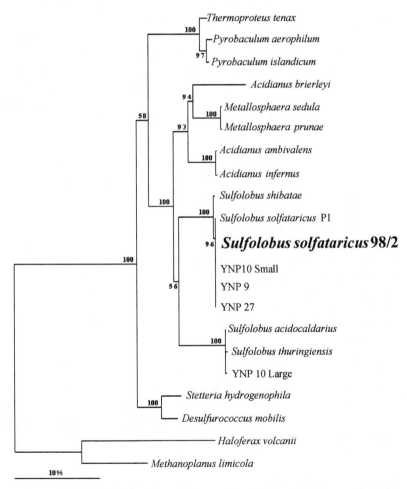

FIG. 1. 16S rRNA phylogenetic tree. The positions of *S. solfataricus* strain 98/2 and related natural isolates relative to other cultured members of the crenarchaea are shown. Evolutionary distance is proportional to branch length; the bar represents 10% nucleotide substitutions. Bootstrap values are indicated at all crenarchaeal nodes.

S. solfataricus strain 98/2 has been used for the studies presented below. Environmental studies suggest that hot springs contain a diversity of hyperthermophilic organisms (Barns *et al.*, 1994). However, the ability to manipulate physiologically most hyperthermophilic archaea remains a major challenge for understanding the metabolic basis for their adaptation to growth at temperature extremes. *S. solfataricus* was selected as a model system to address such questions because it is aerobic and is readily cultivated in liquid and solid defined media. These remain distinguishing characteristics among the hyperthermophilic archaea (Danson, 1993).

S. solfataricus can grow chemoheterotrophically on reduced carbon compounds at an optimal temperature of 80°C (De Rosa *et al.*, 1975; Grogan, 1989). It oxidizes glucose completely to carbon dioxide. Glucose oxidation proceeds via a nonphosphorylated version of the ED pathway (De Rosa *et al.*, 1984; Selig *et al.*, 1997). This pathway lies in contrast to that found in many bacteria, where the intermediates are phosphorylated. Use of nonphosphorylated intermediates in *S. solfataricus* allows the formation of reduced pyridine nucleotides but eliminates ATP formation.

2. The S. solfataricus *Catabolite Repression System*

We reported recently that the crenarchaeote, *S. solfataricus,* harbors a catabolite repression-like system analogous to that observed in members of the bacterial and eukaryotic domains. This system controls the expression of a group of genes required for carbohydrate utilization including the α-glucosidase or maltase (*malA*), the α-amylase (*amyA*), and the β-glycosidase (*lacS*) (Haseltine *et al.*, 1996, 1998, 1999a,b; Rolfsmeier *et al.*, 1996a,b, 1998). The location of these enzymes and their substrates in the context of the archaeal cell is shown in Fig. 2. Discovery of the *S. solfataricus* catabolite repression system occurred in the context of investigations on the catabolism of polysaccharides. Catabolite repression is a paradigm for studies concerned with global and specific gene control mechanisms and has received wide attention in both bacteria and eukaryotes. Our studies, however, are the first to suggest the existence of this system in the archaea. Several hallmarks of catabolite repression have been advanced in other systems. These include (1) transient repression by glucose, (2) a carbon source hierarchy, and (3) a global mode of regulation (Magasanick and Neidhardt, 1987; Saier *et al.*, 1996). All of these have now been detected in *S. solfataricus.*

a. The α-*Glucosidase* (malA). The first CR-regulated gene that was discovered was *malA,* which encodes a previously characterized α-glucosidase essential for hydrolysis of α-1,4-linked polysaccharides

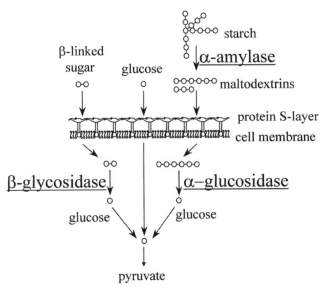

FIG. 2. Location of glycosyl hydrolases. Catabolite repression regulates the levels of all three of these enzymes including two cytoplasmic activities (α-glucosidase and β-glycosidase) and one secreted activity (α-amylase).

(Rolfsmeier et al., 1995). Levels of the enzyme in crude cell extracts varied severalfold in response to growth on different carbon sources, suggesting that its synthesis was subject to gene regulation (Rolfsmeier et al., 1996, 1998). malA was therefore cloned using a reverse genetics strategy. This resulted in the first report of an α-glucosidase gene from archaea. malA is 2083 bp and encodes a protein of 693 amino acids with a calculated mass of 80.5 kDa. It is flanked on the 5' side by an unusual 1-kb intergenic region. Northern blot analysis of the malA region identified transcripts for malA and an upstream ORF located 5' to the 1-kb intergenic region. The malA transcription start site was located by primer extension analysis to a guanine residue eight bp 5' of the malA start codon. The malA species distribution and its close phylogenetic relationship to eukaryotic α-glucosidase were also reported. A schematic summarizing the composition and transcription of the malA locus is shown in Fig. 3A.

b. The β-Glycosidase (lacS). It has become apparent recently that the S. solfataricus catabolite repression system regulates the expression of several genes (Haseltine et al., 1998, 1999a). One of these additional genes is called lacS and was found to be more strongly affected by the catabolite repression system than malA. The lacS gene encodes a

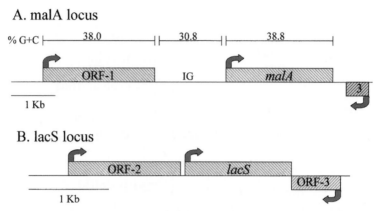

FIG. 3. (A) The *S. solfataricus malA* locus. The location of *malA* and adjacent ORFs are indicated by boxes. The locations of probes for Northern analysis are indicated beneath the boxes. The direction of transcription is indicated by the arrows. Approximate G+C mole percentage compositions for the regions indicated are shown above the boxes. The bar represents 1 kb. (B) The *S. solfataricus lacS* locus. The location of *lacS* and adjacent ORFs are indicated by boxes. The direction of transcription is indicated by the arrows. Restriction sites are indicated by single-letter designations. The bar represents 1 kb.

β-glycosidase which hydrolyzes β-linked oligosaccharides such as cellobiose and lactose (Cubellis *et al.*, 1990; Grogan, 1991). The *lacS* gene was cloned by screening an *S. solfataricus* expression library for thermostable β-glycosidase activity (Haseltine *et al.*, 1999a). The resulting isolate carried a plasmid in which the *lacS* was under the expression of P$_{lac}$. The plasmid contained an insert of 1768 bp and encoded the entire *lacS* coding sequence of 1467 bp as well as 176 bp 5′ to the start codon and 117 bp 3′ to the termination codon. The deduced amino acid sequence exhibited one nonconservative change resulting from three contiguous substitution mutations relative to the *lacS* sequence derived from *S. solfataricus* strain MT-4 (Cubellis *et al.*, 1990). The *lacS* gene was then used as a Southern blot probe to recover genomic clones spanning the *lacS* locus from an *S. solfataricus* 98/2 phage λ library (Rolfsmeier *et al.*, 1998). These fragments were then characterized by restriction analysis and DNA sequencing. A schematic of the *lacS* locus is shown in Fig. 3B.

c. The α-Amylase (amyA). The α-amylase was purified and characterized from culture supernatants of *Sulfolobus solfataricus* strain 98/2 during growth on starch as the sole carbon and energy source (Haseltine *et al.*, 1996). The enzyme is a homodimer with a subunit mass of 120 kDa. It catalyzes the hydrolysis of starch, dextrin, and α-cyclodextrin with similar efficiencies. This enzyme is unrelated to another activity previously characterized from this organism, which has also been referred to as an α-amylase (Miura *et al.*, 1999). The latter activity is actually a cytoplasmic glycosyltrehalose trehalohydrolase whose function remains obscure (Feese *et al.*, 2000). Western blot analysis of *S. solfataricus* α-amylase mutants indicate that the secreted *S. solfataricus* α-amylase is the major amylolytic activity produced by this organism (Haseltine *et al.*, 1996; Montalvo-Rodriguez *et al.*, unpublished).

3. Catabolite Repression Constitutes a Global Gene Regulatory System

Detailed studies have been conducted on the expression of *malA*, *lacS*, and *amyA* (Haseltine *et al.*, 1996, 1999a,b; Rolfsmeier *et al.*, 1998). Enzyme assays and Southern blot analysis revealed that these genes were coordinately expressed. This supported the hypothesis that coordinated gene expression relied on a global gene regulatory system. Since the three target glycosyl hydrolases are located on regions of the *S. solfataricus* genome separated by a minimum of 25 Mb, they are physically and genetically unlinked. Consequently coordinate expression must employ a trans-acting diffusible factor.

a. Steady-State Changes in Catabolite Repression-Regulated Gene Expression. Coordinate expression of the glycosyl hydrolases relied on the use of enzyme assays, specific polyclonal antibodies, and RNA strand-specific probes (riboprobes). The catabolite repression regulatory system was found to act at the level of transcription to coordinate the expression of all three genes (Haseltine *et al.*, 1999a; Montalvo-Rodriguez *et al.*, unpublished). Maximum gene expression occurred during growth on sucrose as the sole carbon and energy source. Minimum gene expression occurred during growth on sucrose supplemented with yeast extract. The effect of added yeast extract results primarily from the presence of asparate and asparagine as noted in earlier studies on the α-amylase (Haseltine *et al.*, 1996). Since the largest repressive carbon source effect was obtained with yeast extract, this supplement was used for subsequent studies. The-fold change in steady-state levels of the measured parameters is shown in Fig. 4.

The α-amylase also exhibits unique forms of regulatory control which are classified as forms of catabolite repression in other systems (Haseltine *et al.*, 1996). In some cases these regulatory phenomena are not evident for the other glycosyl hydrolases. The addition of exogenous glucose represses the production of α-amylase, demonstrating that a classical glucose effect is operative in *S. solfataricus*. In addition, the absolute levels of α-amylase detected in culture supernatants varied greatly with the type of sole carbon source used to support growth. Aspartate was identified as the most repressing sole carbon source for

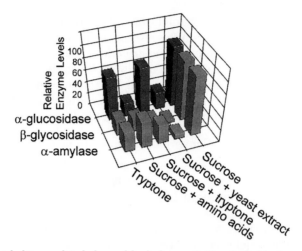

F ɪɢ. 4. Catabolite-regulated glycosyl hydrolase activities. Individual enzyme specific activities detected during growth on various carbon and energy sources are indicated on the *y*-axis in amounts relative to those produced during growth in sucrose minimal medium. Growth media and types of enzymes are indicated.

α-amylase production, while glutamate was the most derepressing. It was hypothesized that the inverse effect of these two amino acids may reflect their routes of consumption after deamination into the TCA cycle.

b. The Kinetics of Catabolite Repression-Regulated Gene Expression. The kinetics of change of regulated gene expression can be used to understand better the nature of the regulatory mechanism used for that gene. In *S. solfataricus,* this strategy has been applied in two cases. Studies on the induction of secretion of the α-amylase by relief of glucose repression occurred within one cell division (Haseltine *et al.,* 1996). Protein pulse radiolabeling revealed a requirement for *de novo* synthesis of protein.

The second instance where regulatory kinetics has been applied *in vivo* consisted of studies on *lacS* expression in response both to the removal (depletion) of repressing carbon sources and to the addition of such carbon sources (Haseltine *et al.,* 1999). Full induction for *lacS* occurred in less than one generation (6 h), while full repression occurred more slowly than the rate of cell dilution (division) and required a total elapsed time of 200 h, nearly 33 generations. Quantitative Northern blot analysis was used to measure the rates of induction and repression for *lacS*. The *lacS* mRNA induction rate was 0.29 pg mRNA/mg/h ($r^2 = 0.92$), while the *lacS* mRNA repression rate was 0.0034 pg mRNA/mg/h ($r^2 = 0.87$). These results are consistent with (but do not prove) a model in which increased transcript synthesis underlies the catabolite repression regulatory response. The delay in *lacS* repression following the addition of repressing carbon sources strongly suggests that a mechanism distinct from that involved in *lacS* induction is employed.

4. Isolation and Characterization of Archaeal Catabolite Repression Mutants

An historic breakthrough in understanding the catabolite repression mechanism of *E. coli* occurred when catabolite repression-regulatory mutants were isolated and characterized. Such mutants exhibited a pleiotropic defect in catabolite repression of gene expression where the affected genes had a reduced or uninducible level of expression. The catabolite repression-regulatory mutants turned out to encode adenyl cyclase (Cya) and the catabolite repression protein (Crp/Cap) (Magasanick and Neidhardt, 1987; Saier *et al.*, 1996). By analogy to the *E. coli* system, similar efforts have been undertaken in *S. solfataricus* in an effort to recover mutants which perturb the archaeal catabolite repression system.

a. Isolation of Archaeal Catabolite Repression Mutants. The *S. solfataricus lacS* gene encodes a β-glycosidase with broad substrate specificity including activity against β-galactosides and their chemical analogues such as the colorimetric indicator X-gal (Cubellis *et al.*, 1990; Grogan, 1991). *S. solfataricus* forms blue colonies on a medium containing X-gal, similar to the behavior of *E. coli*. Disruption of *lacS* by insertion element transposition results in the formation of colorless colonies on indicator medium (Schleper *et al.*, 1994). Since the catabolite repression system in *S. solfataricus* affects the expression of several unlinked genes including *lacS*, an effort was made to isolate *lacS* mutants which were altered in catabolite repression regulation using the *lacS*-dependent blue/white colony phenotype. Two mutant classes, intragenic and extragenic, were recovered. Both types were colorless on indicator medium, however, unlike those observed previously (Schleper *et al.*, 1994), none resulted from insertion element disruption of *lacS*. Only the extragenic class were considered as catabolite repression mutant candidates.

b. The Catabolite Repression Mutation, car, *Results in Pleiotropic Regulatory Defects.* Three independently isolated *lacS* extragenic mutants were analyzed, named *car* for "catabolite repression," and given designations of *car1, car2,* and *car5*. One intragenic *lacS* mutant, called *lacS100,* was also analyzed for comparison. To clarify the reason why

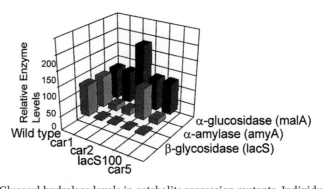

FIG. 5. Glycosyl hydrolase levels in catabolite repression mutants. Individual enzyme specific activities in the mutant strains are indicated on the *y*-axis in amounts relative to those detected in the wild-type strain. Strains and types of enzymes are indicated.

these mutants had become phenotypically *lacS* deficient and therefore formed white colonies on X-gal medium, enzyme assays and Western blot and Northern blot studies were performed (Haseltine *et al.*, 1999b). The *lacS100* mutant was altered only in the levels of *lacS*, while *car1-2*, and *5-* were altered in the expression of all three catabolite repression-regulated genes, *lacS*, *malA*, and *amyA* (Fig. 5). Each *car* mutant is altered in the expression of all three glycosyl hydrolases to the same extent, suggesting that the mutations are likely to be allelic.

These intragenic and extragenic *lacS* mutations were distinguished further using Southern blot analysis and PCR (Haseltine *et al.*, 1999b). In the case of the *car* mutants, Southern blot analysis excluded any gross genomic alterations such as insertion elements or deletions. Southern blot analysis of *lacS100*, however, indicated that this isolate had undergone deletion of the entire *lacS* gene. To exclude further the presence of more subtle alterations in the *lacS* genes of the *car* mutants, PCR was used to amplify separately the *lacS* coding region and the *lacS* promoter region. The coding region was subcloned into an expression plasmid and transformed into *E. coli*, resulting in levels of thermostable β-glycosidase identical to those obtained following identical manipulations with the wild-type *lacS* gene. This result excludes the existence of mutations in the *lacS* coding sequence. In addition, the DNA sequences of the *lacS* promoter region (regions −150 to +25) from *car1* and *car2* were identical to that of the wild-type strain. This excluded the existence of cis-acting mutations in the *lacS* promoter. These results are most likely explained by the hypothesis that the *car* mutants suffered a mutation which compromises the action of a trans-acting factor regulatory factor necessary for the expression of *lacS* and other catabolite repression-regulated genes. The identification of the *car* gene

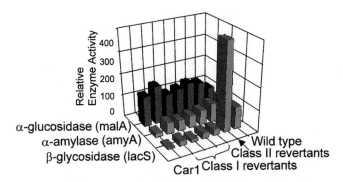

FIG. 6. *Car1 (scr)* revertants. Levels of the there glycosyl hydrolases detected in *car1* and various classes of *car1* revertants are indicated on the *y*-axis in amounts relative to those detected in the wild-type strain.

should provide significant information on the mechanism employed for catabolite repression in archaea.

c. Reversion Analysis of car. To determine if the pleiotropic phenotype of *car* resulted from a single mutation, *car1* revertants were isolated using *lacS*-specific screens and selections (Haseltine *et al.*, 1999b). Revertants were recovered at an unexpectedly high frequency (approx. $1/10^4$) and were named *scr* for "suppressor of catabolite repression." These revertants were pleiotropic and restored levels of the three glycosyl hydrolases as indicated by enzyme assay and Western blot analysis either partially or completely depending on the revertant class (Fig. 6). In one case, an *scr* revertant exhibited levels of the β-glycosidase several times those of the wild-type strain. Attempts to revert the *lacS100* mutation, however, failed, consistent with the finding that the *lacS100* mutation resulted from a deletion of *lacS*. The existence of *car* and its *scr* revertants reveals the presence of a trans-acting transcriptional regulatory system for mediating catabolite repression of glycosyl hydrolase expression in this member of the archaea.

V. Enzymologic, Metabolic, and Genomic Hints about Catabolite Repression in Archaea

A. UNUSUAL PATHWAYS FOR GLUCOSE METABOLISM AND UNIQUE ENZYMES INVOLVED

1. Thermococcus

The thermococci are located on the euryarchaeal branch of the archaea. Primarily marine in habit, they ferment glucose via a modified version of the EMP pathway. Different types of hexose kinases and glyceraldehyde-3-phosphate oxidases distinguish their version from the conventional system: an ADP-dependent hexokinase and 6-phosphofructokinase (versus the ATP-dependent enzymes of bacteria) and a glyceraldehyde-3-phosphate ferredoxin oxidoreductase (GAPOR) in place of GAP-dehydrogenase and phosphoglycerate kinase (van der Oost *et al.*, 1998). In contrast, *Thermococcus zilligi*, a nonmarine species, employs a second glycolytic pathway in parallel with the EMP pathway, which is probably a pentose phosphoketolase (PPK) pathway (Xavier *et al.*, 2000). Thus *T. zilligi* splits carbon flux at the level of glucose-6-phosphate; one branch continues down the EMP pathway, while the other follows the PPK pathway. Interestingly, the two branches were inversely regulated by growth conditions. The EMP pathway is more efficient in the presence of glucose and repressed by tryptone. Perhaps the effect of tryptone, a source of high-quality carbon compounds, results from the carbon source preference and the operation of a catabolite repression-like system in this organism.

2. Pyrococcus

Another example of carbon source preference occurs in the related pyrococci. Like the thermococci, *Pyrococcus furiosus* also uses a modified EMP pathway. In this case, the activities of the two kinases and GAPOR are higher on cells grown on maltose than in cells grown on pyruvate (van der Oost *et al.*, 1998). It should be noted that in the pyrococci, the first three oxidation steps in the EMP pathway are catalyzed by unique enzymes: a glucose ferredoxin oxidoreductase, an aldehyde ferredoxin oxidoreductase (AOR, a tungsten-containing enzyme), and, finally, a pyruvate ferredoxin oxidoreductase (Adams, 1993). The presence of these enzymes in the pathway allows the glucose to be converted to acetate without the need to use NADPH, which is thermolabile. There is no production of ATP during the conversion of glucose to pyruvate, only during the formation of acetate.

3. Thermoproteus

Thermoproteus tenax, like *S. solfataricus,* is a member of the crenarchaea. It is a facultative heterotroph which metabolizes glucose via a modified EMP and the ED pathways (Siebers *et al.*, 1997, 1998). Carbon flux through these pathways merges at the three-carbon stage. At this point an NAD$^+$-dependent glyceraldehyde-3-phosphate dehydrogenase can be allosterically regulated by metabolites and represents one major point of carbon flux control (Brunner *et al.*, 1998). A second point of regulation occurs at the level of pyruvate kinase (PK). In *T. tenax* the activity and the transcript level of this enzyme were compared in heterotrophically and autotrophically grown cells. Both were higher in heterotrophic cells, indicating a regulation of the PK activity at the mRNA synthesis level (Schramm *et al.*, 2000). Such regulation appears to result from a preference for fixed versus oxidized carbon sources, perhaps a sign of catabolite control.

B. Types of Archaeal Carbohydrate Transport Systems

Membrane protein transport systems are essential components of the catabolite repression apparatus in bacteria and eukaryotes. In proteobacteria, the phosphotransferase system (PTS) is critical for signal transduction of carbon source availability. However, transport systems like the bacterial PTS are likely to be absent in archaea or, at least, highly divergent from the bacterial ones. BLAST searches of the archaeal genomes sequences with proteobacterial PTS components gave only high E values, indicating random matches. However, a role for membrane protein transport systems remains a possibility for archaeal catabolite repression. It remains worthwhile therefore to consider how sugars are transported in the archaea.

Genomic analysis of *Sulfolobus solfataricus* indicates the presence of transporters belonging to the major facilitator superfamily (MSF), sugar permeases, and ATP-binding cassette (ABC) transporters. Two systems have been described for glucose transport in *S. solfataricus*. A pH-dependent glucose transport system (Cusdin *et al.*, 1996) and a high-affinity-binding protein-dependent ABC transporter (Albers *et al.*, 1999). The first system has a high affinity for glucose, mannose, and galactose and may be of the proton-symport type because perturbation of the proton gradient affected transport efficiency. The second system is composed of a membrane-bound glycoprotein (GBP) that may be a component of an ABC transport system because the gene encoding for it is adjacent to putative subunits of an ABC transporter including an ATPase and two membrane proteins.

In *Sulfolobus shibatae*, two maltose transport systems are present. Both systems are constitutive and not induced by sugars or repressed by glucose. Interestingly, *S. shibatae* does not utilize glucose preferentially over disaccharides (Yallop and Charalambous, 1996). In *Thermococcus litoralis*, an ABC transporter system has been described with a high affinity for maltose and trehalose which is not inhibited by maltodextrins and is induced by trehalose (Horlacher *et al.*, 1998; Xavier *et al.*, 1996). The affinity for both maltose and trehalose is surprising, as these sugars are structurally distinct. One hypothesis to explain this substrate preference is that maltose and trehalose metabolism may have common enzymes (Horlacher *et al.*, 1998). In the halophile, *Haloferax volcanii*, a glucose-specific ABC transporter has been characterized. There is evidence that this transporter is active during anaerobic growth, but a different transport system must operate under aerobic conditions (Wanner and Soppa, 1999). The system transporting glucose during aerobic growth is inducible and requires a gradient of Na^+ to function, being a sodium/glucose symporter (Tawara and Kamo, 1991).

C. OCCURRENCE AND FUNCTION OF cAMP IN ARCHAEA

cAMP is found in bacterial prokaryotes and eukaryotes, where it functions as a mediator in signal transduction of catabolite repression. In gram-negative bacteria, cAMP controls transcription of specific promoters by interacting with DNA-binding proteins (Epstein *et al.*, 1975). In eukaryotes its mechanism of action is different in that cAMP activates a protein kinase that in turns phosphorylates the factor CREB (CRE binding protein) inducing its interaction with CRE (cAMP responsive element) in the target promoter (Andrisani, 1999; Bullock and Jabener, 1998). In archaea, the presence of cAMP has been reported for species representing the three main biotypes including *Methanobacterium thermoautotrophicum, Halobacterium volcanii,* and *Sulfolobus solfataricus* (Leichtling *et al.*, 1986). It is interesting that in *M. thermoautotrophicum* the level of cAMP changed, increasing, in correspondence with H_2 starvation (Leichtling *et al.*, 1986).

Adenylate cyclase is the enzyme responsible for the formation of cAMP. These enzymes can be subdivided into distinct groups (Sismeiro *et al.*, 1998). Class I enzymes include those of *E. coli* and other proteobacteria, Class II enzymes include those of several bacterial pathogens, while Class III enzymes also include guanylyl cyclases from both bacterial and eukaryotic organisms (Pei and Grishin, 2001). Adenyl cyclase orthologues of any of these three classes are not obvious by BLAST searching in archaeal genomes. If an adenylate cyclase is present in archaea, it must be divergent from the ones belonging to other domains. Interesting, a new class of adenylate cyclase has been discovered in bacteria (Sismeiro *et al.*, 1998). The enzyme, isolated in *Aeromonas hydrophila,* has sequence homology with proteins of unknown function present in various archaeal genomes. Efforts to verify the identity of the *Methanobacterium jannaschii* putative orthologue using recombinant protein failed to support this assignment. Thus, the independent confirmation of cAMP existence in archaea by virtue of the presence of a unique adenyl cyclase orthologue awaits further study.

VI. Future Prospects

We can anticipate that studies on the mechanism of catabolite repression in the archaea will extend our understanding of this regulatory paradigm in all three domains of life. The introduction of archaeal genetic systems and the continued use of purified and semipurified transcription systems will be of particular help in efforts to identify critical cis- and trans-acting components. Perhaps, too, studies on archaeal catabolite repression may help improve our understanding of how early life evolved strategies enabling their survival and proliferation in extreme environments.

REFERENCES

Adams, M. W. (1993). *Annu. Rev. Microbiol.* **47**, 627–658.
Albers, S. V., Elferink, M. G., Charlebois, R. L., Sensen, C. W., Driessen, A. J., and Konings, W. N. (1999). *J. Bacteriol.* **181**, 4285–4291.
Andrisani, O. M. (1999). *Crit. Rev. Eukaryot. Gene. Expr.* **9**, 19–32.
Barns, S. M., Fundyga, R. E., Jeffries, M. W., and Pace, N. R. (1994). *Proc. Natl. Acad. Sci. USA* **91**, 1609–1613.
Baumann, P., Qureshi, S. A., and Jackson, S. P. (1995). *Trends Genet.* **11**, 279–283.
Brunner, N. A., Brinkmann, H., Siebers, B., and Hensel, R. (1998). *J. Biol. Chem.* **273**, 6149–6156.
Bullock, B. P., and Habener, J. F. (1998). *Biochem.* **37**, 3795–3809.
Cheng, W. H., Taliercio, E. W., and Chourey, P. S. (1999). *Proc. Natl. Acad. Sci. USA* **96**, 10512–10517.
Cheung, J., Danna, J. K., O'Connor, E. M., Price, L., and Shand, R. F. (1997). *J. Bacteriol.* **179**, 548–551.
Cohen-Kupiec, R., Blank, C., and Leigh, J. A. (1997). *Proc. Natl. Acad. Sci. USA* **94**, 1316–1320.
Collier, D. N., Hager, P. W., and Phibbs, P. V. (1996). *Res. Microbiol.* **147**, 551–561.
Conti, M. (2000). *Mol. Endocrinol.* **14**, 1317–1327.
Crasnier, M. (1996). *Res. Microbiol.* **147**, 479–482.
Crosson, S. M., and Roesler, W. J. (2000). *J. Biol. Chem.* **275**, 5804–5809.
Cubellis, M. V., Rozzo, C., Montecucchi, P., and Rossi, M. (1990). *Gene* **94**, 89–94.
Cusdin, F. S., Robinson, M. J., Holman, G. D., Hough, D. W., and Danson, M. J. (1996). *FEBS Lett.* **3**, 193–195.
Danson, M. J. (1993). *In* "The Biochemistry of Archaea" (M. Kates, D. J. Kushner, and A. T. Matheson, Eds.). Elsevier, New York.
DeDecker, B. S., O'Brien, R., Fleming, P. J., Geiger, J. H., Jackson, S. P., and Sigler, P. B. (1996). *J. Mol. Biol.* **264**, 1072–1084.
Del Castillo Agudo, L., Nieto Soria, A., and Sentandreu, R. (1992). *Gene* **120**, 59–65.
De Rosa, M., Gambacorta, A., and Bu'lock, J. D. (1975). *J. Gen. Microbiol.* **86**, 156–164.
De Rosa, M., Gambacorta, A., Nicolaus, B., Giardina, P., Poerio, E., and Buonocore, V. (1984). *Biochem. J.* **224**, 407–414.
Deutscher, J., Kuster, E., Bergstedt, U., Charrier, V., and Hillen, W. (1995). *Mol. Microbiol.* **15**, 1049–1053.
Deutscher, J., Fischer, C., Charrier, V., Galinier, A., Lindner, C., Darbon, E., and Dossonnet, V. (1997). *Folia Microbiol. (Praha)* **42**, 171–178.
Ebbole, D. J. (1998). *Fungal Genet. Biol.* **25**, 15–21.
Epstein, W., Rothman-Denes, L. B., and Hesse, J. (1975). *Proc. Natl. Acad. Sci. USA* **72**, 2300–2304.
Farrar, J., Pollock, C., and Gallagher, J. (2000). *Plant Sci.* **154**, 1–11.

Feese, M. D., Kato, Y., Tamada, T., Kato, M., Komeda, T., Miura, Y., Hirose, M., Hondo, K., Kobayashi, K., and Kuroki, R. (2000). *J. Mol. Biol.* **301**, 451–464.

Fischer, F., Zillig, W., Stetter, K.O., and Schreiber, G. (1983). *Nature* **10**, 511–513.

Flores, C. L., Rodriguez, C., Petit, T., and Gancedo, C. (2000). *FEMS Microbiol. Rev.* **24**, 507–529.

Fujita, Y., Miwa, Y., Galinier, A., and Deutscher, J. (1995). *Mol. Microbiol.* **17**, 953–960.

Gancedo, J. M. (1998). *Microbiol. Mol. Biol. Rev.* **62**, 334–361.

Gohl, H. P., Grondahl, B., and Thomm, M. (1995). *Nucleic Acids Res.* **23**, 3837–3841.

Grogan, D. W. (1989). *J. Bacteriol.* **171**, 6710–6719.

Grogan, D. W. (1991). *J. Bacteriol.* **57**, 1644–1649.

Hain, J., Reiter, W.-D., Hudepohl, U., and Zillig, W. (1992). *Nucleic Acids Res.* **20**, 5423–5428.

Hansen, T., and Schonheit, P. (2000). *Arch. Microbiol.* **173**, 103–109.

Hanzelka, B. L., Darcy, T. J., and Reeve, J. N. (2001). *J. Bacteriol.* **183**, 1813–1818.

Haseltine, C., Rolfsmeier, M., and Blum, P. (1996). *J. Bacteriol.* **178**, 945–950.

Haseltine, C., Rolfsmeier, M., Bini, E., Carl, A., Rodriguez-Montalvo, R., Clark, A., and Blum, P. (1998). *American Society for Microbiology, Annual Meeting.*

Haseltine, C., Montalvo-Rodriguez, R., Bini, E., Carl, A., and Blum, P. (1999a). *J. Bacteriol.* **181**, 3920–3927.

Haseltine, C., Montalvo-Rodriguez, R., Carl, A., Bini, E., and Blum, P. (1999b). *Genetics* **152**, 1353–1361.

Hennigan, A. N., and Reeve, J. (1994). *Mol. Microbiol.* **11**, 655–670.

Horlacher, R., Xavier, K. B., Santos, H., DiRuggiero, J., Kossmann, M., and Boos, W. (1998). *J. Bacteriol.* **180**, 680–689.

Hueck, C. J., and Hillen, W. (1995). *Mol. Microbiol.* **15**, 395–401.

Inui, M., Vertes, A. A., and Yukawa, H. (1996). *Res. Microbiol.* **147**, 562–566.

Jones, B. E., Dossonnet, V., Kuster, E., Hillen, W., Deutscher, J., and Klevit, R. E. (1997). *J. Biol. Chem.* **272**, 26530–26535.

Kardinahl, S., Schmidt, C. L., Hansen, T., Anemuller, S., Petersen, A., and Schafer, G. (1999). *Eur. J. Biochem.* **260**, 540–548.

Klenk, H.-P., Palm, P., Lottspeich, F., and Zillig, W. (1992). *Proc. Natl. Acad. Sci. USA* **89**, 407–410.

Koch, K. E., Ying, Z., Wu, Y., and Avigne, W. T. (2000). *J. Exp. Bot.* **51**, 417–427.

Langer, D., Hain, J., Thuriaux, P., and Zillig, W. (1995). *Proc. Natl. Acad. Sci. USA* **92**, 5768–5772.

Leichtling, B. H., Rickenberg, H. V., Seely, R. J., Fahrney, D. E., and Pace, N. R. (1986). *Biochem. Biophys. Res. Commun.* **136**, 1078–1082.

Lutfiyya, L. L., Iyer, V. R., DeRisi, J., DeVit, M. J., Brown, P. O., and Johnston, M. (1998). *Genetics* **150**, 1377–1391.

Magasanik, B., and Neidhardt, F. C. (1987). *In* "*Escherichia coli* and *Salmonella typhimurium* Cellular and Molecular Biology" (F. C. Neidhardt, J. L. Ingrahm, K. Brooks-Low, B. Magasanick, M. Schaechter, and H. E. Umbarger, Eds.). American Society for Microbiology Press, Washington, DC.

Marsh, T. L., Reich, C. I., Whitelock, R. B., and Olsen, G. J. (1994). *Proc. Natl. Acad. Sci. USA* **91**, 4180–4184.

Miura, Y., Kettoku, M., Kato, M., Kobayashi, K., and Kondo, K. (1999). *J. Mol. Microbiol Biotechnol.* **1**, 129–134.

Miwa, Y., Nakata, A., Ogiwara, A., Yamamoto, M., and Fujita, Y. (2000). *Nucleic Acids Res.* **28**, 1206–1210.

Moll, R., and Schäfer, G. (1988). *FEBS Lett.* **232**, 359–363.

Montalvo-Rodriguez, R, Thomas, P., and Blum, P., Unpublished observations.

Movsesian, M. A. (2000). *Expert Opin. Invest. Drugs* **9**, 963–973.

O'Gara, F., Birkenhead, K., Boesten, B., and Fitzmaurice, A. M. (1989). *FEMS Microbiol. Rev.* **5**, 93–101.

Oren, A. (1999). *Microbiol. Mol. Biol. Rev.* **63**, 334–348.

Ozcan, S., Dover, J., and Johnston, M. (1998). *EMBO J.* **17**, 2566–2573.

Pace, N. R. (1991). *Cell* **65**, 531–533.

Palmer, J. R., and Daniels, C. J. (1995). *J. Bacteriol.* **177**, 1844–1849.

Palmer, J. R., and Reeve, J. N. (1993). *In* "The Biochemistry of Archaea" (M. Kates, D. J. Kushner, and A. T. Matheson, Eds.). Elsevier, New York.

Park, E. A., Song, S., Vinson, C., and Roesler, W. J. (1999). *J. Biol. Chem.* **274**, 211–217.

Pei, J., and Grishin, N. V. (2001). *Proteins* **42**, 210–216.

Qureshi, S. A., and Jackson, S. P. (1998). *Mol. Cell* **1**, 389–400.

Qureshi, S. A., Baumann, P., Rowlands, T., Khoo, B., and Jackson, S. P. (1995a). *Nucleic Acids Res.* **23**, 1775–1781.

Qureshi, S. A., Khoo, B., Baumann, P., and Jackson, S. P. (1995b). *Proc. Natl. Acad. Sci. USA* **92**, 6077–6081.

Ramseier, T. M. (1996). *Res. Microbiol.* **147**, 489–493.

Reiter, W.-D., Palm, P., and Zillig, W. (1988). *Nucleic Acids Res.* **16**, 1–19.

Reiter, W.-D., Hudepohl, U., and Zillig, W. (1990). *Proc. Natl. Acad. Sci. USA* **87**, 9509–9513.

Roder, R., and Pfeifer, F. (1996). *Microbiology* **142**, 1715–1723.

Roesler, W. J. (2000). *Mol. Cell Endocrinol.* **162**, 1–7.

Roitsch, T. (1999). *Curr. Opin. Plant Biol.* **2**, 198–206.

Rolfsmeier, M., and Blum, P. (1995). *J. Bacteriol.* **177**, 482–485.

Rolfsmeier, M., Haseltine, C., and Blum, P. (1996a). *American Society for Microbiology, Abstracts of the Annual Meeting.*

Rolfsmeier, M., Haseltine, C., Bini, E., Clark, A., and Blum, P. (1996b). *Thermophiles '96 International Meeting,* Athens, GA.

Rolfsmeier, M., Haseltine, C., Bini, E., Clark, A., and Blum, P. (1998). *J. Bacteriol.* **180**, 1287–1295.

Rolland, F., De Winde, J. H., Lemaire, K., Boles, E., Thevelein, J. M., and Winderickx, J. (2000). *Mol. Microbiol.* **38**, 348–358.

Ronne, H. (1995). *Trends Genet.* **11**, 12–17.

Rowlands, T., Baumann, P., and Jackson, S. P. (1994). *Science* **264**, 1326–1329.

Saier, M. H. (1989). *Microbiol. Rev.* **53**, 109–120.

Saier, M. H. (1996). *FEMS Microbiol. Lett.* **138**, 97–103.

Saier, M. H. (1998). *Biotechnol. Bioeng.* **58**, 170–174.

Saier, M. H., Jr., Ramseier, T. M., and Reizer, J. (1996). *In* "Escherichia coli and Salmonella" (F. C. Neidhardt, Eds.). American Society for Microbiology Press, Washington, DC.

Saier, M. H., Chauvaux, S., Cook, G. M., Deutscher, J., Paulsen, I. T., Reizer, J., and Ye, J. J. (1996). *Microbiol.* **142**, 217–230.

Sandman, K., Gralying, R. A., Dobrinski, B., Lurz, R., and Reeve, J. N. (1994). *Proc. Natl. Acad. Sci. USA* **91**, 12624–12628.

Schleper, C., Roder, R., Singer, T., and Zillig, W. (1994). *Mol. Gen. Genet.* **243**, 91–96.

Schramm, A., Siebers, B., Tjaden, B., Brinkmann, H., and Hensel, R. (2000). *J. Bacteriol.* **182**, 2001–2009.

Selig, M., Xavier, K. B., Santos, H., and Schonheit, P. (1997). *Arch. Microbiol.* **167**, 217–232.

Shand, R.F., and Betlach, M. (1991). *J. Bacteriol.* **173**, 4692–4699.

Siebers, B., Wendisch, V. F., and Hensel, R. (1997). *Arch. Microbiol.* **168**, 120–127.

Siebers, B., Klenk, H. P., and Hensel, R. (1998). *J. Bacteriol.* **180**, 2137–2143.

Sismeiro, O., Trotot, P., Biville, F., Vivares, C., and Danchin, A. (1998). *J. Bacteriol.* **180**, 3339–3344.

Skalhegg, B. S., and Tasken, K. (2000). *Front. Biosci.* **5**, D678–D693.

Sowers, K., Thai, T., and Gunsalus, R. (1993). *J. Biol. Chem.* **268**, 23172–23178.

Stulke, J., and Hillen, W. (1998). *Naturwissenschaften* **85**, 583–592.

Stulke, J., Arnaud, M., Rapoport, G., and Martin-Verstraete, I. (1998). *Mol. Microbiol.* **28**, 865–874.

Stulke, J., and Hillen, W. (2000). *Annu. Rev. Microbiol.* **54**, 849–880.

Sugden, C., Donaghy, P. G., Halford, N. G., and Hardie, D. G. (1999). *Plant Physiol.* **120**, 257–274.

Tawara, E., and Kamo, N. (1991). *Biochim. Biophys. Acta* **1070**, 293–299.

Terauchi, Y., Sakura, H., Yasuda, K., Iwamoto, K., Takahashi, N., Ito, K., Kasai, H., Suzuki, H., Ueda, O., Kamada, N., Jishage, K., Komeda, K., Noda, M., Kanazawa, Y., Taniguchi, S., Miwa, I., Akanuma, Y., Kodama, T., Yazaki, Y., and Kadowaki, T. (1995). *J. Biol. Chem.* **270**, 30253–30256.

Thompson, D., and Daniels, C. (1998). *Mol. Microbiol.* **27**, 541–551.

Trumbly, R. J. (1992). *Mol. Microbiol.* **6**, 15–21.

Tyler, B., Loomis, W. F., and Magasanik, B. (1967). *J. Bacteriol.* **94**, 2001–2011.

van der Oost, J., Schut, G., Kengen, S. W., Hagen, W. R., Thomm, M., and de Vos, W. M. (1998). *J. Biol. Chem.* **273**, 28149–28154.

Wanner, C., and Soppa, J. (1999). *Genetics* **152**, 1417–1428.

Wettach, J., Gohl, H. P., Tschochner, H., and Thomm, M. (1995). *Proc. Natl. Acad. Sci. USA* **92**, 472–476.

Wills, C. (1996). *Res. Microbiol.* **147**, 566–572.

Woese, C., Kandler, O., and Wheelis, M. L. (1990). *Proc. Natl. Acad. Sci. USA* **87**, 4576–4579.

Xavier, K. B., Martins, L. O., Peist, R., Kossmann, M., Boos, W., and Santos, H. (1996). *J. Bacteriol.* **178**, 4773–4777.

Xavier, K. B., da Costa, M. S., and Santos, H. (2000). *J. Bacteriol.* **182**, 4632–4636.

Yallop, C. A., and Charalambous, B. M. (1996). *Microbiology* **142**, 3373–3380.

Yang, C. F., and DasSarma, S. (1990). *J. Bacteriol.* **172**, 4118–4121.

Zhu, W., Zeng, Q., Colangelo, C. M., Lewis, M., Summers, M. F., and Scott, R. A. (1996). *Nat. Struct. Biol.* **3**, 122–124.

Zillig, W., Palm, P., Klenk, J., Langer, D., Hudepohl, U., Hain, J., Lanzendorfer, M., and Holz, I. (1993). *In* "The Biochemistry of Archaea" (M. Kates, D. J. Kushner, and A. T. Matheson, Eds.). Elsevier, New York.

SUBJECT INDEX

A

AAA$^+$ superfamily, *see* ATPases
    associated with various cellular
    activities
Amino-terminal hydrolases, 20S
    proteasomes as, 290–291
α-Amylase, *Sulfolobus solfataricus*
    catabolite repression, 354–355
Ancestral proteins
    minimum number, 64–66
    present-day protein comparison, 66–67
Animals, catabolite repression, 346
AP endonucleases, 155–156
Archaea
    catabolite repression, 348–350, 359–362
    cyclic AMP, 362
    DNA recombination, 116–117, 121
    DNA repair pathways, 146–162
    DNA strand exchange, 126–127
    energy-dependent proteases, 324–325
    glucose metabolism, 359–361
    histones, 83–86, 93–94, 201–202
    histones, *Methanothermus fervidus*
        variants, 86–93
    Hsp70, 237
    methanogenic, chemofossil evidence,
        29–30
    molecular chaperones, 220–223, 227,
        231–233
    MRE11 protein, 117, 121
    nascent polypeptide-assisted complex,
        232–233
    nucleosome DNA, 79–82
    nucleosomes, 79, 85–86, 201–202, 206
    peptidyl-prolyl isomerase, 221–222
    prefoldin, 238, 242–244
    proteasomes
        active sites, 285
        additional protein binding, 284–285
        assembly and processing, 294–298
        characteristics, 323–324
        chemical inhibitors, 298–299
        composition, 281–282

distribution, 285–288
energy-dependent proteases,
    324–325
GroEL similarity, 282
non-AAA$^+$ regulators, 308–310
peptide bond hydrolysis mechanism,
    288–294
subunit distribution, 285–288
Thermoplasma acidophilum, 282
ubiquitin-dependent degradation
    pathway, 310–314
ubiquitin-independent degradation
    pathway, 314–318
yeast similarity, 283–284
protein disulfide isomerase, 221–222
protein folding, 230–231, 234–237,
    265–269
RAD50 protein, 116–117, 121
RadA protein, 126–127
RadA protein paralogues, 134–136
Sgs1 helicase, 116
single-stranded DNA binding proteins,
    130–131
small heat-shock proteins, 255–256
Spo11 protein, 110–111
transcription
    bioinformatic view, 207–209
    catabolite repression apparatus,
        349–350
    elongation, 191–193
    gene regulation, 203–207
    gene-specific regulation, 209–212
    histones, 201–202
    nucleosomes, 201–202
    nucleosome templates, 81–82
    polarity, 188–190
    preinitiation, 191
    preinitiation complex formation,
        188–190
    promoter elements, 174–179
    RNA polymerase, 194–201
    termination, 193–194
transcription initiation factors,
    179–188, 204

367

CONTENTS OF PREVIOUS VOLUMES

379